ALLERGY ANALYSIS:

ALTERNATIVE METHODS

Stephen G Perry, Ph.D.

outskirts press

ALLERGY ANALYSIS: Alternative Methods

v10.1

Outskirts Press, Inc.
http://www.outskirtspress.com

ISBN: 978-1-9772-6222-6

PRINTED IN THE UNITED STATES OF AMERICA

Contents

Preface

It has been estimated that a significant percentage of the population exhibit clinical symptoms due directly to food sensitivities with the possibility that many more suffer from other allergies or ill-defined medical problems that may at least be aggravated by their diets. Then, as now, a number of clinical tests aimed at aiding the clinician in the diagnosis of allergy problems are not definitive and do not provide for effective screening and isolation of unfavorable symptomatology. In some cases, sophisticated and highly advanced procedures have not benefited from systematic application and follow-up study to appreciate fully the advances made. In the mid 1970's we reported that the *in vitro* Cytotoxic Test correlated with *in vivo* food challenges and was an excellent screening technique to identify allergens that affect the skin, and potentially other organ systems such as the cardiovascular, gastrointestinal, respiratory, and central nervous. In this book "ALLERGY ANALYSIS ALTERNATIVE METHODS", we offer a follow-up study for the results of systematic Cytotoxic Testing. Data is presented that correlates Cytotoxic Test results with clinical findings, familial distributions, and population studies. This book is written in detail to clarify and further substantiate our earlier reports (Ulett, GA, and SG Perry 1974; 1975) and ongoing studies. The detail should also permit other scholars to analyze the data according to their fields of expertise. I also recommend this book to students and practicing clinicians as a guide for interpreting single diagnostic tests and to plan testing and treatment regimens to better serve the patient's well-being and preventative health programs.

I dedicate this book to George A. Ulett, Ph.D. M.D. whose vision created the Missouri Institute of Psychiatry (MIP) and provided inspired leadership to MIP and its many programs, some of which achieved international recognition.

Sharon Authenreith-Netherly deserves considerable recognition for her skills as a technician and her dedication to the patients and volunteers that participated in Dr Ulett's programs.

Of particular recognition are the many that participated in the program and volunteered of themselves and their families then spread the word in the community for others to come and seek treatment from complaints of discomfort that they suffered and yet had not found relief. If you found them today, they are still probably talking about the commonsense benefit that they and family members received.

It has taken all too long to tell this story!

CHAPTER 1

Historical

BLACK, M.D. (1956), an El Paso, Texas Pediatrician that described the in vitro (in glass) effects of allergens A New Diagnostic Method in Allergic Disease was a benchmark publication by Arthur P. on white blood cells of allergic children. The defining clinical event that led Dr. Black to investigate and conceptualize his diagnostic methodology came in 1928 when a blood transfusion he administered led to acute agranulocytosis and subsequent death of a child. Black reported that prior to transfusion the child had marked leukocytosis (rising WBC) with a high percentage of polymorph nuclear cells (segmented neutrophils)". To understand, the adverse reaction Dr. Black used donor and recipient bloods remaining from cross matching to prepare warm stage slides with supravital staining. Under the microscope, he observed the destruction of the recipient's polymorphonuclear leukocytes in 30 - 40 minutes while control cell preparations remained viable and active. Dr. Black found no explanation in 1928; however, in 1956 he speculated that the phenomenon might have been an allergic one..."in which the donor's plasma contains an allergen, the child's plasma, a homologous antibody, and the leukocytes act as the shock organ". His search of the literature for evidence describing the in vitro behavior of leukocytes in the presence of both allergen and reaginic plasma from a sensitized individual yielded only papers reporting pieces of that puzzle. He noted that Hektoen (1906) reported in a study of bacterial phagocytosis that the variable factor is the serum and not the leukocytes. Hektoen also demonstrated that antibody can be eluted from leukocytes in salt solutions. Studying bacterial hypersensitivity Blatt and Nantz (1955) used washed cells relatively void of plasma and antibody in experiments employing leukocytes and plasma with no leukocytotoxic effect or leukocytes and antigen with no leukocytotoxic effect, but not leukocytes, plasma, and allergen together in a single preparation. Prausnitz and Kustner (1921) demonstrated this three-way requirement in dramatic fashion by transferring skin sensitizing antibodies from an allergic person to the skin of a nonallergic person, then using these inoculated leukocyte containing sites for skin tests with specific antigens. Described as a highly reliable diagnostic test the P-K test is impractical for many reasons and therefore this passive transfer testing procedure applies only in exceptional circumstances. These observations laid the foundation for Black (1956) to develop a new method, referred to as the Cytotoxic Test, with ten supporting clinical case studies.

The Cytotoxic Test: Black (1956) provided a list of recommended materials for Cytotoxic Testing. The leukocytes were concentrated and separated from erythrocytes with the use of acid-citrate-dextrose anticoagulant and centrifugation. The original procedure recommended only heparinized hematocrit tubes with anticoagulant taken up by capillary action to 6-8 mm followed by a finger or heel stick sample to within a cm of the end of the tube. To ensure the admixture of blood and anticoagulant a rubber bulb attached to the tube and its contents expelled into a hollow ground slide and reaspirated several times. The crit tube(s) were refilled, sealed, and centrifuged 15 - 30 minutes at 1200 rpm. Scratch each tube at the bottom of the buffy coat and discard the erythrocytes. The contents of the tube(s) are admixed in a hollow ground slide then taken up in plain capillary tube. Two tubes provided enough plasma-leukocyte cell suspension for six tests. Allergens were prepared by placing a drop (~ 0.1 ml) of distilled water on a slide and mixing an amount of dried allergen (Holister-Stier Labs, same as used for skin testing) transferred on the narrow end of a toothpick then allowed to dry in an area less than a circular cover slip. If neutral red supravital staining (25 mgs/100 mls ETOH) is going to be used place a drop in the center of the cover slip and allowed to dry. At the time of testing select the appropriate slide(s), add a drop (approx. 0.1 ml) of plasma-leukocyte cell suspension on the allergen matrix of the slide then cover slipped and sealed with a Vaseline-mineral oil sealant. Incubate at 37 C observing under a warm stage microscope at 30 min intervals. Carry controls, which are viable and stable under these conditions for hours through the same procedures. The Cytotoxic effects manifest as progressive changes seen in the polymorphonuclear leukocytes: as a loss in ameboid activity, a rounding of cell contour, a decrease and cessation of cytoplasmic movement, and increased staining. Erythrocytes and other cellular elements of the blood are generally not affected. Black developed this procedure as a diagnostic tool to support his practice. His case reports (Black, 1956) indicated modest testimonial success.

The Bryans (1960) employed the Cytotoxic food allergy test, varying only slightly from the method of Black (1956), as one of many diagnostic tools in their global assessment of patients presenting themselves to their otolaryngology practice associated with Barnes Hospital/Washington University Medical School, St. Louis, Missouri. They improved the techniques for isolating viable leukocytes in such quantities to support large-scale screening of patients for a broader spectrum of potential allergens. They standardized the buffy coat/plasma mixture to ensure 15 - 40 leukocytes per microscopic (60X apochromatic high dry objective) field. Since cells settle rapidly precautions are required in their procedures to gain consistency of handling and distributing the buffy coat cell suspensions. The standardized antigen preparations 0.1% (1.0mg/1.0 ml) solutions are matched with the dried matrix on the slide being slightly larger than the cover slip for more uniformity with batches of test slides made in advance. Clean "Silicad" glassware, slides and cover slips were a prerequisite for performing the test especially improving the controls where unfavorable protein binding to glass may occur. Performing the test at room temperature (80 - 85 F) at determined intervals proved acceptable and permitted more uniformity with batch screening. The Bryan's (1960) evaluated the degree of cytotoxic reaction as negative, or positive (slight, moderate, or marked). The Bryan's described the Cytotoxic reactions of the neutrophils as presenting the following changes: 1) loss of amoeboid movement and cessation of streaming of cytoplasmic granules; 2) rounding of the cell contour with or without spine-like projections; 3) vacuolization and spreading of the cytoplasm where the cell may become thin and flat; and 4) fragmentation of the cell with rupture of the membranes and dispersion of the granules. A negative cytotoxic reaction shows no changes as described above and are consistent with the no changes seen in the controls carried with each experiment (i.e., nonreactive foods, or distilled water). A slight positive shows half of the leukocytes losing their amoeboid activity and displaying a rounded contour upon scanning 10 - 15 fields. A moderate positive cytotoxic reaction shows all of the leukocytes lose their amoeboid activity and display a rounded contour with a few cells disintegrating. A marked positive cytotoxic reaction shows the majority of leukocytes disintegrated and in some cases this happens immediately. Some patients suffering symptoms of the ear, nose and throat were found to also present clinical evidence for an allergic etiology involving food and based on the

Bryans (1960) follow-up Cytotoxic testing of 107 of their patients this possibility was tested. They also presented a table that ranked the frequency that specific allergens were tested for and from which the percentage of positive or negative occurrences could be derived for their patient population of 107 (i.e., 122 tests for pork were 63.9% negative and 36.1% positive). Through diet management some form of relief for symptoms and syndromes studied in their population of 107 patients was achieved for 78.2%, no improvement for 12.8%, and for 9.0% it was unknown if any benefit was accomplished. The Bryans (1967, 1969, 1971, and 1972) continued to refine the description of their testing techniques and offered more detailed analyses of their findings from over 2500 patients tested.

The Cytotoxic Test Validated: Ulett M.D., Ph.D., (1973, 1974) while Director of the Missouri Institute of Psychiatry, a cooperative effort between the Missouri University Medical School and the Missouri Division of Mental Health, introduced a Cerebral Allergy program. This program included the use of Cytotoxic Testing as a diagnostic tool used to assist in the assessment of mental and psychological disorders. The Institute's computer analyzed EEG's and psychopharmacology programs had already received international recognition. The Bryans provided training and valuable consultation that expedited the establishment and early success for this program. The first studies followed the Cytotoxic Testing procedure of Black (1956) as described by the Bryans (1960, 1969, and 1971). Using the results of Cytotoxic Testing to design strict diet management programs Dr. Ulett built a highly successful program, which he continued until his retirement in the late 1990s, treating a broad range of clinical problems presented by patients seeking mental health services. Ulett and Perry (1974, 1975) made some significant observations while continuing to seek ways to improve and validate the Cytotoxic Testing procedure. Initial Cytotoxic Tests on subjects fasted since the previous evening were followed on a subsequent day with a second fasting Cytotoxic Test then a third challenge Cytotoxic Test at 1.5 - 2.0 hours after consuming foods known to be Cytotoxic Test positive based on the initial Cytotoxic Test results. Ulett and Perry (1974) reported that when subjects were challenged with cytotoxic (in vitro) positive tested foods, they showed a systemic (in vivo) leukocytosis that gave a direct correlation between the two different testing methods. That is, Cytotoxic Test positive food(s) when consumed following carefully managed procedures with proper controls caused the (in vivo) WBC to rise, and cytotoxic test negative food(s) caused no change in the in vivo WBC providing a parallel, independent confirmation and validation of the accuracy of Cytotoxic Testing results. Further evidence documenting the clinical impact that modest amounts of cytotoxic test positive foods can have been presented in the form of in vivo dose-response data. The in vivo leukocytosis response to measured amounts of cytotoxic test positive foods when plotted against the log of the amount of food consumed yielded classical sigmoid dose-response curves and zero response curves for cytotoxic test negative foods (see Chapter 3 Single Sensitizing Foods). At the very least this finding should make fasting a specimen requirement for a WBC as it would result in a dramatic shift in the accepted WBC normal ranges used to interpret one of the most common diagnostic tests in all of clinical medicine. WBCs monitored at 30-minute intervals during the food challenge were the basis for establishing in vivo leukocytosis maxima. Cytotoxic tests performed on blood specimens taken at 1.5 - 2.0 hours into a food challenge revealed additional cytotoxic test positive foods that were not detected during fasting cytotoxic tests thus identifying a significant number of false negative food allergens. The evidence for selecting this testing regime will be discussed in full detail in Chapter 2. Ulett and Perry (1975) analyzed the frequency that food allergen(s) display as a function of age and took a closer look at coffee as a single food allergen(s), while exploring the oral-respiratory route of entry with tobacco as offending allergen(s). The results served to validate further the excellent correlation between in vitro cytotoxic testing and in vivo WBC response test while revealing an incredible vision into potential mechanisms of food allergy and human molecular biology. Several models were advanced (Ulett and Perry, 1975) as a basis for correlating proven sensitivities with known clinical problems where allergic etiologies are suspected.

The Cytotoxic Test Questioned: Most that have tried the cytotoxic test validate its use as it applies to clinical practice. If they master the test's capricious, disciplined, and demanding nature and it serves to provide any semblance of diagnostic value in their practice of medicine then, in the past, there was a semblance of validation. Some did or did not succeed with the test, then failed to interpret properly the results or held up unrealistic expectations for what the results should mean. What should a positive cytotoxic test result mean if there is no outward or apparent disorder? If there is a disorder, how does one link any diagnostic test directly or indirectly to an offending allergen or allergic process?

Chambers, et al (1958) were among the first to test and conclude that the validity of the cytotoxic test was open to question. The 24 clinically hypersensitive subjects that Chambers tested presented clear-cut clinical sensitivities with 16 out of 24 (66.7%) subjects being skin test positive for the allergens tested (subject(s)/allergen; 9/ragweed, 1/cow's milk, 6/house dust, 2/dog, 2/peanut, 1/egg, 1/cat, 1/mixed feathers, and 1/wheat). The 36 not clinically sensitive subjects that Chambers tested had 11 out of 36 (47.8%) subjects being skin test positive for the allergens tested (subject(s)/allergen; 4/ragweed, 9/cow's milk, 3/house dust, 3/dog, 3/peanut, 4/egg, 0/cat, 7/mixed feathers, and 3/wheat,). Only one cytotoxic tested allergen gave an active or positive test in the two groups tested.

Lieberman, et al (1975) in a collaborative study out of the departments of medicine, allergy-immunology, and pediatrics at the University Of Tennessee College Of Medicine, Memphis undertook to corroborate the cytotoxic test findings as reported by Black (1956) and the Bryans (1960, 1967, 1969, 1971, and 1972). They tested a broad range of food allergens adhering closely to the procedures described by the Bryans. According to Lieberman's group, the study failed to confirm previous claims for the validity of the cytotoxic food test. They reported that the test had little diagnostic accuracy in the evaluation of atopic patients with well-established allergic reactions to foods and they were concerned about a substantial incidence of false-positive reactions. Although they reported the test to be subjective with poor correspondence between examiners, no criticism was reported concerning the cytotoxic reaction or the principles of the cytotoxic test per se.

The Cytotoxic Test has come under evaluation in several books and reviews. Golbert (1975) published "A Review of Controversial and Therapeutic Techniques Employed in Allergy" that included a discussion of the cytotoxic test procedure. Terr (1985) wrote an editorial entitled The Cytotoxic Test again highlighting this procedure. In 1991, Goldberg and Kaplan published "A Review of Controversial Concepts and Techniques in the Diagnosis and Management of Food Allergies" that included a discussion of the cytotoxic test citing Ulett and Perry (1974) as an interesting and confusing reference. Middleton et. al., Editors (1993) cited Ulett and Perry (1974; 1975) commenting that..."proponents of the cytotoxic test have claimed efficacy only in the form of case reports or descriptions of a series of patients claiming that symptomatic improvements occur when foods are eliminated from the diet based on the results of the cytotoxic test". None of these reports relied on a definitive assessment using a protocol with a placebo-controlled double-blind food challenge (PCDBFC)." Other investigators have failed to confirm that the cytotoxic food test results correlate with clinical illness (Lieberman, 1975; Bensen and Arkins, 1976. John Anderson (1994) published a review highlighting what he considered the major contributions to understanding adverse reactions to foods. In this review he clearly outlines what he believes should be the standards to guide research and clinical practice relating to the field of food allergy. Bernstein and Storms (1995) in a publication "Practice Parameters for Allergy Diagnostic Testing" list criteria for unproven diagnostic tests to include: 1) Procedures that are invalid for any purpose (the cytotoxic test was named as an example), 2) Procedures capable of valid measurement, but not appropriate for diagnostic use, in IgE mediated allergy, and 3) Procedures of theoretical value, but not practical (expense, sensitivity, specificity, or general availability). In this review they state, "Controlled studies for the cytotoxic test demonstrated that the results are not reproducible and do not correlate with clinical evidence of allergy". Clearly, a high standard has been set for clinical studies and practices that is not yet the standard for the development of diagnostic and investigational procedures.

Over the years, the death of the cytotoxic test has been greatly exaggerated. Ulett and Perry (1974; 1975) made a significant contribution concerning the utility of the cytotoxic test that has been overlooked or misinterpreted as reflected in the reports cited here. One might have to finish reading and studying the remaining contents of this book to appreciate fully what the accomplishments of their research and clinical studies. If Ulett and Perry could reduce the results of cytotoxic testing to a gold standard test, the WBC, then costly time consuming impractical double-blind placebo-controlled food challenge (DBPCFC) "gold standard" testing protocols may not be required in all cases. Sampson and Ho (1997) identified a "subset of patients" that they maintained are likely (>95%) to experience clinical reactions to egg, milk, peanut, or fish as determined by using CAP Systems FEIA (Pharmacia) food specific antibody detection procedures. They went on to maintain that this is an example where one might eliminate the need to perform DBPCFC test protocols in a significant number of patients suspected of having IgE mediated food allergy thus saving time and money in the pursuit of practical treatment. In examples like this 30 - 40% of food, specific IgE positive individuals experience clinical symptoms when food was ingested (Sampson and Albergor, 1984). As an example, Sampson's group has published evidence that children with atopic dermatitis associated with cow's milk, egg or peanut allergy correlated with high concentrations of allergen specific IgE antibodies, which were also predictive of food induced clinical symptoms (James and Sampson, 1992; Bernhisel-Broadbent et al, 1994). Where research and learning flourish the frontiers of knowledge in medicine will move forward. Considering what is happening in all disciplines of science those that embrace the broader perspective will find new avenues to investigate the specific mechanisms underlying this clinical enigma we know as allergy.

REFERENCES

Anderson JA: Milestones marking the knowledge of adverse reactions to food in the decade of the 1980s. Annals Allergy 72:143-154. 1994.

Benson TE and Arkins JA: Cytotoxic testing for food allergy evaluation of reproducibility and correlation. J Allergy Clin Immunol, 58:471, 1976.

Bernhisel-Broadbent J, Dentzes HM, Dentzes RZ and Sampson HA: Allergenicity and antigenicity of chicken egg ovomucoid (Gal d III) compared with ovalbumin (Gal d I) in children with egg allergy. J Allergy Clin Immunol, 93:1047-59, 1994.

Bernstein IL and Storms WW: Practice parameters for allergy diagnostic testing. Ann Allergy, 75:543-625, 1995.

Black AP: A New Diagnostic method in allergy disease. Pediat 17:716-724, 1956.

Blatt H and Nantz FA: Further studies on use of tissue culture of blood leukocytes in clinical evaluation of bacterial hypersensitivity of tuberculin type. Annals Allergy, 8:622, 1955.

Bryan WTK and Bryan MP: Clinical examples of resolution of some idiopathic and other chronic disease by careful allergic management. Laryngoscope, 82:1231-38, 1972.

Bryan WTK and Bryan MP: Cytotoxic Reactions in the diagnosis of food allergy. Otolarnyngol Clinics of N Am, 4:523-534, 1971.

Bryan WTK and Bryan MP: Cytotoxic Reactions in the diagnosis of food allergy. Laryngoscope, 79:1453-1472, 1969.

Bryan WTK and Bryan MP: Diagnosis of food allergy by cytotoxic reactions. Trans Am Soc Opthal Otolarnyngol Allergy 8:14 1967

Bryan WTK and Bryan MP: The application of in vitro cytotoxic reactions to clinical diagnosis of food allergy. Laryngoscope 70:810-824, 1960.

Chambers VV, Hudson BH, and Glaser JH: A Study of the reactions of human polymorphonuclear leukocytes to various allergens. J Allergy 29:93-102, 1958.

Goldberg BJ and Kaplan MS: Controversial Concepts and Techniques in the Diagnosis and Management of Food Allergies. In: Food Allergy: Immunology and Allergy Clinics of North America, Anderson JA (Ed) Vol II, 1994.

Goldbert, TM: A review of controversial and therapeutic techniques employed in allergy. J Allergy Clin Immunol 56:170-190, 1975.

Hektoen L, Phagocytosis and oponins. JAMA, 46:1407 1906.

James JM and Sampson HA: Immunological changes associated with the development of tolerance in children with cow's milk allergy. J Pediatr, 121:371-7, 1992.

Lieberman P, Crawford L, Bjelland J, Connell B and Rice M: Controlled study of the cytotoxic food test. JAMA, 231:728-730, 1975.

Middleton E, Reed CE, Ellis EF, Adkinson NF, Yunginger JW and Busse WW, Editors: Allergy Principles and

Practice Vols I & II: Chapter 71 Unconventional theories and unproven methods in allergy, Mosby, 1993.

Prausnitz C and Kustner H: Studien uber die Ueberempfindlichkeit. Centralbl. f. Bakt. I Abt Originale 86:160-169, 1921.

Sampson HA and Ho DG: CAP System FEIA (quantitative assay) to determine the potential utility of quantitative food specific IgE values in the diagnosis of IgE mediated food hypersensitivity. J Allergy Clin Immunol 100:444-51, 1997.

Sampson HA and Albergo R: Comparison of results of skin tests, RAST and double-blind placebo-controlled food challenges in children with atopic dermatitis. J Allergy Clin Immunol, 74:26-33, 1984.

Terr AI: The cytotoxic test. West J Medicine, 139:702-703, 1983.

Ulett GA and Perry SG: Cytotoxic testing and leukocyte increase as an index to food sensitivity. II. Coffee and Tobacco. Annals Allergy 34:150-160, 1975.

Ulett GA and Perry SG: Cytotoxic testing and leukocyte increase as an index to food sensitivity. Annals Allergy 33:23-32, 1974.

Ulett GA, Itil E and Perry SG: Cytotoxic food testing in alcoholics. Quart J Stud Alc. 35:930-942, 1974.

Ulett GA and Itil E: Alcoholism and allergy (Presented at the 115th annual meeting of the American Psychiatric Association, Honolulu, Hawaii, 1973).

CHAPTER 2

First Analysis

THE PUBLICATIONS OF Ulett and Perry (1975 and 1974) shed new light on two lines of food allergen diagnostic testing that were dependent on the principles of in vitro leukocyte cytotoxic reaction (Bryans, 1971, 1969, 1960; Black, 1956) and in vivo leukocytosis (Vaughn, 1939, 1936a. 1936b, 1934a, 1934b; Sabin, 1925). Both lines of testing had ardent proponents and years of published reports, that led in many directions. Both lines of testing were often dominated by clinical interpretations that were; subjective, anecdotal, lacking good controls, and were nonquantitative yet endured by some benefit derived in clinical practice for doctor and/or patient. Both lines of testing were perfectly suited for rigorous quantitation and detailed morphological identification under well-controlled, carefully chosen conditions and time frames.

Having been cytotoxic food tested repeatedly with reasonable accuracy over a period of several months, testing for evidence of an in vivo response to cytotoxic test positive foods and the prospect of quantitating and even automating the procedure became high priorities. A detailed description of our systematic initial discoveries will be shared with you in this chapter followed in subsequent chapters with detailed analysis of our findings. The emphasis will be on the analytical aspects of developing a useful diagnostic procedure. You will be amazed at what we discovered. You should marvel at the simplicity of our investigations. You should be outraged that the clinically useful normal range (~4.4-11 x $10^3/mm^3$) for the widely and frequently used WBC has been so wrong for so long with fasting not required! Physicians do require fasting samples when ordering laboratory tests but turn around and interpret the WBC results with the wrong normal range. What happens between ~7.7 -11.5 $10^3/mm^3$ when the WBC count goes up (with the immune system) good, bad, or indifferent? It is our hope that isolated allergy diagnostic test results may become more meaningful when interpreted in the backdrop of our findings and that this alternative method for allergy analysis has opened some new doors and avenues for preventative medicine and improved healthcare procedures.

Leukocytosis "Under Control": Based on earlier prefasted cytotoxic tests (Ulett and Perry, 1974) "allergic" or sensitizing (chicken, chocolate, peanut, and tea) and "nonallergic" or nonsensitizing (apple, beef, corn, and licorice) in vivo challenge meals were selected and test periods in the 8:00am - 12:00 noon time slot were reserved. Specimens, for a 70 food prefasted control cytotoxic test and control WBC or control WBC/differentials, were obtained. The clock was then started on the "allergenic" meal, "nonallergenic" meal, or sham-meal in vivo control sessions (see Figure 2.1). During the experimental challenge period, WBCs or WBC/differentials were determined at 30-minute intervals over 4.0 - 5.0 hours. Initially, a specimen for a 70 food in vitro cytotoxic test was taken at the 4.0-hour mark of the "allergenic" in vivo challenge meal. The test results, detailed in table 2.1, were spectacular. Three turns at bat and three home runs! Even the slight rise in WBC for the "nonallergenic" meal was viewed positively as we had hoped to see differences that at least could be dealt with statistically. Upon reviewing the results of cytotoxic testing done at 4.0 hours, in the wake of the leukocytosis maximum (see Ulett and Perry, 1974 Table III), six new cytotoxic test positive foods appeared (SGP), including apple and corn that had been picked for the first "nonallergenic" meal. Based on this information a new "nonallergenic" meal (cherry, egg, orange, and pork) was selected and a follow-up test period was scheduled.

The second allergenic test series was completed eight days later with the same remarkable pattern of results (Table 2.2), thus giving confirmation to the first experiments. In the original test series, the meal period lasted nearly 30 minutes with tea and chocolate taken in the last few minutes. Adjusting the challenge meal periods to 10 minutes appeared to sharpen the response in both time and magnitudes (85% vs. 40% WBC increase) illustrating how critically controlling the conditions of the procedures are to the outcome and proper interpretation of these kinds of test results. The "nonallergenic" control meal showed no WBC increase over its control WBC level giving an absolute black and white comparison between "allergenic" and "nonallergenic" meal WBC responses!

TABLE 02.01 CYTOTOXIC TESTED FOOD CHALLENGE WITH ANALYSIS OF *IN VIVO* HEMATOLOGY RESULTS.

	TIME HOUR		DIFFERENTIAL				
			**				
ANALYTE		WBC	SEG	LYMPH	MONO	EOSIN	BASO
RANGE		4.8-10.8	58-66	20-40	2-8	2-4	0-1
UNITS		10(3)	%	%	%	%	%
CASE/AGE/SEX							
SGP/35/M	**0.0**	5.55					
"ALLERGENIC"	0.5	**5.85**					
MEAL (30 MIN)	1.0	**6.30**					
(CHICKEN)	1.5	**6.35**					
(CHOCOLATE)	2.0	**6.50**					
(PEANUT)	2.5	**7.80**					
(TEA)	3.0	**7.00**					
5/22/73	3.5	**6.35**					
8:45AM	4.0	**6.25**					
	5.0	**5.70**					
========							
AVERAGE		**6.46**					
SGP/35/M	**0.0**	5.80	57	42		1	0
FASTING	0.5						
PERIOD	1.0	**5.75**	49	49		2	0
5/23/73	1.5	**5.80**	55	43		2	0
8:45AM	2.0	**5.75**	56	43		1	0
	2.5	**5.80**	63	35		2	0
	3.0	**5.75**	55	42		2	1
	3.5	**5.80**	69	30		1	0
	4.0	**5.75**	59	39		2	0
========							
AVERAGE		**5.77**	58	40		2	0
SGP/35/M	**0.0**	5.55	53	48	2	0	0
"NONALLERGENIC"	0.5	**5.65**	56	42	1	1	0
MEAL (30 MIN)	1.0	**5.65**	50	48	0	2	0
(APPLE)	1.5	**6.00**	47	45	5	3	0
(BEEF)	2.0	**6.05**	51	49	0	0	0
(CORN)	2.5	**6.00**	54	44	1	1	0
(LICORICE)	3.0	**5.90**	52	48	0	0	0
5/24/73	3.5	**5.75**	55	42	0	3	0
9:45AM	4.0	**5.55**	56	42	0	2	0
========							
AVERAGE		**5.82**	53	45	1	2	0

These results represent parallel, independent, and corroborating lines of evidence; one from an *in vitro* bioassay (Cytotoxic Test) and the other from an *in vivo* quantitative systemic response (measured *in vivo* leukocytosis). Positive responses as measured by these distinctly separate techniques are interpreted as meaning that the individual is or has been sensitized by the food antigen(s) that elicit the positive response, and that these are diagnostic tests for determining which potential sensitizer(s) can affect the individual being tested.

TABLE 02.02 CYTOTOXIC TESTED FOOD CHALLENGE WITH ANALYSIS OF *IN VIVO* HEMATOLOGY RESULTS.

	TIME		DIFFERENTIAL					
	HOUR		***					
ANALYTE		WBC	SEG	LYMPH	MONO	EOSIN	BASO	PULSE
RANGE		4.8-10.8	58-66	20-40	2-8	2-4	0-1	
UNITS		10(3)	%	%	%	%	%	
CASE/AGE/SEX								
SGP/35/M	**0.0**	5.95	65	30	5	0	0	64
"ALLERGENIC"	0.5	**6.05**	59	38	1	2	0	68
MEAL (10 MIN)	1.0	**6.50**	70	25	3	2	0	64
(CHICKEN)	1.5	**6.85**	68	24	5	3	0	64
(CHOCOLATE)	2.0	**11.00**	72	24	1	3	0	63
(PEANUT)	2.5	**10.00**	70	29		1	0	63
(TEA)	3.0	**8.00**	68	30	1	1	0	64
5/30/73	3.5	**6.15**	68	28	2	2	0	61
9:30AM	4.0	**5.80**	65	32	1	2	0	60
	5.0	**5.70**	68	29	1	1	0	57
========								
AVERAGE		**7.34**	68	29	2	2	0	63
SGP/35/M	**0.0**	5.45	54	48	2	1	1	59
"NONALLERGENIC"	0.5	**5.55**	58	42	1	1	0	60
MEAL, (10 MIN)	1.0	**5.50**	53	48	3	0	0	60
(CHERRY)	1.5	**5.55**	56	45	1	0	0	64
(EGG)	2.0	**5.45**	54	49	2	1	0	65
(ORANGE)	2.5	**5.50**	55	44	1	0	1	65
(PORK)	3.0	**5.45**	59	48	2	0	0	63
6/5/73	3.5	**5.55**	56	42	1	1	0	60
9:00AM	4.0	**5.50**	54	42	0	1	0	60
========								
AVERAGE		**5.51**	56	45	1	1	0	62

The first test (Table 02.01) was most encouraging as we observed a 40% increased WBC response to foods that had been determined cytotoxic test positive (Figure 02.01 dotted line) which then returned to control level at 5.0 hours. The test period continued to be a period of strict fasting and protected exposure from "other" environmental sensitizers (i.e., smoke, pollen, etc.) that might further unstablize the WBC and bias the results during the course of the test. This was the first time that we looked for the possibility of an *in vivo* response to foods tested previously as cytotoxic test positive. The results were exciting and as with most newfound results suggested may more experiments to test the validity of these initial findings. On the following day a second test (Table 02.01) was set up to monitor a control test period when no food was taken (Figure 02.01 double open circles) to see what the WBC pattern would be during the same experimental time slot. Under the carefully controlled time-period, we were even more excited about the previous day's results. This gave us two-way controls for comparison with experimentals. The first control was the 0.0-time control that provided a fixed point in time reference for comparison of the experimental 30-minute time intervals (0.5 - 5.0 hrs.). The more dynamic controls were gained by comparison of each point in time (control vs. experimental), allergic foods vs. non-allergic foods (0.0-time control through the 5.0-time interval) and allergic vs. sham meals (no food or only water) over the entire experimental period. After false negatives were identified, the "nonallergenic" meal provided a real time internal control (Figure 02.01 small open circles) to validate the WBC response that permitted us to be excited about the *in vivo* WBC response! Critical documented answers that we never hear from patients.

The temporal patterns of these first *in vivo* WBC responses (figure 02.01) were displayed as a percent of their 0.0 hr. control values vs. time and graphically show the dramatic comparisons of results from our first test series. Expressing the data as a percent of the controls normalized all observations to a common point of reference. In these examples, the experimental intervals were compared to the levels measured before the test challenge began. Eventually we were able to compare fasting or nonsensitizing controls vs. experimental at the timed intervals in real time and with astonishing clarity. As clinical research goes, we would have been ecstatic to see differences that would eventually become apparent after many analyses sorted out statistically. I am reminded of an old advisor's admonishment (Oliver Lowery, M.D.); "If you have to rely on statistics, then you didn't design the experiment right". Thank you, Ollie, for your training and your strict standards on experiment design!

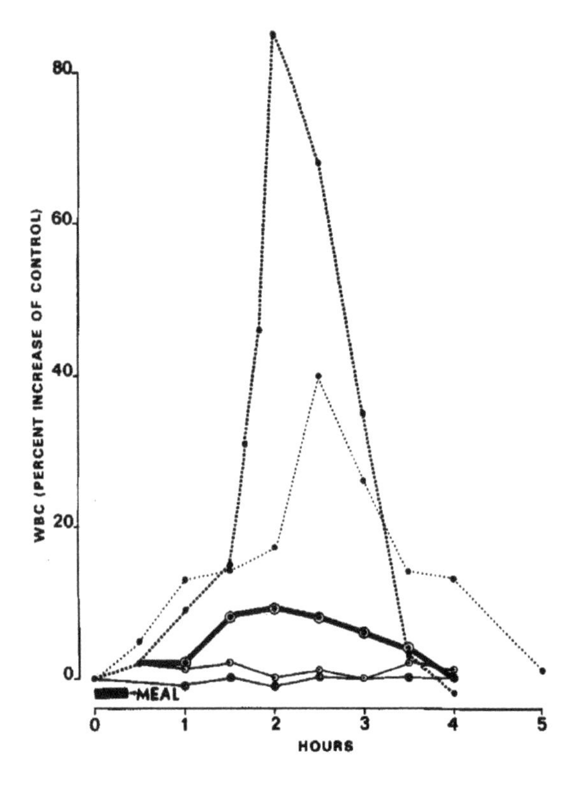

Figure 02.00 Systemic circulatory response (leukocytosis) to sensitizing and nonsensitizing cytotoxic tested foods; a) response to a sensitizing meal (chicken, chocolate, peanut, and tea - dotted line), b) same as (a) eight days later (hatched line), nonsensitizing foods (beef, licorice with false negatives corn and apple - large circles), nonsensitizing foods (cherry, egg, orange, and pork - small circles), and e) sham meal control (two cups of water - double circles).

The subject was fasted from 6:00 PM the previous evening striving to achieve the cleanest controls possible. The meals were managed for content and duration further cleaning up the controls and focusing in on the systemic WBC response (selective margination?). Bochner (1997) defined margination as a dynamic and reversible process involving the formed elements of the blood and the endothelial lining of the blood vessels. Bochner went on to describe margination as involving the tethering and rolling of leukocytes on the endothelium where induction of firm adhesion influenced by chemokines, platelet activating factor, leukotrines and other chemotractant's act through transmembrane receptors on the leukocyte. Clinically, cytokines and chemokines (Freri, 1999; 1993) have been linked to asthma and leukotrines B4 (Koro et al) has been linked to atopic dermatitis showing the powerful chemical mediator and chemotractant regulatory system affects triggered by leukocytes during allergic symptomatology. Very interesting!

That allergenic food or other environmental antigen(s) in the circulation may be signaling their presence resulting in the proportional recruitment of "reserve" leukocytes to meet the "external invasion" is an intriguing concept. In the experiments described above and detailed in the first three chapters of this book the allergenic exposure was a ten-minute pulse and a measured amount of food. In the real world, the exposure may be brief or chronic extracting varying degrees bodily resources to cope with the consequences of exposure. Chronic long-term exposures even at "subclinical" levels could place demands on the one's health and well-being with an ultimate selective advantage going to those not as affected as others. This may be manifested subtlety in the classroom, office, or athletic field at 3:00 PM, or in life threatening circumstances during a transfusion as may have been the case with Dr Black's patient as related in chapter one.

The stability of the controls was again tested (Tables 02.03 and 02.04) revealing that two fasted subjects-maintained WBC values that did not vary over a 24-hour test period beginning at midnight and testing at 1.0-hour intervals all the way through to midnight the following day. Critical to understanding the WBC response is the value of each individuals "normal" WBC (Bessman and David, 1986). Our experience indicated that "healthy subjects" generally had relatively low WBCs (5 – 6 thousand per cubic mm) that varied over a very narrow range unless unstabilized for known or unknown clinical reasons or set in motion by a host of environmental sensitizers. As we have shown sensitizing foods, with only a brief exposure can unstablize the WBC for up to five hours.

These 24-hour studies, and the control test periods detailed above (Tables 02.03 and 02.04), led us to conclude very early on that "allergenic" foods and probably other airborne environmental sensitizers are largely responsible for the unstable WBC. Considering meal patterns and inadvertent contact with airborne sensitizers such as pollens, aromatics, or tobacco smoke chronically unstablize the WBC, the speculation that circadian rhythms (Haus, 1992) may be the cause of unexplained WBC fluctuations must be re-evaluated. It is significant that Cytotoxic tested food challenges now can be directly linked to fundamental diagnostic parameters such as the WBC and differential and probably beyond to specific biological processes involving the immune defense system. The white count increase includes plasma white cells that recognize antigens then respond with monoclonal antibodies good bad , or indifferent, and survive while other white cells are disintegrating.

In Chapter 4.0, we report the results of computer analyzed EEGs that were recorded while we were measuring the changes in the WBC. What we saw in real time at a subconscious level was exciting. In our 1975 publication (Ulett and Perry, 1975), we proposed a model where in phase I the white cell had the genetic capacity to recognize antigen(s), then in phase I initiate events at the molecular level, and finally these events have the potential to manifest an allergic response in an end organ that may also be genetically determined. The computer analyzed alpha, beta, and theta waves in the EEG from the frontal lobes appear to synchronize with the WBC and lymphocyte patterns while the parietal lobe EEG activity appeared to be independent from the WBC and lymphocyte pattern for challenge and fasting control test periods. The parietal and temporal lobe patterns were quite different as we will see in detail in Chapter 4. These real time communications at the molecular level opens some interesting lines for continued research and lend support for the proposed model.

NOTES:

____

____

____

____

____

TABLE 02.03 ANALYSIS OF FASTING 24 HOUR WBC/DIFFERENTIAL RESULTS IN AN INDIVIDUAL (SSA).

			DIFFERENTIAL					
	TIME	WBC	***					
CASE	HOUR	10(3)	SEG	LYMPH	MONO	EOSIN	BASO	PULSE
SSA	**0.0**	6.25	46.0	52.0	1.0	0.0	0.0	56.0
AGE 24	1.0	**6.15**	49.0	50.0	1.0	0.0	0.0	56.0
SEX F	2.0	**6.30**	45.0	53.0	1.0	1.0	0.0	48.0
	3.0	**6.20**	49.0	50.0	0.0	1.0	0.0	56.0
	4.0	**6.30**	46.0	52.0	1.0	1.0	0.0	50.7
	5.0	**6.20**	49.0	50.0	1.0	0.0	0.0	54.0
	6.0	**6.10**	45.0	52.0	2.0	1.0	0.0	56.0
	7.0	**6.05**	47.0	53.0	0.0	0.0	0.0	54.7
	8.0	**6.00**	42.0	55.0	2.0	1.0	0.0	56.0
	9.0	**6.10**	45.0	52.0	2.0	1.0	0.0	68.0
	10.0	**6.00**	46.0	53.0	1.0	0.0	0.0	64.0
	11.0	**6.00**	49.0	50.0	0.0	1.0	0.0	66.7
	12.0	**6.15**	45.0	53.0	2.0	0.0	0.0	68.0
	13.0	**6.20**	49.0	51.0	0.0	0.0	0.0	56.0
	14.0	**6.00**	46.0	52.0	1.0	1.0	0.0	68.0
	15.0	**6.05**	48.0	50.0	1.0	1.0	0.0	69.3
	16.0	**6.20**	46.0	50.0	2.0	2.0	0.0	68.0
	17.0	**6.10**	42.0	54.0	2.0	2.0	0.0	60.0
	18.0	**6.25**	49.0	51.0	0.0	0.0	0.0	56.0
	19.0	**6.10**	44.0	52.0	2.0	2.0	0.0	60.0
	20.0	**6.20**	46.0	53.0	1.0	0.0	0.0	57.3
	21.0	**6.05**	50.0	49.0	1.0	0.0	0.0	56.0
	22.0	**6.05**	46.0	52.0	2.0	0.0	0.0	52.0
	23.0	**6.15**	41.0	56.0	2.0	1.0	0.0	56.0
	24.0	**6.05**	40.0	58.0	1.0	1.0	0.0	58.7
===============								
AVERAGE		**6.1**	**46.0**	**52.1**	**1.2**	**0.7**	**0.0**	**59.0**
SUM		147.0	1104.0	1251.0	28.0	17.0	0.0	1415.4
SUM OF SQS		900.0	50960.0	65313.0	46.0	23.0	0.0	84346.9
NUMBER (N)		24.0	24.0	24.0	24.0	24.0	24.0	24.0
SQRT OF N		4.9	4.9	4.9	4.9	4.9	4.9	4.9
MAXIMUM		6.3	50.0	58.0	2.0	2.0	0.0	69.3
MINIMUM		6.0	40.0	49.0	0.0	0.0	0.0	48.0
RANGE		0.3	10.0	9.0	2.0	2.0	0.0	21.3
STD ERROR (SEM)		0.0	0.4	0.4	0.1	0.1	0.0	0.9
SEM %		0.2	0.9	0.7	7.1	11.8	0.0	1.5
STD DEV (SD)		0.1	2.8	2.1	0.8	0.7	0.0	6.2
SD % COEF VAR		1.5	6.0	4.1	65.3	97.4	0.0	10.5

TABLE 02.04 ANALYSIS OF FASTING 24 HOUR WBC/DIFFERENTIAL RESULTS IN AN INDIVIDUAL (SSA).

			DIFFERENTIAL					
	TIME	WBC	**					
CASE	HOUR	10(3)	SEG	LYMPH	MONO	EOSIN	BASO	PULSE
TDS	**0.0**	5.95	54.0	40.0	3.0	3.0	0.0	48.0
AGE 25	1.0	**6.05**	60.0	35.0	2.0	3.0	0.0	50.7
SEX M	2.0	**6.10**	60.0	38.0	1.0	1.0	0.0	52.0
	3.0	**5.95**	65.0	32.0	0.0	3.0	0.0	48.0
	4.0	**6.05**	64.0	34.0	1.0	1.0	0.0	48.0
	5.0	**6.15**	70.0	28.0	1.0	1.0	0.0	56.0
	6.0	**6.05**	68.0	30.0	1.0	1.0	0.0	48.0
	7.0	**6.20**	61.0	35.0	2.0	2.0	0.0	49.3
	8.0	**6.05**	63.0	35.0	0.0	2.0	0.0	60.0
	9.0	**6.20**	65.0	30.0	2.0	3.0	0.0	76.0
	10.0	**6.25**	61.0	37.0	1.0	1.0	0.0	76.0
	11.0	**6.20**	69.0	25.0	2.0	4.0	0.0	74.7
	12.0	**6.50**	65.0	32.0	1.0	2.0	0.0	76.0
	13.0	**6.15**	64.0	35.0	0.0	1.0	0.0	73.3
	14.0	**6.25**	61.0	37.0	1.0	1.0	0.0	61.3
	15.0	**6.20**	60.0	38.0	0.0	1.0	1.0	60.0
	16.0	**6.10**	65.0	32.0	1.0	2.0	0.0	57.3
	17.0	**6.50**	69.0	31.0	0.0	0.0	0.0	60.0
	18.0	**6.15**	60.0	37.0	1.0	1.0	1.0	58.7
	19.0	**6.25**	65.0	32.0	0.0	3.0	0.0	57.3
	20.0	**6.20**	66.0	34.0	0.0	0.0	0.0	64.0
	21.0	**6.20**	69.0	30.0	1.0	0.0	0.0	61.3
	22.0	**6.15**	62.0	35.0	1.0	2.0	0.0	60.0
	23.0	**6.10**	66.0	31.0	1.0	2.0	0.0	53.3
	24.0	**6.00**	69.0	30.0	0.0	1.0	0.0	52.0
===============								
AVERAGE		**6.2**	**64.5**	**33.0**	**0.8**	**1.6**	**0.1**	**59.7**
SUM		148.0	1547.0	793.0	20.0	38.0	2.0	1433.2
SUM OF SQS		913.1	99973.0	26459.0	28.0	86.0	2.0	87602.5
NUMBER (N)		24.0	24.0	24.0	24.0	24.0	24.0	24.0
SQRT OF N		4.9	4.9	4.9	4.9	4.9	4.9	4.9
MAXIMUM		6.5	70.0	38.0	2.0	4.0	1.0	76.0
MINIMUM		6.0	60.0	25.0	0.0	0.0	0.0	48.0
RANGE		0.6	10.0	13.0	2.0	4.0	1.0	28.0
STD ERROR (SEM)		0.0	0.4	0.5	0.1	0.2	0.0	1.2
SEM %		0.4	0.6	1.6	10.0	10.5	0.0	2.0
STD DEV (SD)		0.1	3.3	3.3	0.7	1.1	0.3	9.4
SD % COEF VAR		2.1	5.2	10.1	84.2	66.9	0.0	15.7

Leukocytosis in ten subjects: The next step was to test additional subjects using the combination of in vitro cytotoxic tests with the in vivo food challenged monitoring of the WBC. Our initial observations were confirmed in each of four additional male and five female subjects (Ulett and Perry, 1974). The in vivo WBC response (figure 2.2) was consistent for male or female subjects of widely varying ages (20 - 59 years).

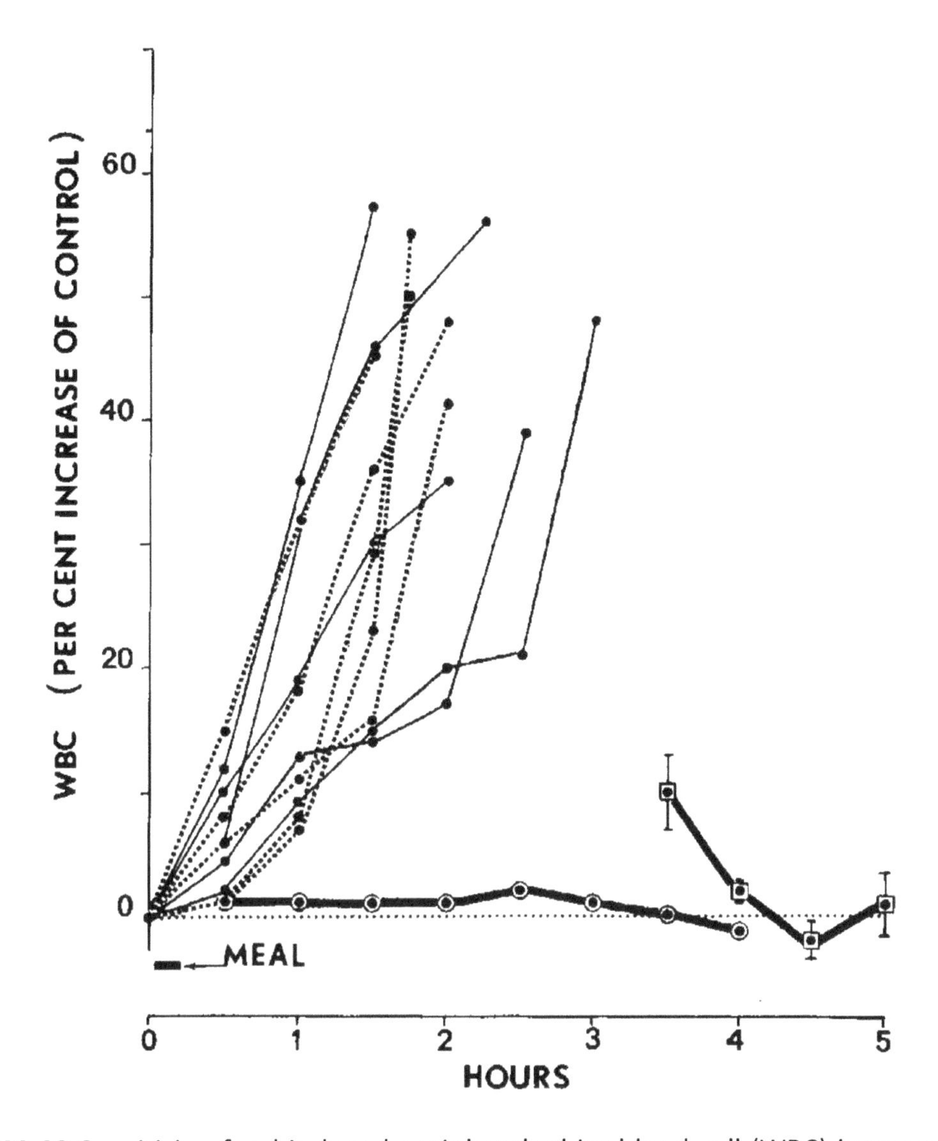

Figure 02.02 Sensitizing food induced peripheral white blood cell (WBC) increase in five males (solid lines) and five female (hatched lines) prefasted subjects, with return to control values (open squares) plotted as an average for the group +/-S.E.M. Nonsensitizing food tested controls (open circles) were run on the same subjects on the morning following their allergic test.

When male vs. female populations were analyzed 71 females age span newborn to 82 years in age averaging 25.2+/-9.7 SD (38.6%CV) with an overall sensitivity of 36.5+/-8.0 SD (21.8%CV). The 59 males age span newborn to 72 years in age averaging 30.9+/-9.4 SD (30.3%CV) with an overall sensitivity of 38.3+/-8.2 SD (21.4%CV). Although the difference in age between the males and female was not significant (P< 0.10) there was no significant difference (P<0.30) in the average sensitivity (percent positives) between the male and female populations. The data for male vs. female test positive sensitivities can be visualized below (Figure 06.01). It appears that sensitivities are manifest sequentially as antigen exposure triggers immune system prenatal and early postnatal genetic potential independent of sex or age. If you are not stung by a bee, "or a food", until late in life then the same fundamental process kicks in with a proportional reaction base on Phases I. II., and III of your genetic capacity. We have proposed three phases that add up to an allergic manifestation. Phase I we identify (*in vitro* Cytotoxic Test or *in vivo* Food Challenge Test positive) as an immune cell recognition step, Phase II: initiated immune defense steps, and Phase III: end organ(s) allergic response (i.e., skin, gastrointestinal, cardiovascular, central nervous, etc. system(s). We are measuring Phase I. with the *in vitro* Cytotoxic Test, or the *in vivo* Food Challenge reported the **Leukocytosis and Controls:** The graphic data displayed in figure 02.02 was derived from the raw data in Tables 02.05 and 02.06 and expressed as a percent of the respective controls reducing all subjects to a common point for convenient comparison. When the raw WBC data were analyzed, a number of exciting findings were revealed. The control data (Table 02.05) represents the peripheral WBC response to the nonsensitizing ("nonallergenic") meal. The control WBC averaged in the 6.7 (x1000) range, the standard deviation was in the 1.2 range, and the percent coefficient of variation was consistently in the 17% range on average across the test period and the WBC showed no significant variation among any of the control intervals. The expected range (plus or minus 2 SD) was consistently within the 4.4 - 9.1 range that is considerably tighter than the 4.0 - 10 range reported by many clinical laboratories or published as clinical laboratory standards (CLR, through 1998-1999). The last time that I, in a clinical lab setting and in response to a CLIA inspection challenge, randomly sampled 100 specimens submitted for a complete blood count the WBC was 7.3 +/-1.52 SD (20.8% CV) and the "normal range" was 4.26 - 10.33 for "fasting not required" WBCs completed within 24 hours.

TABLE 02.05 WBC (x1000) COMPARISONS AT HOURS AFTER EXPOSURE TO NONSENSITIZING FOODS

CASE	AGE	SEX	CTL/WBC 0.0	WBC 0.5	WBC 1.0	WBC 1.5	WBC 2.0	WBC 2.5	WBC 3.0	WBC 3.5	WBC 4.0
SGP	35	M	**5.55**	5.65	5.65	**6.00**	6.05	6.00	5.90	5.75	5.55
GAU	55	M	**5.90**	6.00	6.05	**5.95**	6.00	5.90	6.00	5.90	5.90
JMH	32	F	**5.75**	5.80	5.95	**5.80**	5.90	5.75	5.80	5.85	5.85
CAP	33	F	**8.45**	8.60	8.50	**8.50**	8.55	8.45	8.55	8.65	8.50
PGG	20	F	**8.80**	9.00	8.65	**9.00**	8.90	8.85	8.90	8.60	8.75
RCH	23	M	**5.10**	5.22	5.20	**5.25**	5.15	5.20	5.20	5.05	5.05
TRS	27	M	**5.65**	5.70	5.70	**5.75**	5.70	5.70	5.80	5.60	5.60
SSS	24	F	**6.25**	6.25	6.25	**6.25**	6.20	6.15	6.30	6.25	6.25
CAH	23	F	**8.10**	8.00	7.85	**7.70**	7.80	7.90	7.85	7.85	7.80
LEM	59	M	**7.30**	7.30	7.30	**7.10**	7.25	7.20	7.30	7.25	7.35
=============	====										
AVERAGE	33.1		**6.69**	**6.75**	**6.71**	**6.73**	**6.75**	**6.71**	**6.76**	**6.68**	**6.66**
WBC(SUM)			66.9	67.5	67.1	67.3	67.5	67.1	67.6	66.8	66.6
WBCSQ(SUM)			463.4	472.6	464.5	467.7	470.6	465.2	472.0	461.0	459.4
NONSENS/SENS			**P<.488**	**P<.35**	**P<.05**	**P<.0005**	**P<.005**	**P<.01**	**P<.05**	**P<.20**	**P<.45**
NUMBER (N)	10.0		10.00	10.00	10.00	10.00	10.00	10.00	10.00	10.00	10.00
SQRT OF N	3.2		3.2	3.2	3.2	3.2	3.2	3.2	3.2	3.2	3.2
MAXIMUM	59.0		8.80	9.00	8.65	9.00	8.90	8.85	8.90	8.65	8.75
MINIMUM	20.0		5.10	5.22	5.20	5.25	5.15	5.20	5.20	5.05	5.05
RANGE	39.0		3.7	3.8	3.5	3.8	3.8	3.7	3.7	3.6	3.7
STD ERROR (SEM)	3.9		0.37	0.38	0.35	0.38	0.38	0.37	0.37	0.36	0.37
SEM %	11.8		5.53	5.60	5.14	5.57	5.56	5.44	5.47	5.39	5.56
STD DEV (SD)	12.3		**1.17**	**1.20**	**1.09**	**1.19**	**1.19**	**1.15**	**1.17**	**1.14**	**1.17**
SD % COEF VAR	37.3		**17.5**	**17.7**	**16.3**	**17.6**	**17.6**	**17.2**	**17.3**	**17.1**	**17.6**
EXPECTED -			**4.3**	**4.4**	**4.5**	**4.4**	**4.4**	**4.4**	**4.4**	**4.4**	**4.3**
RANGE (2SD)			**9.0**	**9.1**	**8.9**	**9.1**	**9.1**	**9.0**	**9.1**	**9.0**	**9.0**

When a statistical comparison (Student t-test) was made between the 0.0-time nonsensitizing food controls (Table 02.05) and the 0.0-time sensitizing food controls (Table 02.06) no significant differences were seen even though the tests on individuals were done on successive days. The statistical comparisons between ten nonsensitized and the same ten sensitized subjects are reported in Table 02.05 above with the results labeled as (NONSENS/SENS). No differences were seen at the 0.5, 3.5 and 4.0 testing intervals, however the increased WBC (leukocytosis) response seen at hours 1.0 - 3.0 (Table 02.06) proved to be highly significant. When a statistical comparison was made between the 0.0-hour controls and 0.5 - 4.0-hour average WBC results for ten sensitizing foods challenged subjects' significant differences were seen. The statistical comparisons are reported in Table 02.06 below with the averaged WBC levels and labeled (CTL/SENS). For example, following a 10-minute pulse of food the average WBC at 30 minutes shifted from a 0.0-time control value of 6.66 to 7.07 a difference that proved not significant (P<0.30). The difference between the 0.5-hour nonsensitizing WBC of 6.75 compared to the 0.5-hour sensitizing WBC of 7.07 was likewise not significant (P<0.35). This two-way comparison proved to be a significant difference (P<0.05) at 1.0 hour as were similar two-way comparison at the 1.5 - 3.0-time intervals (Tables 2.5 and 2.6).

TABLE 02.06 WBC(X1000) COMPARISONS AT HOURS AFTER EXPOSURE TO SENSITIZING FOODS.

CASE	AGE	SEX	CTL/WBC 0.0	WBC 0.5	WBC 1.0	WBC 1.5	WBC 2.0	WBC 2.5	WBC 3.0	WBC 3.5	WBC 4.0
SGP	35	M	**5.55**	5.85	6.30	**6.35**	6.50	7.80	7.00	6.35	6.25
GAU	55	M	**5.85**	5.95	6.25	**6.75**	7.00	7.05	8.65	6.85	6.00
JMH	32	F	**5.75**	5.80	6.20	**7.40**	8.45	8.10	8.00	7.30	5.80
CAP	33	F	**8.85**	8.95	9.45	**10.85**	10.20	9.60	8.75	8.50	8.40
PGG	20	F	**8.65**	9.00	9.65	**10.20**	12.15	12.00	11.10	9.65	8.70
RCH	23	M	**5.05**	5.67	6.80	**7.92**	6.82	5.80	5.53	5.32	5.13
TRS	27	M	**5.60**	5.95	7.40	**8.15**	8.47	8.70	7.50	6.70	6.20
SSS	24	F	**6.35**	7.30	8.35	**9.20**	8.45	7.40	6.55	6.30	6.15
CAH	23	F	**7.70**	8.30	9.10	**10.50**	11.40	9.70	8.50	8.00	7.70
LEM	59	M	**7.25**	7.95	8.60	**9.40**	9.80	8.95	8.05	7.45	7.05
=============	====										
AVERAGE	33.1		**6.66**	**7.07**	**7.81**	**8.67**	**8.92**	**8.51**	**7.96**	**7.24**	**6.74**
WBC(SUM)			66.6	70.7	78.1	86.7	89.2	85.1	79.6	72.4	67.4
WBCSQ(SUM)			460.3	517.3	627.1	774.8	830.0	750.6	654.4	538.3	466.5
CTL/SENS				**P<.30**	**P<.05**	**P<.0005**	**P<.005**	**P<.01**	**P<.05**	**P<.20**	**P<.45**
NUMBER (N)	10.0		10.00	10.00	10.00	10.00	10.00	10.00	10.00	10.00	10.00
SQRT OF N	3.2		3.2	3.2	3.2	3.2	3.2	3.2	3.2	3.2	3.2
MAXIMUM	59.0		8.85	9.00	9.65	10.85	12.15	12.00	11.10	9.65	8.70
MINIMUM	20.0		5.05	5.67	6.20	6.35	6.50	5.80	5.53	5.32	5.13
RANGE	39.0		3.8	3.3	3.5	4.5	5.7	6.2	5.6	4.3	3.6
STD ERROR (SEM)	3.9		0.38	0.33	0.35	0.45	0.57	0.62	0.56	0.43	0.36
SEM %	11.8		5.71	4.71	4.42	5.19	6.33	7.29	6.99	5.98	5.30
STD DEV (SD)	12.3		**1.20**	**1.05**	**1.09**	**1.42**	**1.79**	**1.96**	**1.76**	**1.37**	**1.13**
SD % COEF VAR	37.3		**18.0**	**14.9**	**14.0**	**16.4**	**20.0**	**23.0**	**22.1**	**18.9**	**16.8**
EXPECTED -			**4.3**	**5.0**	**5.6**	**5.8**	**5.4**	**4.6**	**4.4**	**4.5**	**4.5**
RANGE (2SD)			**9.1**	**9.2**	**10.0**	**11.5**	**12.5**	**12.4**	**11.5**	**10.0**	**9.0**

By comparison, follow the average WBC across the test period. Now go back, match the WBC values with their standard deviations and coefficient of variations, and note the pattern of this variation as they relate to the time intervals. Then finally look at the expected ranges across the test period remembering that a ten-minute exposure to normal amounts of "allergenic" food triggered some changes that appear to be of "subclinical" consequence based on the published WBC "normal ranges".

The WBC may be the most commonly ordered test in the history of medicine. The current WBC normal range is probably the sloppiest in all of Clinical Chemistry and seriously needs to be addressed. At the very least, a six-hour fasting specimen requirement that includes all drinks (except water), and tobaccos should be required. In the following chapter, we will look more closely at single sensitizing food.

REFERENCES

Bessman MD and David J: Automated blood counts and differentials: A practical guide, White Cells, Chapter 4 pgs. 84-132, The Johns Hopkins University Press Baltimore & London, 1986.

Black AP: A New Diagnostic method in allergy disease. Pediat 17:716-724, 1956.

Bochner BS: Update on cells and cytokines cellular adhesion and its antagonism. J. Allergy Clin Immunol 100:581-5, 1997.

Bryan WTK and Bryan M: Cytotoxic Reactions in the diagnosis of food allergy. Otolarnyngol Clinics of N Am 4:523-533, 1971.

Bryan WTK and Bryan M: Cytotoxic Reactions in the diagnosis of food allergy. Laryngoscope 79:1453-1472, 1969.

Bryan WTK and Bryan M: The application of in vitro cytotoxic reactions to clinical diagnosis of food allergy. Laryngoscope 70:810-824, 1960.

Freri M: Corticosteroid effects on cytokines and chemokines. Allergy and Asthma Proc 20:147-159, 1999.

Freri M: The role of cytokines in asthma. Asthma Management 2:25-31, 1993.

Koro O, Furutani K, Hide M, Yamada S and Yamamoto S: Chemical mediators in atopic dermatitis: Involvement of leukotrines B4 released by a type I allergic reaction in the pathogenesis of atopic dermatitis. J Allergy Clin Immunol 103:663-70, 1999.

Sabin FR, Cunningham RS, Doan CA, and Kindwall JA: The normal rhythm of the white blood cells. Bull Johns Hopkins Hosp 37:14-67, 1925.

Ulett GA, and SG Perry: Cytotoxic testing and leukocyte increase as an index to food sensitivity. II. Coffee and Tobacco. Annals Allergy 34:150-160, 1975.

Ulett GA, and SG Perry: Cytotoxic testing and leukocyte increase as an index to food sensitivity. Annals Allergy 33:23-32, 1974.

Vaughan WT: Practice of Allergy. St. Louis MO CV Mosby Company, 1939.

Vaughan WT: Theory concerning mechanism and significance of allergic response. J Lab & Clin Med 21:629-49, 1936a.

Vaughan WT: Index as a diagnostic method in the study of food allergy. With a discussion of its reliability. J Lab & Clin Med 21:1278-88, 1936b.

Vaughan WT: Food allergens III. The Leucopenic Index, Preliminary report. J Allergy 5:601-5: 1934a.

Vaughan WT: Further studies on the Leucoxenic Index in food allergy. J Allergy 6:78-85, 1934b.

NOTES:

CHAPTER 3

Single Sensitizing Foods

FOODS WERE ONE of the "Stepping-Stones in Allergy" for Unger and Harris (1974) as they reviewed the history, contributors, and progress in the field of Allergy in a series of publications for Annals of Allergy. Anderson (1994) used "Milestones" to give his perspective to advances made in the understanding of adverse reactions to foods. Textbooks by Metcalfe et al (1991) and Allergy Principals and Practice (1988; 1993) have become often cited references that chronicle the progress of Allergy as a clinical science and the role that "allergenic" foods play in this still emerging medical discipline. The European perspective on the immunological, chemical, and clinical problems of food allergy was discussed by Orttolani and Clzedik-Eysenberg (1998). The publications of Sampson et al (1989, 1988, 1984 and 1983) are examples where a food allergen and related reactions have been directly linked to allergic atopic dermatitis meeting strict standards for clinical studies and diagnostic procedures.

Although it is not a one step process, or even a simple process, investigators continue the search to identify offending foods and the paths they follow or influence to exact biological or clinical effect(s). A single food is not a single antigen and especially coffee is not only caffeine and tobacco are not just nicotine. Each food, assuming that they are pure and uncontaminated, potentially contains multiple antigens. Aside from the strictly pharmacological effects of the various components in coffee, tobacco, and some foods in general, there are probably many little known and very specific food effects that hold some really important secrets about our health and well-being. Building on our initial discoveries (Ulett and Perry, 1974; 1975) we continue here to detail and try to understand a very specific cytotoxic test positive food initiated *in vivo* WBC response phenomenon.

Leukocytosis and Dose Response: Up to this point we had tested combinations of "allergenic" and "nonallergenic" foods and the next step was to look at single foods. We went back to our initial studies to identify individual foods and test for response to measured amounts (varying doses) of these foods. The original protocol was modified to permit quantitative amounts of chicken, chocolate, and peanut to be consumed in a randomized order for each series (Figure 3.1). In Figure 3.1a we show the data for two identical quantity and composition "allergenic" meals of chicken, chocolate, peanut, and tea detailed in Chapter 2 (Figure 2.1). As discussed earlier the difference in WBC response could be due in large part to duration of the meal consumption. Limiting the meal to 10 minutes resulted in an increased WBC response that occurred earlier in time. Figure 3.1b graphs the WBC response to varying amounts of chicken. Chocolate (figure 3.1c) and peanut (Figure 3.1d) are comparable corroborating graphs clearly displaying that varying amounts of single cytotoxic test positive foods elicit a dose related WBC *in vivo* response. The data supporting figure 3.1 are tabulated and analyzed in detail in tables 3.1 - 3.3. Incredibly, when these results were analyzed statistically (Student t Test), even as a very small sample, the *in vivo* WBC response tested as highly significantly different, compared to the controls, in spite of a worst-case analysis of averaging all the varying amounts of each single sensitizing food consumed over the days and time periods studied!

The chicken WBC response shown in Figure 3.1b was derived as a per cent of the 0.0-time control vs. the 0.5 - 4.0 test interval data reported in Table 3.1 and illustrates clear differences directly related to the amount of cytotoxic test positive food consumed. Note also that the amount of food taken on any given day in the series was not in direct ascending or descending order ruling out progressive effects that might bias the results. The 25 - 270 grams cover realistic amounts of food that could be consumed in a meal. When the average response for this individual was calculated the 0.5 - 2.5 interval WBC response increases proved to be significantly (95% confidence level or better/Student t-Test) different when compared to 0.0-time control WBC levels. Especially note that at 1.5 hours the average WBC response of 7.64 +/-0.89 SD (11.7% CV) had an expected range of 5.9 - 9.4 in this sample. At 1.5 hours these were statistically significant (99.95% confidence level) different from the ranges at the beginning or end of this test period as compared to the 0.0-time control.

TABLE 03.01 CHICKEN (SKINLESS BREAST) DOSE RESPONSE WITH TEMPORAL WBC ANALYSIS.

ANALYTE	CHICKEN	CTL/WBC	WBC	WBC	WBC	WBC	WBC	WBC	WBC	WBC
UNITS	GMS	$x10^3mm^3$	$x10^3mm^3$	$x10^3mm^3$	$x10^3mm^3$	$x10^3mm^3$	$x10^3mm^3$	$x10^3mm^3$	$x10^3mm^3$	$x10^3mm^3$
HOURS		**0.0**	0.5	1.0	1.5	2.0	2.5	3.0	3.5	4.0
CASE/AGE/SEX										
SGP/35/M										
(7/5)(12:30PM)	25.0	**5.15**	5.25	5.45	7.70	5.40	5.10	5.15		
(7/5)(8:30PM)	50.0	**5.15**	5.65	5.95	6.45	5.90	5.50	5.20	5.25	5.15
(7/6)(8:25AM)	75.0	**5.25**	5.70	6.30	7.20	6.55	6.00	5.45	5.15	5.10
(7/2)(9:20AM)	139.0	**5.35**	5.85	7.20	8.40	7.60	6.56	5.75	5.40	5.30
(7/3)(8:30AM)	270.0	**5.30**	5.95	6.95	8.45	7.80	6.85	6.00	5.40	5.30
======										
AVERAGE		**5.24**	**5.68**	**6.37**	**7.64**	**6.65**	**6.00**	**5.51**	**5.30**	**5.21**
WBC(SUM)		26.2	28.4	31.9	38.2	33.3	30.0	27.6	21.2	20.9
WBCSQ(SUM)		137.3	161.6	204.9	294.7	225.5	182.2	152.3	112.4	108.7
CTL0.0/0.5-4.0			**P<.005**	**P<.0025**	**P<.0005**	**P<0.01**	**P<0.05**	**P<0.10**	**P<0.45**	**P<0.45**
NUMBER (N)		5.00	5.00	5.00	5.00	5.00	5.00	5.00	4.00	4.00
SQRT OF N		2.2	2.2	2.2	2.2	2.2	2.2	2.2	2.0	2.0
MAXIMUM		5.35	5.95	7.20	8.45	7.80	6.85	6.00	5.40	5.30
MINIMUM		5.15	5.25	5.45	6.45	5.40	5.10	5.15	5.15	5.10
RANGE		0.2	0.7	1.8	2.0	2.4	1.8	0.9	0.3	0.2
STD ERROR(SEM)		0.04	0.14	0.35	0.40	0.48	0.35	0.17	0.06	0.05
SEM %		0.76	2.46	5.49	5.24	7.22	5.83	3.09	1.18	0.96
STD DEV (SD)		**0.09**	**0.31**	**0.78**	**0.89**	**1.07**	**0.78**	**0.38**	**0.13**	**0.10**
SD % COEF VAR		**1.7**	**5.5**	**12.3**	**11.7**	**16.1**	**13.0**	**6.9**	**2.4**	**1.9**
VARIANCE		0.01	0.06	0.41	0.57	0.87	0.42	0.11	0.01	0.01
EXPECTED -		5.1	5.1	4.8	5.9	4.5	4.4	4.7	5.1	5.0
RANGE (2SD)		5.4	6.3	7.9	9.4	8.8	7.6	6.3	5.6	5.4

The chocolate WBC maximum response (Figure 03.01c) was almost double that of its control level for the 142.9 grams (4 Hershey's Bar) amount. As with chicken the amounts consumed are in the range of amounts that might realistically be consumed. The calculated average WBC response closely matches the one Hershey's Bar (35.7 grams) response and proved to be significantly different (Table 3.2) from its 0.0-time control at the 0.5 - 4.0 intervals. At 1.5 hours the average WBC response was 8.44 +/- 1.57 SD (19.9% CV) with an expected range of 5.3 - 11.6 that proved significantly different at the 99.75 confidence level compared to the 0.0-time control. Although this is a complex food with many potential antigens it is also a fairly refined food concentrating these antigens and showing how relatively small amounts of sensitizing food can elicit a clear WBC response. Even single foods present a complex result leaving many questions to be answered, however at this point in our testing we were very pleased to see many questions being answered and confirmed. Consider that a person Cytotoxic Test negative for chocolate would have tested negative for all concentrations of chocolate. For those Cytotoxic Test positive (~50+%, Ulett, and Perry, 1975) what has been put in motion and what are the effects. Which component(s) in chocolate are responsible? This is where the fun and reward begin.

TABLE 03.02 CHOCOLATE (HERSHEY'S, 35.7 GMS) DOSE RESPONSE WITH TEMPORAL WBC ANALYSIS.

ANALYTE	CHOCOLATE	CTL/WBC	WBC	WBC	WBC	WBC	WBC	WBC	WBC	WBC
UNITS	GMS	x10³mm³	x10³mm³	x10³mm³	x10³mm³	x10³mm³	x10³mm³	x10³mm³	x10³mm³	x10³mm³
HOURS		**0.0**	0.5	1.0	1.5	2.0	2.5	3.0	3.5	4.0
CASE/AGE/SEX										
SGP/35/M										
(6/27)(8:50AM)	8.9	**5.30**	5.50	5.80	6.45	5.55	5.25	5.25	5.15	5.20
(6/26)(8:30AM)	17.9	**5.43**	5.88	6.53	7.30	6.48	5.75	5.45	5.28	5.42
(6/28)(8:20AM)	35.7	**5.25**	5.90	7.10	8.75	7.15	6.25	5.85	5.40	5.25
(6/25)(8:30AM)	71.4	**5.35**	6.83	8.15	9.70	7.97	6.82	6.22	5.92	5.65
(6/29)(8:30AM)	142.9	**5.20**	6.85	7.80	10.00	8.75	7.25	6.05	5.25	5.25
======										
AVERAGE		**5.31**	**6.19**	**7.08**	**8.44**	**7.18**	**6.26**	**5.76**	**5.40**	**5.35**
WBC(SUM)		26.5	31.0	35.4	42.2	35.9	31.3	28.8	27.0	26.8
WBCSQ(SUM)		140.8	193.2	254.0	365.5	264.0	198.8	166.8	146.2	143.5
CTL0.0/0.5-4.0			**P<.005**	**P<.0025**	**P<.0025**	**P<.010**	**P<.025**	**P<.050**	**P<.350**	**P<.400**
NUMBER (N)		5.00	5.00	5.00	5.00	5.00	5.00	5.00	5.00	5.00
SQRT OF N		2.2	2.2	2.2	2.2	2.2	2.2	2.2	2.2	2.2
MAXIMUM		5.43	6.85	8.15	10.00	8.75	7.25	6.22	5.92	5.65
MINIMUM		5.20	5.50	5.80	6.45	5.55	5.25	5.25	5.15	5.20
RANGE		0.2	1.4	2.4	3.6	3.2	2.0	1.0	0.8	0.5
STDERROR(SEM)		0.05	0.27	0.47	0.71	0.64	0.40	0.19	0.15	0.09
SEM %		0.87	4.36	6.64	8.41	8.91	6.39	3.37	2.85	1.68
STD DEV(SD)		**0.10**	**0.60**	**1.05**	**1.59**	**1.43**	**0.89**	**0.43**	**0.34**	**0.20**
SD % COEF VAR		**1.9**	**9.8**	**14.9**	**18.8**	**19.9**	**14.3**	**7.5**	**6.4**	**3.8**
VARIANCE		0.01	0.30	0.72	1.88	1.25	0.51	0.13	0.07	0.03
EXPECTED-		5.1	5.0	5.0	5.3	4.3	4.5	4.9	4.7	5.0
RANGE(2SD)		5.5	7.4	9.2	11.6	10.0	8.1	6.6	6.1	5.8

The peanut WBC response (Figure 03.01d) followed the pattern of chicken and chocolate with the average WBC response for the 0.5 - 2.5-hour intervals proving to be significantly different from their respective controls at no less than the 95% confidence level (Table 03.03). All three of these foods are high on most lists of offending foods. Peanut has been investigated extensively as a proven true allergenic food with a well-documented history of being linked to known clinical allergies and to allergic anaphylaxis (Sampson, 1998, 1990; Sampson et al, 1996, 1992; Keating et al, 1990). Peanut has also been fractionated with very specific antigens being identified and characterized (Stanley et al, 1996; Burks et al, 1996; and Bannon et al, 1996). Here we have some of the many pieces of a puzzle laid out in a logical pattern in position to be linked together in at least one of the possible peanut allergy scenarios.

Does the WBC response play a role in this scenario? With what has been presented to this point and what you will read in subsequent chapters about the familial relationships of cytotoxic test positive and WBC response specificity seems to reveal a possible lymphocyte component and a genetic element of the allergic response.

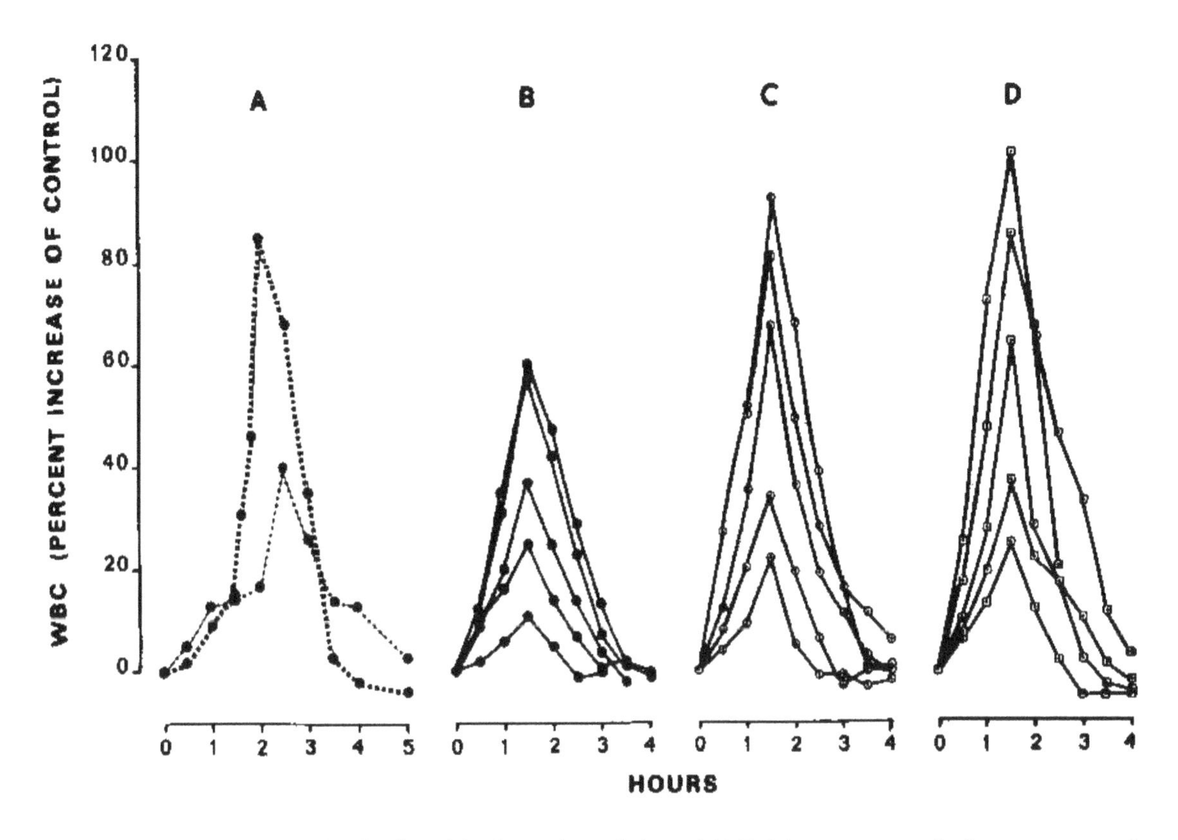

Figure 3.1 Sensitizing single food induced peripheral WBC increases relative to controls in the same individual under various test conditions; a) Initial response to a sensitizing meal of chicken, chocolate, peanut, and tea (dotted line) and a second response the same meal eight days later (hatched line); b) a dose response series testing 50, 75, 100, and 270 grams of chicken breast (solid circles); c) a dose-response series testing 9.0, 17.9, 35.7, 71.4, and 142.9 grams of milk chocolate (open circles); and d) a dose-response series testing 50, 75, 100, 150, and 200 grams of peanut (open squares). Nonsensitizing foods (orange, egg, and cherry in 50-, 100-, and 200-gram amounts) not shown in this figure, for this individual, were run as controls and did not vary from the 0.0-time control values.

The WBC dose response series was reduced to single graphic display (Figure 03.02) first published by Ulett and Perry (1974). When the WBC response maxima (Figure 03.01) were plotted against the log of the amount of food consumed a classic sigmoid dose response was revealed. Distinctly different shaped curves for each food tested became apparent.

Curves displaced up and to the left would reflect a greater sensitivity and could prove a useful way of indexing shifts with time or treatment within an individual or for comparisons between individuals of known clinical status.

In pharmacological terms an LD50 (lethal dose for 50% of the population), TD50 (toxic dose for 50% of the population), or ED50 (effective dose for 50% of the population) could be extrapolated from the sigmoid midpoint, or have we discovered a potential AD50 (allergic dose 50% for an individual)?

TABLE 03.03 PEANUT DOSE RESPONSES WITH TEMPORAL WBC ANALYSIS.

ANALYTE	PEANUT	CTL/WBC	WBC	WBC	WBC	WBC	WBC	WBC	WBC	WBC
UNITS	GMS	$x10^3mm^3$	$x10^3mm^3$	$x10^3mm^3$	$x10^3mm^3$	$x10^3mm^3$	$x10^3mm^3$	$x10^3mm^3$	$x10^3mm^3$	$x10^3mm^3$
HOURS		**0.0**	0.5	1.0	1.5	2.0	2.5	3.0	3.5	4.0
CASE/AGE/SEX										
SGP/35/M										
(7/11)(8:15AM)	50.0	**5.40**	5.70	6.10	6.75	6.05	5.50	5.15	5.10	5.15
(7/9)(8:30AM)	75.0	**5.40**	5.75	6.45	7.45	6.60	6.30	5.50	5.25	5.15
(7/10)(8:15AM)	100.0	**5.20**	5.75	6.65	8.55	6.65	6.10	5.70	5.25	5.10
(7/11)(12:35PM)	150.0	**5.15**	6.00	7.55	9.55	8.85	6.20			
(7/12)(8:30AM)	200.0	**5.15**	6.45	8.85	10.35	8.50	7.50	6.85	5.70	5.30
======										
AVERAGE		**5.26**	**5.93**	**7.12**	**8.53**	**7.33**	**6.32**	**5.80**	**5.33**	**5.18**
WBC(SUM)		26.3	29.7	35.6	42.7	36.7	31.6	23.2	21.3	20.7
WBCSQ(SUM)		138.4	176.2	258.4	372.5	275.0	201.8	136.2	113.6	107.1
CTL 0.0/0.5-4.0			**P<.0025**	**P<.005**	**P<.0005**	**P<.0025**	**P<0.01**	**P<0.10**	**P<0.30**	**P<0.25**
NUMBER (N)		5.00	5.00	5.00	5.00	5.00	5.00	4.00	4.00	4.00
SQRT OF N		2.2	2.2	2.2	2.2	2.2	2.2	2.0	2.0	2.0
MAXIMUM		5.40	6.45	8.85	10.35	8.85	7.50	6.85	5.70	5.30
MINIMUM		5.15	5.70	6.10	6.75	6.05	5.50	5.15	5.10	5.10
RANGE		0.3	0.8	2.8	3.6	2.8	2.0	1.7	0.6	0.2
STD ERROR (SEM)		0.05	0.15	0.55	0.72	0.56	0.40	0.43	0.15	0.05
SEM %		0.95	2.53	7.72	8.44	7.64	6.33	7.33	2.82	0.97
STD DEV (SD)		**0.11**	**0.34**	**1.23**	**1.61**	**1.25**	**0.89**	**0.85**	**0.30**	**0.10**
SD % COEF VAR		**2.1**	**5.7**	**17.3**	**18.9**	**17.1**	**14.2**	**14.7**	**5.6**	**1.9**
VARIANCE		0.01	0.08	0.98	1.74	1.26	0.43	0.41	0.05	0.01
EXPECTED -		5.0	5.3	4.7	5.3	4.8	4.5	4.1	4.7	5.0
RANGE (2SD)		5.5	6.6	9.6	11.7	9.8	8.1	7.5	5.9	5.4

Distinctly shaped curves for the three nonsensitizing foods (orange, egg, and cherry) were also evident in figure 3.2 and are recorded in detail in tables 3.4 – 3.6 below. Not only do we have a zero-time control for reference, but we also have a dynamic real time control for reference. It is these nonsensitizing foods with an AD50 of zero that provide the basis of diet management plans that Dr. Ulett and his wife Dr. Ulett used with considerable success for their patients that presented themselves for a wide range of clinical complaints. We will report on a number of these cases in Chapter 5.0. Interesting, patients with neurological/psychological complaints received relief and also enjoyed secondary benefits in the form of significant weight loss and loss of facial acne.

Please visualize with me, in quantitative terms, using the WBC and differential a widely used and understood clinical test what is happening here. We are seeing an *in vivo* WBC response on its own merit, that happens to correlate separate and distinctly with the results of *in vitro* Cytotoxic Testing. Clearly when test positive foods are eliminated the patient benefited.

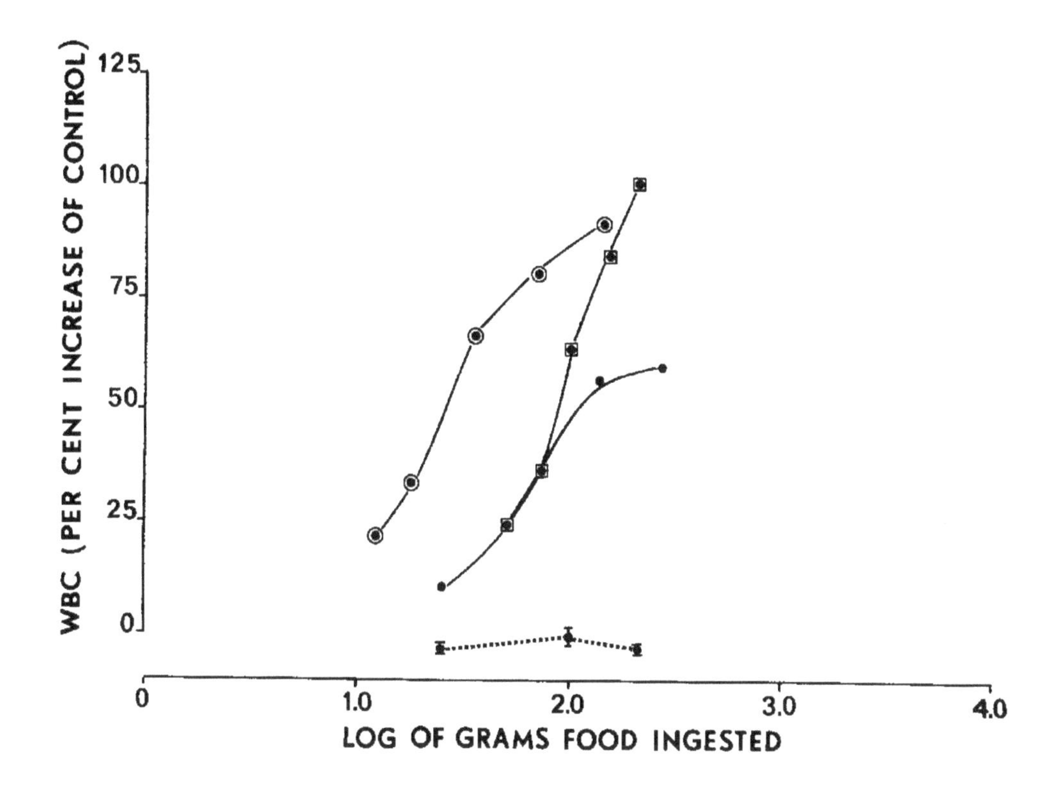

Figure 3.2 Dose response curves for single sensitizing foods: chicken (small solid circles), chocolate (open circles) and peanut (open squares), with three Nonsensitizing foods (orange, egg, and cherry) as controls plotted (hatched lines) as an average for the group +/- S.E.M. All experiments were done on the same prefasted subject (S.G.P.) who was previously evaluated for food sensitivities using cytotoxic testing. The response was measured at leukocytosis maxima, relative to an initial control level, that occurred at 1.5 hours for the foods tested.

WBC Response and Other Control Parameters: Although there are good controls for the positive WBC, RBCs and pulse rates provided additional clinically established parameters that were measured and tracked parallel to the WBC.

The chicken dose response challenge (Table 03.07) data did not reveal any significant patterns for elements of the differential or for the pulse rate that was comparable to the temporal pattern of the WBC response

The chocolate dose response challenge (Table 03.08) data did not reveal any significant pattern for the differential or for the pulse rate corroborating the picture seen for chicken.

While the WBC response was manifest these related parameters were stable and unremarkable relative to the changes seen for the WBC

It should be noted that the fluctuation in the WBC is completely within the "currently accepted" normal range for the "not required to be fasted" WBC.

If the specimen requirement for the WBC (Differential) required fasting, then the higher WBC counts should be raising some flags that should be accounted for. This has not happened for all too long

TABLE 03.04 ORANGE DOSE RESPONSE WITH ANALYSIS OF TEMPORAL WBC PATTERNS.

ANALYTE	ORANGE	CTL/WBC	WBC	WBC	WBC	WBC	WBC	WBC
UNITS	GMS	x10^{3}mm^3	x10^{3}mm^3	x10^{3}mm^3	x10^{3}mm^3	x10^{3}mm^3	x10^{3}mm^3	x10^{3}mm^3
HOURS		**0.0**	0.5	1.0	1.5	2.0	2.5	3.0
CASE/AGE/SEX								
SGP/35/M								
(7/25)(11:35AM)	50.0	**6.30**	6.25	6.15	6.35	6.25	6.05	6.30
(7/25)(8:35AM)	100.0	**6.30**	6.20	6.30	6.25	6.25		
(7/24)(11:55AM)	200.0	**6.90**	6.85	6.85	6.80	6.85	6.80	6.85
======								
AVERAGE		**6.50**	**6.43**	**6.43**	**6.47**	**6.45**	**6.43**	**6.58**
WBC(SUM)		19.5	19.3	19.3	19.4	19.4	12.9	13.2
WBCSQ(SUM)		127.0	124.4	124.4	125.6	125.0	82.8	86.6
NUMBER (N)		3.00	3.00	3.00	3.00	3.00	2.00	2.00
SQRT OF N		1.7	1.7	1.7	1.7	1.7	1.4	1.4
MAXIMUM		6.90	6.85	6.85	6.80	6.85	6.80	6.85
MINIMUM		6.30	6.20	6.15	6.25	6.25	6.05	6.30
RANGE		0.6	0.6	0.7	0.6	0.6	0.8	0.6
STD ERROR (SEM)		0.20	0.22	0.23	0.18	0.20	0.38	0.28
SEM %		3.08	3.37	3.63	2.84	3.10	5.84	4.18
STD DEV (SD)		**0.35**	**0.38**	**0.40**	**0.32**	**0.35**	**0.53**	**0.39**
SD % COEF VAR		**5.3**	**5.8**	**6.3**	**4.9**	**5.4**	**8.3**	**5.9**
EXPECTED -		5.8	5.7	5.6	5.8	5.8	5.4	5.8
RANGE (2SD)		7.2	7.2	7.2	7.1	7.1	7.5	7.4

TABLE 03.05 EGG (BOILED) DOSE RESPONSE WITH ANALYSIS OF TEMPORAL WBC PATTERNS.

ANALYTE	EGG	CTL/WBC	WBC	WBC	WBC	WBC	WBC	WBC
UNITS	GMS	x10³mm³	x10³mm³	x10³mm³	x10³mm³	x10³mm³	x10³mm³	x10³mm³
HOURS		**0.0**	0.5	1.0	1.5	2.0	2.5	3.0
CASE/AGE/SEX								
SGP/35/M								
(7/24)(8:55AM)	50.0	**6.90**	6.80		6.85	6.90	6.85	6.90
(7/22)(9:05AM)	100.0	**9.40**	9.30	9.45	9.35	9.30	9.35	9.30
(7/17)(11:56AM)	200.0	**8.30**	8.25	8.30	8.25	8.20	8.25	8.15
======								
AVERAGE		**8.20**	**8.12**	**8.88**	**8.15**	**8.13**	**8.15**	**8.12**
WBC(SUM)		24.6	24.4	17.8	24.5	24.4	24.5	24.4
WBCSQ(SUM)		204.9	200.8	158.2	202.4	201.3	202.4	200.5
NUMBER (N)		3.00	3.00	2.00	3.00	3.00	3.00	3.00
SQRT OF N		1.7	1.7	1.4	1.7	1.7	1.7	1.7
MAXIMUM		9.40	9.30	9.45	9.35	9.30	9.35	9.30
MINIMUM		6.90	6.80	8.30	6.85	6.90	6.85	6.90
RANGE		2.5	2.5	1.2	2.5	2.4	2.5	2.4
STD ERROR (SEM)		0.83	0.83	0.57	0.83	0.80	0.83	0.80
SEM %		10.16	10.27	6.48	10.22	9.84	10.22	9.86
STD DEV (SD)		**1.44**	**1.44**	**0.81**	**1.44**	**1.39**	**1.44**	**1.39**
SD % COEF VAR		**17.6**	**17.8**	**9.2**	**17.7**	**17.0**	**17.7**	**17.1**
RANGE (2SD)		5.3-11.1	5.2-11.0	7.2-10.5	5.3-11.0	5.4-10.9	5.3-11.0	5.3-10.9

The peanuts dose response challenge (Table 3.9) data for the differential and the RBC did not reveal any significant patterns comparable to the temporal pattern seen for the WBC response. Questions were raised to us that haemo-concentration might account for the fluctuations seen in the WBC. The RBC has been very stable with little or no fluctuation in all cases where we have monitored the RBC (see tables 3.9 and 3.27 - 3.29) as a control for the WBC response.

The chicken and peanut tests also demonstrated the WBC response in same day serial (piggy backed) challenges. When the initial WBC response returned to control levels a second challenge with a different amount of cytotoxic test positive food gave a WBC response that was proportional to the initial as well as other measured food challenges in the series.

NOTES:

TABLE O3.O6 CHERRY DOSE RESPONSE WITH ANALYSIS OF TEMPOROA WBC PATTERNS

ANALYTE	CHERRY	CTL/WBC	WBC	WBC	WBC	WBC	WBC	WBC	WBC	WBC
UNITS	GMS	x10³mm³	x10³mm³	x10³mm³	x10³mm³	x10³mm³	x10³mm³	x10³mm³	x10³mm³	x10³mm³
HOURS		**0.0**	0.5	1.0	1.5	2.0	2.5	3.0	3.5	4.0
CASE/AGE/SEX										
SGP/35/M										
(7/17)(7:55AM)	50.0	**8.15**	8.25	8.10	8.00	8.25	8.15	8.30	8.40	8.30
(7/16)(12:15PM)	100.0	**7.75**	7.80	7.70	7.75	7.75	7.80	7.70		
(7/16)(9:15AM)	200.0	**7.90**	7.90	7.95	7.75	7.65	7.70	7.75		
======										
AVERAGE		**7.93**	**7.98**	**7.92**	**7.83**	**7.88**	**7.88**	**7.92**	**8.40**	**8.30**
WBC(SUM)		23.8	24.0	23.8	23.5	23.7	23.7	23.8	8.4	8.3
WBCSQ(SUM)		188.9	191.3	188.1	184.1	186.6	186.6	188.2	70.6	68.9
NUMBER (N)		3.00	3.00	3.00	3.00	3.00	3.00	3.00	1.00	1.00
SQRT OF N		1.7	1.7	1.7	1.7	1.7	1.7	1.7	1.0	1.0
MAXIMUM		8.15	8.25	8.10	8.00	8.25	8.15	8.30	8.40	8.30
MINIMUM		7.75	7.80	7.70	7.75	7.65	7.70	7.70	8.40	8.30
RANGE		0.4	0.5	0.4	0.3	0.6	0.5	0.6	0.0	0.0
STD ERROR (SEM)		0.13	0.15	0.13	0.08	0.20	0.15	0.20	0.00	0.00
SEM %		1.68	1.88	1.68	1.06	2.54	1.90	2.53	0.00	0.00
STD DEV (SD)		**0.23**	**0.26**	**0.23**	**0.14**	**0.35**	**0.26**	**0.35**	**0.00**	**0.00**
SD % COEF VAR		**2.9**	**3.3**	**2.9**	**1.8**	**4.4**	**3.3**	**4.4**	**0.0**	**0.0**
EXPECTED -		7.5	7.5	7.5	7.5	7.2	7.4	7.2	8.4	8.3
RANGE (2SD)		8.4	8.5	8.4	8.1	8.6	8.4	8.6	8.4	8.3

NOTES:

TABLE 03.07 CHICKEN DOSE RESPONSE CHALLENGE WITH ANALYSIS OF HEMATOLOGY RESULTS.

	TIME	DIFFERENTIAL						
	HOUR	**						
ANALYTE		WBC	SEG	LYMPH	MONO	EOSIN	BASO	PULSE
RANGE		4.8-10.8	58-66	20-40	2-8	2-4	0-1	
UNITS		x10³/mm³	%	%	%	%	%	RATE
CASE/AGE/SEX								
SGP/35/M	**0.0**	**5.15**	**55**	**39**	**4**	**2**		
CHICKEN	0.5	5.65	51	49				
(50 GMS)	1.0	5.95	54	44	1	1		
7/5	1.5	6.45	61	34	4	1		
8:30AM	2.0	5.90	58	40	2			
	2.5	5.50	58	40	1	1		
	3.0	5.20	51	41	2	1		
PIGGY-BACKED	3.5	5.25	57	40	3			
=======	4.0	5.15	58	38	2	2		
CHICKEN	4.5	5.25	56	38	3	3		
(25 GMS)	5.0	5.45	54	40	4	2		
7/5	5.5	5.70	58	40	1	1		
12:30PM	6.0	5.40	61	36	2	1		
	6.5	5.10	54	42	2	2		
	7.0	5.15	57	40	2	1		
EXP AVERAGE		5.51	56	40	2	1		
=======	**0.0**	**5.25**	**56**	**40**	**3**	**1**		
CHICKEN	0.5	5.70	60	38	1	1		
(75 GMS)	1.0	6.30	65	31	2	2		
7/6	1.5	7.20	58	40	1	1		
8:25AM	2.0	6.55	51	47	1	1		
	2.5	6.00	59	40		1		
	3.0	5.45	56	42	1	1		
	3.5	5.15	53	43	2	2		
	4.0	5.10	60	37	2	1		
EXP AVERAGE		5.93	58	40	1	1		
=======	**0.0**	**5.35**	**58**	**40**	**1**	**1**		**73**
CHICKEN	0.5	5.85	54	43	2	1		72
(139 GMS)	1.0	7.20	56	42	1	1		70
7/2	1.5	8.40	59	40	1			68
9:20AM	2.0	7.60	53	46	1			73
	2.5	6.56	57	40	2	1		73
	3.0	5.75	54	44	1	1		72
	3.5	5.40	56	40	2	2		
	4.0	5.30	61	35	2	2		
EXP AVERAGE		6.51	56	41	2	1		71
=======	**0.0**	**5.30**	**57**	**40**	**2**	**1**		**81**
CHICKEN	0.5	5.95	50	48	1	1		80
(270 GMS)	1.0	6.95	57	41	1	1		79
7/3	1.5	8.45	52	46	1	1		85
8:30AM	2.0	7.80	55	40	1			80
	2.5	6.85	59	40	1			81
	3.0	6.00	56	40	3	1		
	3.5	5.40	51	46	2	1		85
	4.0	5.30	56	41	2	1		86
EXP AVERAGE		6.59	55	43	2	1		82

NOTES:

TABLE 03.08 CHOCOLATE DOSE RESPONSE CHALLENGE WITH ANALYSIS OF HEMATOLOGY RESULTS.

ANALYTE	TIME HOUR	WBC	SEG	LYMPH	MONO	EOSIN	BASO	PULSE
		DIFFERENTIAL ******						
RANGE		4.8-10.8	58-66	20-40	2-8	2-4	0-1	
UNITS		$x10^3/mm^3$	%	%	%	%	%	RATE
CASE/AGE/SEX								
SGP/35/M	**0.0**	**5.30**	**57**	**40**	**2**	**1**		**76**
1/4 HERSHEY BAR	0.5	5.50	52	43	3	1	1	73
(8.93 GMS)	1.0	5.80	56	41	2	1		74
6/27	1.5	6.45	59	39	1	1		80
8:50AM	2.0	5.55	55	40	3	2		76
	2.5	5.25	60	37	2	1		69
	3.0	5.25	57	40	3			64
	3.5	5.15	51	45	2	2		74
	4.0	5.20	56	40	2	1	1	76
EXP AVERAGE		5.52	56	41	2	1	1	73
=======	**0.0**	**5.43**	**52**	**43**	**3**	**2**		**67**
1/2 HERSHEY BAR	0.5	5.88	54	45	1			75
(17.86 GMS)	1.0	6.53	67	31	1	1		73
6/26	1.5	7.30	50	47	2	1		64
8:30AM	2.0	6.48	62	35	2	1		72
	2.5	5.75	56	41	2	1		69
	3.0	5.45	50	46	2	2		65
	3.5	5.28	56	40	2	2		75
	4.0	5.42	56	40	2	1	1	68
EXP AVERAGE		6.01	56	41	2	1	1	70
=======	**0.0**	**5.25**	**51**	**46**	**2**	**1**		**77**
1 HERSHEY BAR	0.5	5.90	55	42	1	2		77
(35.72 GMS)	1.0	7.10	53	46	1			76
6/28	1.5	8.75	56	42	2			79
8:20AM	2.0	7.15	55	40	2	2	1	75
	2.5	6.25	57	41	1	1		71
	3.0	5.85	55	40	3	2		69
	3.5	5.40	57	41	1	1		77
	4.0	5.25	59	40	1			73
EXP AVERAGE		6.46	56	42	2	2	1	75
=======	**0.0**	**5.35**	**55**	**40**	**3**	**2**		**77**
2 HERSHEY BARS	0.5	6.83	58	40	1	1		
(71.44 GMS)	1.0	8.15	53	42	3	2		81
6/25	1.5	9.70	58	40	1	1		79
8:30AM	2.0	7.97	58	39	2	1		76
	2.5	6.82	57	41	1	1		81
	3.0	6.22	55	41	2	2		80
	3.5	5.92	54	40	3	3		79
	4.0	5.65	58	40	1	1		72
	5.0	5.45	54	42	2	2		81
EXP AVERAGE		6.97	56	41	2	2		79
=======	**0.0**	**5.20**	**55**	**40**	**2**	**2**	**1**	**76**
4 HERSHEY BARS	0.5	6.85	52	46	1	1		65
(142.88 GMS)	1.0	7.80	50	47	2	1		81
6/29	1.5	10.00	56	40	2	2		73
8:30AM	2.0	8.75	59	39	1	1		79
	2.5	7.25	52	48				67
	3.0	6.05	56	40	2	2		71
	3.5	5.25	53	45	1	1		72
	4.0	5.25	57	40	1	2		68
EXP AVERAGE		7.15	54	43	1	1		72

TABLE 03.09 PEANUT DOSE RESPONSE CHALLENGE WITH ANALYSIS OF HEMATOLOGY RESULTS.

	TIME							
	HOUR		DIFFERENTIAL ***					
ANALYTE		WBC	SEG	LYMPH	MONO	EOSIN	BASO	RBC
RANGE		4.8-10.8	58-66	20-40	2-8	2-4	0-1	4.7-6.1
UNITS		$x10^3/mm^3$	%	%	%	%	%	$x10^6/mm^3$
CASE/AGE/SEX								
SGP/35/M	**0.0**	**5.40**	**55**	**38**	**3**	**2**		**5.33**
PEANUT	0.5	5.70	59	43	2	1		5.32
(50 GMS)	1.0	6.10	54	38	1	2		5.39
7/11	1.5	6.75	60	40	1	1		5.41
8:15AM	2.0	6.05	57	42	2	1		5.36
	2.5	5.50	55	40	1	2		5.39
	3.0	5.15	58	40	1	1		5.38
	3.5	5.10						5.37
PIGGY-BACK	4.0	5.15	54	42	2	2		5.36
=======	0.0	5.15	58	40	2			5.38
PEANUT	0.5	6.00	61	38		1		5.36
(150 GMS)	1.0	7.55	57	40	2	1		5.40
7/11	1.5	9.55	60	36	2	2		5.39
12:35PM	2.0	8.85	55	43	1		1	5.38
	2.5	6.20	59	39	1	1		5.32
EXP AVERAGE		6.34	57	40	2	1	1	5.37
=======	**0.0**	**5.40**	**57**	**40**	**1**	**1**		
PEANUT	0.5	5.75	59	38		2		
(75 GMS)	1.0	6.45	55	42	3			
7/10	1.5	7.45	58	40	1			
8:30AM	2.0	6.60	60	38		2		
	2.5	6.30	52	45	1	1		
	3.0	5.50	56	40	2	2		
	3.5	5.25	61	37	1	1		
	4.0	5.15	57	40				
EXP AVERAGE		6.06	57	40	2	2		
=======	**0.0**	**5.20**	**55**	**42**	**2**	**1**		
PEANUT	0.5	5.75	57	40	2	1		
(100 GMS)	1.0	6.65	59	39	1	1		
7/10	1.5	8.55	60	38	2			
8:15AM	2.0	6.65	57	40	2	1		
	2.5	6.10	55	43	1	1		
	3.0	5.70	58	40	1	1		
	3.5	5.25	62	35	3			
	4.0	5.10	60	35	3	2		
EXP AVERAGE		6.22	59	39	2	1		
=======	**0.0**	**5.15**	**55**	**41**	**2**	**2**		**5.48**
PEANUT	0.5	6.45	58	41	1			5.45
(200 GMS)	1.0	8.85	56	42	2			5.43
7/12	1.5	10.35	59	38	2	1		5.48
8:30AM	2.0	8.50	60	37	2	1		5.43
	2.5	7.50	60	38	1	1		5.46
	3.0	6.85	55	43	1	1		5.42
	3.5	5.70	58	40	1	1		5.46
	4.0	5.30	54	44	2			5.44
EXP AVERAGE		7.44	58	40	2	1		5.45

Coffee as a Single Sensitizer: Coffee was used as a single sensitizer (Ulett and Perry, 1975) to further illustrate and confirm the *in vivo* WBC dose response to an *in vitro* Cytotoxic Test positive food. Increased amounts of freeze-dried coffee in a volume of 500 mls resulted in an increased WBC response (Table 03.10). The differential appeared stable and remained proportional within the WBC temporal pattern. Using coffee, we again could demonstrate sequential and proportional WBC responses by piggybacking

(Table 03.10), a second challenge test immediately following the WBC returning to control levels after the initial WBC response. When do you reach for that second cup of coffee? For the habituated it's probably different for all, yet somewhere on the descending slope of its WBC response.

When the coffee WBC response maxima (Figure 03.30b) were plotted as the log of the coffee concentration consumed versus the WBC response the linear portion of a sigmoid dose response was revealed (Figure 03.03b inset). As before with the solid foods chicken, chocolate and peanut this graphic presentation provides a classical framework to interpret the degree of sensitivity and make comparisons to other times or other subjects especially for known immediate or slow reacting clinical conditions.

TABLE 3.10 COFFEE DOSE RESPONSE CHALLENGE WITH ANALYSIS OF HEMATOLOGY RESULTS.

	TIME		DIFFERENTIAL				
	HOUR		**				
ANALYTE		WBC	SEG	LYMPH	MONO	EOSIN	BASO
RANGE		4.8-10.8	58-66	20-40	2-8	2-4	0-1
UNITS		$x10^3/mm^3$	%	%	%	%	%
CASE/AGE/SEX							
SGP/35/M	**0.0**	**7.00**	**54**	**42**	**3**	**1**	
COFFEE	0.5	8.25	52	46	1	1	
(20 MGMS/ML)	1.0	9.75	56	41	2	1	
9/24	1.5	11.05	54	44	1	1	
9:10AM	2.0	8.85	51	47	2		
	2.5	8.10	57	41	1	1	
	3.0	7.25	52	45	2	1	
	3.5	6.95	53	45	1	1	
PIGGY-BACK	4.0	6.90	50	46	2	2	
=======	4.5	7.40	54	44	2		
COFFEE	5.0	7.95	57	40	2	1	
(10 MGMS/ML)	5.5	8.65	56	41	1	2	
9/24	6.0	7.80	57	41	2		
1:10PM	6.5	7.20	52	45	2	1	
	7.0	6.90	54	44	1	1	
EXP AVERAGE		8.07	54	44	2	1	
=======	**0.0**	**7.25**	**58**	**41**	**1**	**1**	
COFFEE	0.5	11.15	53	44	2	1	
(30 MGMS/ML)	1.0	12.40	59	38	2	1	
10/3	1.5	13.70	54	44	1	1	
9:00AM	2.0	12.05	56	42	1	1	
	2.5	10.45	58	40	1	1	
	3.0	9.50	55	42	2	1	
	3.5	8.30	57	40	2	1	
	4.0	7.40	54	45	1		
	4.5					1	
	5.0	7.20	51	46	2		
EXP AVERAGE		10.24	55	42	2	1	

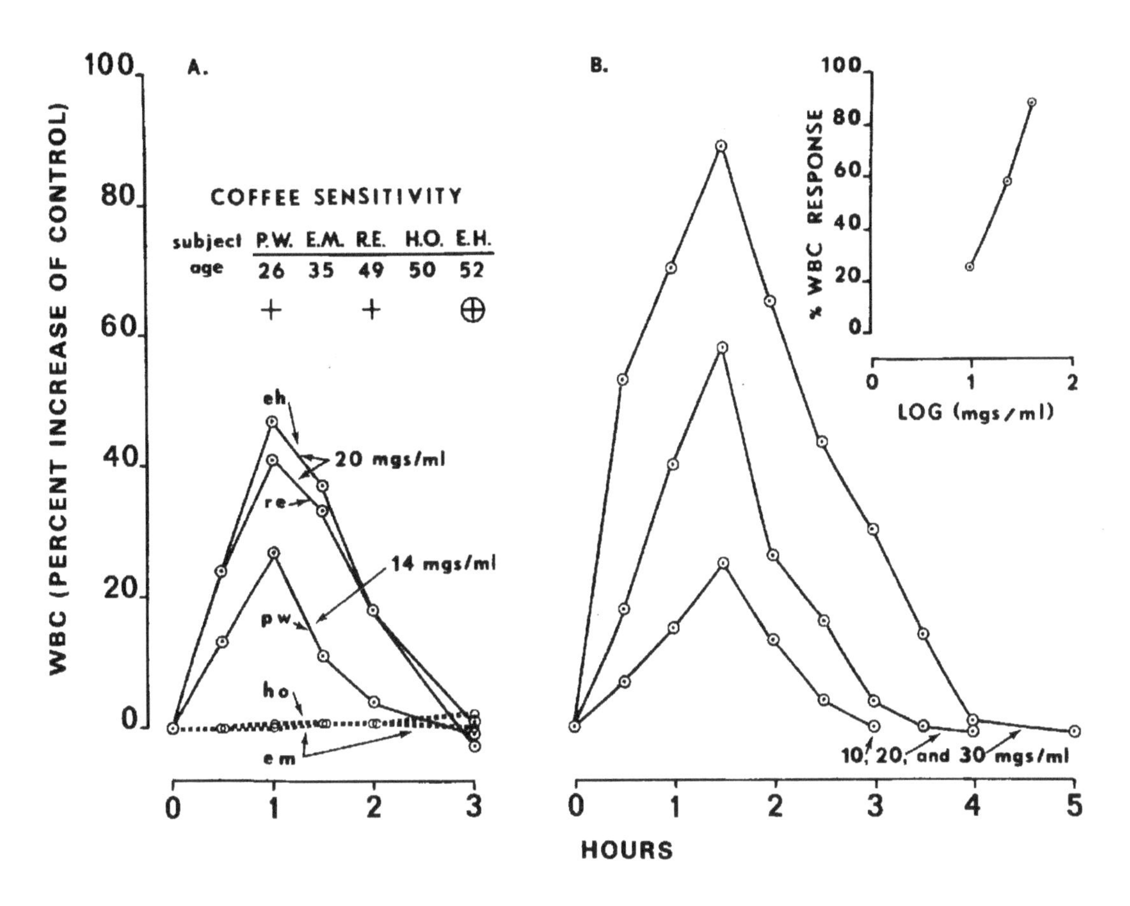

Figure 3.3 Peripheral_WBC responses to coffee: A) in vivo WBC response test and in vitro cytotoxic test results (table inset); B) coffee dose-response data indicating magnitude and time course of the WBC response, also expressed as a dose response curve (Figure inset), following a 10-minute period during which a measured amount of coffee was consumed.

Coffee *in vivo* WBC Response Blind Study: As the result of informal in-service discussions five subjects from the dietary department volunteered to be *in vitro* Cytotoxic Tested and also submit themselves to the *in vivo* WBC food challenge. There were still a few skeptics that felt just knowing that they were consuming something that they were positive for might be enough to unstabilize one's WBC that prompted a blind study. This study also provided an opportunity to repeat and corroborate the findings of our original study (Ulett and Perry, 1974).

Coffee was selected as a single sensitizing food to test for the food challenge WBC response (Tables 3.11 and 3.12, see also (Figure 03.03a). The data was gathered in a series of tests conducted prior to the *in vitro* Cytotoxic Testing (Tables 03.13 -03.18) of the five dietary department volunteers. Three of the five subjects showed positive coffee *in vivo* WBC responses (PJW, RKE and EOH) while two subjects tested negative with no coffee WBC response (EWM and HSO). The *in vivo* WBC response test results correlate exactly with the *in vitro* Cytotoxic Test results (Table 3.18). The data is also summarized in Figure 03.03a and inset as published by Ulett and Perry (1975).

TABLE 03.11 COFFEE BLIND DOSE RESPONSE CHALLENGE WITH ANALYSIS OF HEMATOLOGY RESULTS.

	TIME		DIFFERENTIAL				
	HOUR			**			
ANALYTE		WBC	SEG	LYMPH	MONO	EOSIN	BASO
RANGE		4.8-10.8	58-66	20-40	2-8	2-4	0-1
UNITS		x10³/mm³	%	%	%	%	%
CASE/AGE/SEX							
PJW/26/F	**0.0**	**7.20**	**70**	**28**	**2**		
COFFEE	0.5	8.10	72	24	2	2	
(14 MGM/ML)	1.0	9.15	69	29	1	1	
12/3	1.5	8.00	70	27	1	2	
8:55AM	2.0	7.50	73	25	1	1	
	3.0	7.15	71	26	1	2	
========							
EXP AVERAGE		7.98	71	26	1	2	
EWM/35/F	**0.0**	**6.25**	**70**	**28**	**1**	**1**	
COFFEE	0.5	6.25	68	30	1	1	
(14 MGM/ML)	1.0	6.30	71	26	2	1	
12/4	1.5	6.30	70	26	2	2	
8:55AM	2.0	6.30	72	26	1	1	
	3.0	6.25	69	30	1		
========							
EXP AVERAGE		6.28	70	28	1	1	
RKE/49/F	**0.0**	**6.80**	**60**	**38**	**1**	**1**	
COFFEE	0.5	8.40	61	36	1	2	
(20 MGM/ML)	1.0	10.00	64	34	1	1	
11/30	1.5	9.30	62	36		2	
8:37AM	2.0	8.00	60	37	1	2	
	3.0	6.60	59	38	1	2	
========							
EXP AVERAGE		8.46	61	36	1	2	
HSO/50/F	**0.0**	**6.55**	**65**	**33**	**1**	**1**	
COFFEE	0.5	6.55	63	34	1	2	
(14 MGM/ML)	1.0	6.55	60	37	1	2	
12/3	1.5	6.60	64	34	1	1	
8:50AM	2.0	6.60	66	31		3	
	3.0	6.65	64	34	1	1	
========							
EXP AVERAGE		6.59	63	34	1	2	
EOH/52/F	**0.0**	**6.60**	**55**	**45**			
COFFEE	0.5	8.15	55	44		1	
(20 MGM/ML)	1.0	9.30	53	44	1	2	
11/30	1.5	8.75	57	40	1	2	
8:35AM	2.0	7.80	55	43	1	1	
	3.0	6.65	54	45		1	
========							
EXP AVERAGE		8.13	55	43	1	1	

TABLE 03.12 COFFEE CHALLENGES (BLIND STUDY) WITH ANALYSIS OF TEMPORAL WBC PATTERNS.

ANALYTE				CTL/WBC	WBC	WBC	WBC	WBC	WBC
UNITS				$x10^3mm^3$	$x10^3mm^3$	$x10^3mm^3$	$x10^3mm^3$	$x10^3mm^3$	$x10^3mm^3$
HOURS				**0.0**	0.5	1.0	1.5	2.0	3.0
CASE	AGE	SEX	COFFEE						
COFFEE			MG/ML						
SENSITIZED									
PJW	26	F	14	**7.20**	8.10	9.15	8.00	7.50	7.15
RKE	49	F	20	**6.80**	8.40	10.00	9.30	8.00	6.60
EOH	52	F	20	**6.60**	8.15	9.30	8.75	7.80	6.65
======									
COFFEE									
NONSENSITIZED									
EWM	35	F	14	**6.25**	6.25	6.30	6.30	6.30	6.25
HSO	50	F	14	**6.55**	6.55	6.55	6.60	6.60	6.65
======									
SENSITIZED									
AVERAGE				**6.87**	**8.22**	**9.48**	**8.68**	**7.77**	**6.80**
WBC(SUM)				20.6	24.7	28.5	26.1	23.3	20.4
WBCSQ(SUM)				141.6	202.6	270.2	227.1	181.1	138.9
0 CTL/0.5-4.0					**P<.0025**	**P<.0005**	**P<.0005**	**P<.01**	**P<0.40**
NUMBER (N)				3.00	3.00	3.00	3.00	3.00	3.00
SQRT OF N				1.7	1.7	1.7	1.7	1.7	1.7
MAXIMUM				7.20	8.40	10.00	9.30	8.00	7.15
MINIMUM				6.60	8.10	9.15	8.00	7.50	6.60
RANGE				0.6	0.3	0.9	1.3	0.5	0.6
STD ERROR (SEM)				0.20	0.10	0.28	0.43	0.17	0.18
SEM %				2.91	1.22	2.99	4.99	2.15	2.70
STD DEV (SD)				0.35	0.17	0.49	0.75	0.29	0.32
SD % COEF VAR				5.0	2.1	5.2	8.6	3.7	4.7
EXPECTED -				6.2	7.9	8.5	7.2	7.2	6.2
RANGE(2SD)				7.6	8.6	10.5	10.2	8.3	7.4
NONSENSITIZED									
AVERAGE				**6.40**	**6.40**	**6.43**	**6.45**	**6.45**	**6.45**
WBC(SUM)				12.8	12.8	12.9	12.9	12.9	12.9
WBCSQ(SUM)				82.0	82.0	82.6	83.3	83.3	83.3
VS SENS 0.0-3.0				**P<0.10**	**P<.0025**	**P<.0025**	**P<.0025**	**P<0.01**	**P<0.15**
NUMBER (N)				2.00	2.00	2.00	2.00	2.00	2.00
SQRT OF N				1.4	1.4	1.4	1.4	1.4	1.4
MAXIMUM				6.55	6.55	6.55	6.60	6.60	6.65
MINIMUM				6.25	6.25	6.30	6.30	6.30	6.25
RANGE				0.3	0.3	0.3	0.3	0.3	0.4
STD ERROR (SEM)				0.15	0.15	0.13	0.15	0.15	0.20
SEM %				2.34	2.34	1.95	2.33	2.33	3.10
STD DEV (SD)				**0.21**	**0.21**	**0.18**	**0.21**	**0.21**	**0.28**
SD % COEF VAR				**3.3**	**3.3**	**2.8**	**3.3**	**3.3**	**4.4**
EXPECTED				6.0	6.0	6.1	6.0	6.0	5.9
RANGE(2SD)				6.8	6.8	6.8	6.9	6.9	7.0

Zero-time control specimens were obtained prior to all five of the subjects consuming two cups (500 mls) of a known concentration of freeze-dried coffee within ten minutes following with their WBC/ differentials determined at 30-minute intervals (Table 3.11). These results are displayed graphically in Figure 03.03a as the WBC values, after an overnight fasting period, just prior to consuming the coffee. The 0.0-time specimen results provide a valuable control and reference point for comparison while following the WBC response with time. The timed specimen results provide serial determinations to help track and validate the WBC response changes. The negative WBC response from the cytotoxic test negative specimens provides a very valuable additional control that follows the same time course as the positive WBC coffee response. As data from multiple food WBC responses described in Chapter 2 (Tables 02.05 and 02.06) revealed, those exciting findings are again revealed and corroborated here with coffee as a single sensitizing food (Table 03.11). Statistical comparisons were made between controls and the WBC response coffee challenge test intervals (Table 03.12). The comparative differences between the averaged coffee sensitized subject's 0.0-time controls with their 0.5 - 2.5 test intervals proved significant at the 99% confidence level. The 3.0 hr. test level was not significantly different from the 0.0-time control indicating the time course for the response. The differences between the 0.0 - 3.0-time interval coffee sensitized subjects and the 0.0 - 3.0 times interval coffee nonsensitized subject controls showed the same pattern of significance as when compared to the internal 0.0-time control. With only five subjects, three coffee sensitized subject's 0.5 - 2.5 test intervals proved significantly (99% confidence level) different from the comparable intervals for the control coffee nonsensitized. No significant difference between the 0.0 controls or the 3.0-time intervals could be detected between the coffee sensitized and nonsensitized controls. This remained a Blind Study while the *in vitro* Cytotoxic Tests were conducted.

Cytotoxic Test *in* vitro Screening Blind Study: Five dietary department volunteers were screened for possible sensitivities to 70 items (Tables 03.13 - 03.18) using the method of *in vitro* Cytotoxic Testing described by Black (1956), the Bryan's (1960, 1969 and 1971) and Ulett and Perry (1974). The Cytotoxic Tests were run on successive days in the 8:00 - 9:00 a.m. time slot with the initial prefasting test followed on the second day with a challenge allergenic meal and cytotoxic test performed near the leukocytosis maximum. This second cytotoxic test reveals the false negatives described by Ulett and Perry (1974).

Based on the results of first *in vitro* Cytotoxic Tests challenge, sensitizing meals were formulated. A specimen was taken for a challenge *in vitro* Cytotoxic Test at the *in vivo* WBC maximum to reveal the additional positives (false negatives) missed on the initial prefasted Cytotoxic Test. Note these false negatives are not missed with the *in vivo* WBC Response Challenge Test (Ulett and Perry, 1974). Using a blind study protocol the following meals were administered: PJW: coffee (270 mls), bacon (six stripes), lettuce (2/3 cup-shredded); EMW: egg (two boiled), tomato (one large), potato (one small-boiled); RKE: egg (two boiled), chocolate (1.75 Hershey's Bars), coffee (350 mls); HSO: peach (two medium halves in water), tuna (3.25 oz.), radish (four small), carrot (one six inches); EOH: coffee (16 ozs.), sugar (two teaspoons), banana (one medium), whole-wheat bread (two slices), tomato (two medium). These meals represent modest amounts comparable to the average person's daily per meal intake. We previously reported (Ulett and Perry, 1974) that of the cytotoxic test positive foods 34% could be detected only by *in vitro* Cytotoxic Tests done near the time of maximum WBC response (leukocytosis). Today this mystery might find answers using the *in vivo* WBC response model (Ulett and Perry, 1975) to look for molecular entities, such as the cytokines (Fishman et. al., 1996; Horowitz, 1996) that initiate or control this response.

The results of Cytotoxic Testing for the blind study are summarized in Tables 03.19 An index of food sensitivity for each food group can be derived from the vertical columns of data, while the frequency among the five dietitians for each of the foods listed can be derived from the horizontal lines of data as first described in Ulett and Perry (1974). The per cent food sensitivity gives a useful index for each individual but note (Table 03.13) that the specific sensitivities to various fruits permit a powerful ability to discriminate between (i.e., identify) individuals using food sensitivities as potential genetic markers. This will be even more apparent in Chapter 5 when familial relationships are presented.

TABLE 03.13 DISTRIBUTION OF CYTOTOXIC TEST POSITIVE SENSITIVITY TO FRUITS.

DIETARY DEPT COFFEE BLIND STUDY VOLUNTEERS

SUBJECT	PJW	EWM	RKE	HSO	EOH
AGE/TESTED	26	35	49	50	52
APPLE	0	0	0	0	1
BANANA	0	0	0	0	0
CANTALOUPE	0	0	1	0	0
CHERRY	1	0	1	0	0
HONEYDEW	0	0	0	0	1
ORANGE	1	0	1	0	1
PEACH	1	1	0	1	0
PEAR	0	0	0	0	0
PINEAPPLE	0	1	1	1	1
STRAWBERRY	0	1	0	0	0
TOMATO	0	1	0	0	1
VANILLA	0	0	1	1	0
WATERMELON	1	0	0	0	0
POSITIVES	**4**	**4**	**5**	**3**	**5**
NUMBER (N)	13	13	13	13	13
% POSITIVE	**30.8**	**30.8**	**38.5**	**23.1**	**38.5**

TABLE 03.14 DISTRIBUTION OF CYTOTOXIC TEST POSITIVE SENSITIVITY TO GRAINS.

DIETARY DEPT COFFEE BLIND STUDY VOLUNTEERS

SUBJECT	PJW	EWM	RKE	HSO	EOH
AGE/TESTED	26	35	49	50	52
BARLEY	0	0	1	1	0
CORN	0	0	1	1	1
MALT	0	0	0	0	0
OAT	1	1	1	0	0
RICE	0	0	0	1	0
RYE	1	0	0	1	0
WHEAT	0	0	1	1	1
POSITIVES	**2**	**1**	**4**	**5**	**2**
NUMBER (N)	7	7	7	7	7
% POSITIVE	**28.6**	**14.3**	**57.1**	**71.4**	**28.6**

TABLE 03.15 DISTRIBUTION OF CYTOTOXIC TEST POSITIVE SENSITIVITY TO VEGETABLES.

	DIETARY DEPT COFFEE BLIND STUDY VOLUNTEERS				
SUBJECT	PJW	EWM	RKE	HSO	EOH
AGE/TESTED	26	35	49	50	52
BROCCOLI	0	1	0	0	0
CABBAGE	1	0	1	1	1
CARROT	0	0	0	1	1
CUCUMBER	0	0	0	0	1
HOPS	0	1	1	0	0
HORSERADISH	0	0	0	0	0
LETTUCE	1	0	0	0	0
ONION	0	1	0	1	0
PEA	0	0	1	0	0
PEANUT	0	0	0	0	1
POTATO	0	1	1	0	1
RADISH	0	0	0	1	0
SOYBEAN	0	1	0	1	1
SPINACH	0	0	0	0	0
STRINGBEAN	0	0	0	0	0
SUGAR, BEET	0	0	0	1	0
SUGAR, CANE	0	0	1	0	1
POSITIVES	**2**	**5**	**5**	**6**	**7**
NUMBER (N)	17	17	17	17	17
% POSITIVE	**11.8**	**29.4**	**29.4**	**35.3**	**41.2**

TABLE 03.16 DISTRIBUTION OF CYTOTOXIC TEST POSITIVE SENSITIVITY TO FISH/SEAFOOD.

	DIETARY DEPT COFFEE BLIND STUDY VOLUNTEERS				
SUBJECT	PJW	EWM	RKE	HSO	EOH
AGE/TESTED	26	35	49	50	52
CARP	0	0	0	0	0
CRAB	1	0	0	0	0
LOBSTER	0	0	0	1	0
SHRIMP	1	1	1	0	0
SWORDFISH	1	0	0	0	0
TROUT	0	0	0	0	0
TUNA	1	0	1	1	0
POSITIVES	**4**	**1**	**2**	**2**	**0**
NUMBER (N)	7	7	7	7	7
% POSITIVE	**57.1**	**14.3**	**28.6**	**28.6**	**0.0**

The vertical columns reflect the overall sensitivities to all foods tested for in each individual (Tables 03.13 - 03.18) with a summary of the total sensitivity to all foods tested listed in Table 03.19. This data adds to our earlier published observations (Ulett and Perry, 1974). If a patient presented a picture serious enough to suspect that diet was a significant factor affecting their clinical well-being the tests defined in Tables 03.13 – 03.18 were used to diet-manage the patient to avoid all test positive foods and consume only test negative foods. As reported by Ulett and Perry (1974, 1975) this approach resulted in many interesting and favorable outcomes. In Tables 03.13 - 03.18 the horizontal rows reveal the frequency that various *in vitro* Cytotoxic Test positive foods appeared in the five Dietary department volunteers. Note that coffee appeared positive in three of the five participants (Table 03.18). Not detected in table 03.18, but shown in the Figure 03.03a, EOH was positive in the *in vitro* Cytotoxic Test (Ulett and Perry, 1975) done on a specimen taken at the height of the WBC response to a sensitizing food challenge. The *in vivo* Food Challenge provided a Double-Blind alternative and independent confirmation that EOH's immune-defense system recognized coffee as a potential sensitizer or allergen. In another subject (Figure 03.03b) is an example of using the *in vivo* Food Challenge to further analyze the degree that an "allergenic food" affected an individual by testing varying amounts of coffee as a suspect food. Note that increased amounts of food, which were administered in random order, result in proportional increases in the WBC response. When this was plotted as the log of the amount of offending food consumed vs. the WBC response a classic sigmoid dose response curve was revealed. When the curve is displaced to the left and up what might be call an AD_{50} (allergic dose-fifty percent) might signify how susceptible this individual might be to a given food especially as this curve gets displaced further to the left and up in the graphic presentation (Figure 03.03b). The *in vivo* Food Challenge can be repeated following clinical treatment to measure the effectiveness of treatments administered. Successful treatment plans should displace the curve down and to the right and probably help us understand how the immune-defense system and the WBCs participate in this fascinating process.

TABLE 03.17 DISTRIBUTION OF CYTOTOXIC TEST POSITIVE SENSITIVITY TO MEAT/FOWL/EGG/MILK.

	DIETARY DEPT COFFEE BLIND STUDY VOLUNTEERS				
SUBJECT	PJW	EWM	RKE	HSO	EOH
AGE/TESTED	26	35	49	50	52
BEEF	0	0	0	0	0
LAMB	0	0	0	0	0
PORK	1	0	1	0	0
VEAL	1	0	0	0	0
CHICKEN	1	1	0	1	1
TURKEY, Lt.	0	0	0	0	1
EGG	1	1	1	0	0
MILK	1	1	0	0	0
CHEESE/AMERICAN	0	0	0	0	0
CHEESE/ROQUEFORT	0	0	0	0	0
CHEESE/SWISS	0	1	0	0	0
POSITIVES	**5**	**4**	**2**	**1**	**2**
NUMBER (N)	11	11	11	11	11
% POSITIVE	**45.5**	**36.4**	**18.2**	**9.1**	**18.2**

TABLE 03.18 DISTRIBUTION OF CYTOTOXIC TEST POSITIVE SENSITIVITY TO MISC ITEMS.

	DIETARY DEPT COFFEE BLIND STUDY VOLUNTEERS				
SUBJECT	PJW	EWM	RKE	HSO	**EOH**
AGE/TESTED	26	35	49	50	52
CHOCOLATE	0	0	1	0	0
COFFEE	**1**	**0**	**1**	**0**	**1**
COTTONSEED	0	1	0	0	0
GARLIC	0	0	1	1	0
GASOLINE	0	0	0	0	0
MUSHROOM	0	0	0	0	0
MUSTARD	1	0	0	1	1
PEPPERMINT	0	1	1	0	0
TEA	0	0	0	0	0
TOBACCO	0	0	0	0	0
TURPENTINE	0	1	0	0	0
YEAST, BAKER'S	0	0	1	0	1
YEAST, BREWER'S	1	0	1	0	0
ALCOHOL, 1%	0	0	0	0	0
ALCOHOL, 2%	0	0	0	0	0
POSITIVES	**3**	**3**	**6**	**2**	**3**
NUMBER (N)	15	15	15	15	15
% POSITIVE	**20.0**	**20.0**	**40.0**	**13.3**	**20.0**

TABLE 03.19 SUMMARY OF CYTOTOXIC TEST POSITIVE COFFEE BLIND STUDY VOLUNTEERS.

	DIETARY DEPT COFFEE BLIND STUDY VOLUNTEERS				
SUBJECT	PJW	EWM	RKE	HSO	**EOH**
AGE/TESTED	26	35	49	50	52
POSITIVES	**17**	**11**	**19**	**17**	**17**
NUMBER (N)	70	70	70	70	70
% POSITIVE	**24.3**	**15.7**	**27.1**	**24.3**	**24.3**

Tobacco as a Single Sensitizer: Tobacco, a single sensitizer that followed an oral-respiratory as opposed to a gastro-intestinal route of entry, was tested using the tobacco challenge *in vivo* WBC response test. WBC, differential and RBC tests were performed prior to ten minutes of smoking two popular brand filter king-sized cigarettes followed by these determinations at 15-minute intervals for the first hour then at 30-minute intervals until control values were reestablished at three hours (Ulett and Perry, 1975). Subjects, that in documented histories had indicated themselves as regular smokers (ten or more cigarettes per day) or not, were selected from among a population that had earlier been screened using the method of *in vitro* Cytotoxic Testing. In our initial report (Ulett and Perry, 1975) we provided a graphic presentation (see Figure 03.04) of the tobacco challenge WBC response that illustrates the clear-cut difference between tobacco sensitized and tobacco nonsensitized subjects. The tobacco picture is much the same as that seen for coffee earlier in this chapter and that reported in Chapter 2 (Figure 02.02) for multiple food challenges. The *in vitro* Cytotoxic Test and the *in vivo* WBC Response Test continued to correlate with each other with near perfect precision.

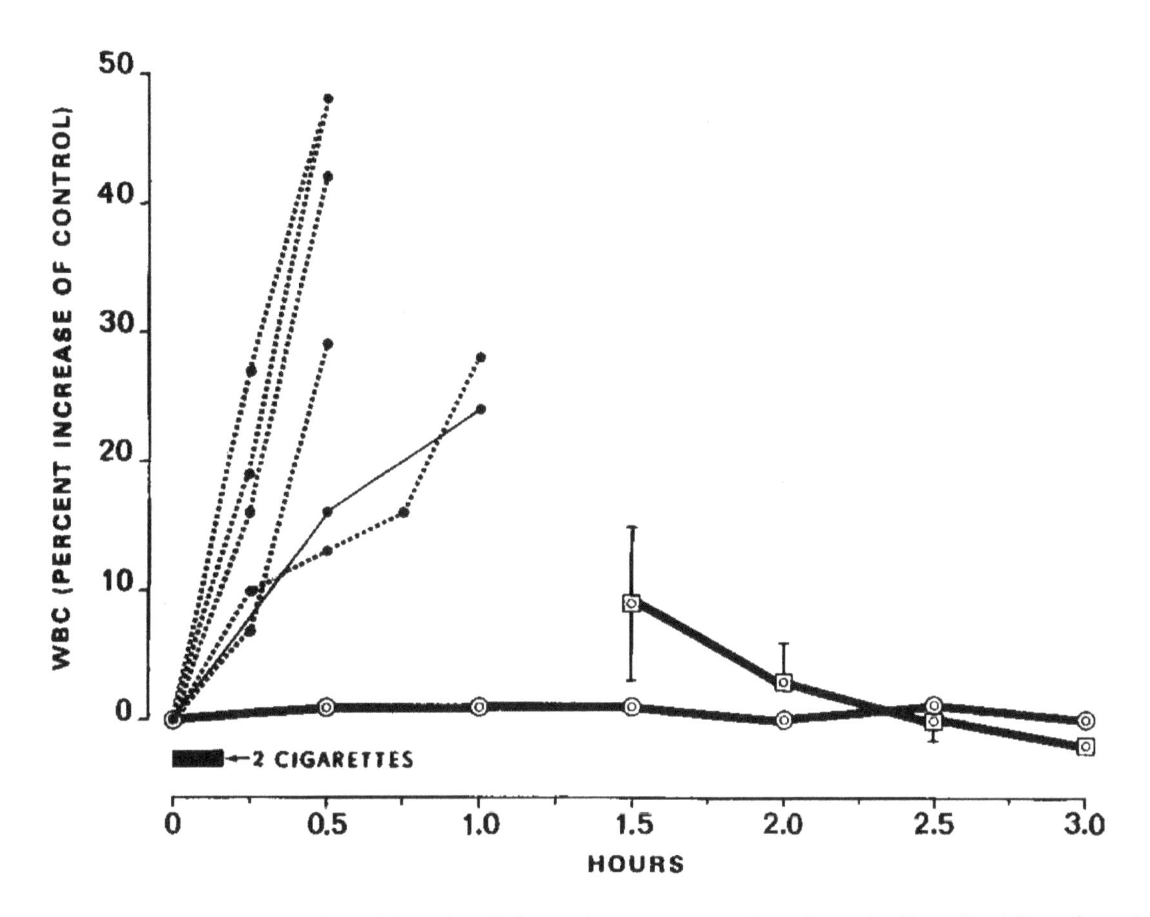

Figure 3.4 Peripheral white blood cell (WBC) increase in five female (hatched lines) and one male (solid line) prefasted subjects in response to smoking two filter king-sized cigarettes within 10 minutes. The return to control values (open squares) was plotted as an average for the six subjects +/-SEM. Three prefasted regular smokers determined tobacco insensitive by the cytotoxic test procedure failed to show a peripheral WBC response (open circles).

By example, regular smokers that were Cytotoxic Test positive for tobacco (sensitized) showed a significant WBC response with no comparable changes in the differential or RBC. Although with PAD (Table 03.20) there was a significant change in the WBC values (23.1% increase over the 0.0-time control at 1.0 hour) the 0.0-time control was high (8.45 $x10^3/mm^3$) compared to other fasted subjects in these studies (see Tables 03.26 and 03.27). A second regular smoker (CAP) that was cytotoxic test positive for tobacco (sensitized) showed a significant WBC response with no comparable changes in the differential or RBC. Again, there was a significant change in the WBC values (22.4% increase over the 0.0-time control at 0.5 hour with the 0.0-time control initially high (8.25 x10^3/mm^3) compared even to all other fasted subjects in our studies (Chapter 2, Tables 02.05 and 02.06).

TABLE 03.20 ANALYSIS OF RBC, WBC, AND DIFFERENTIAL COUNT RESULTS IN TOBACCO SENSITIZED SMOKERS.

CASE	AGE	SEX	TIME HOUR	RBC 10^6mm^3	WBC 10^3mm^3	SEG	LYMPH	MONO	EOSIN	BASO
						DIFFERENTIAL				
PAD	25	F	**0.00**	**5.53**	**8.45**	**60.0**	**38.0**	**1.0**	**1.0**	**0.0**
			0.25	5.55	9.30	62.0	36.0	1.0	1.0	0.0
			0.50	5.45	9.55	64.0	35.0	1.0	1.0	0.0
			0.75	5.35	9.80	65.0	34.0	0.0	1.0	0.0
			1.00	5.52	10.40	61.0	35.0	2.0	2.0	0.0
			1.50	5.54	9.15	64.0	34.0	1.0	1.0	0.0
			2.00	5.55	8.70	60.0	38.0	0.0	2.0	0.0
			2.50	5.55	8.50	58.0	41.0	0.0	1.0	0.0
			3.00	5.54	8.35	64.0	34.0	0.0	2.0	0.0
==============										
AVERAGE				**5.5**	**9.2**	**62.3**	**35.9**	**0.6**	**1.4**	**0.0**
SUM				44.0	73.8	498.0	287.0	5.0	11.0	0.0
SUM OF SQS				242.5	683.3	31042.0	10339.0	7.0	17.0	0.0
NUMBER (N)				8.0	8.0	8.0	8.0	8.0	8.0	8.0
SQRT OF N				2.8	2.8	2.8	2.8	2.8	2.8	2.8
MAXIMUM				5.6	10.4	65.0	41.0	2.0	2.0	0.0
MINIMUM				5.4	8.4	58.0	34.0	0.0	1.0	0.0
RANGE				0.2	2.1	7.0	7.0	2.0	1.0	0.0
STD ERROR (SEM)				0.0	0.3	0.9	0.9	0.3	0.1	0.0
SEM %				0.5	2.8	1.4	2.4	40.0	9.1	0.0
STD DEV (SD)				**0.1**	**0.7**	**2.5**	**2.5**	**0.7**	**0.4**	**0.0**
SD % COEF VAR				**1.3**	**7.9**	**4.0**	**6.9**	**113.1**	**25.7**	**0.0**

This elevated 0.0-time WBC appears to be characteristic of many regular cigarette smokers that we have worked with in these tobacco studies. In an epidemiological study involving 4264 subjects Corre et al (1971) described a leukocyte increase of about 30% in heavy smokers who inhale compared to nonsmokers. Howell (1970) reported a striking difference between the WBCs of smokers and nonsmokers as well as erythrocyte sedimentation rates (E.S.R.) that ranged 10% higher in smokers compared to nonsmokers. Winkle and Statland (1981), Winkel, et al (1981) have reported on the effects of smoking on the leukocyte count. These studies were haunted in their design and interpretation by the prospect of circadian variations as have others in the literature (Haus, 1991) that report WBC fluctuations as diurnal or circadian. Our studies refute these fluctuations given fasting and meaningful controls.

Tobacco smoke alone is known to contain more than 4000 components, although nicotine and carbon monoxide has received the most attention (Lakier, 1992; Krupski, 1991). One mg of nicotine from a cigarette provides a convenient marker for tracking the appearance of a tobacco component into the body as monitored in the circulating plasma concentrations (ng/ml) of nicotine (Pomerleau, 1992). When smoked normally low-dose cigarettes reach plasma levels of 5-6 ng/ml and high-dose cigarettes reach 20-25 ng/ml. High-dose cigarettes smoked deeply can reach nicotine concentrations of 60 ng/ml with maximal concentrations reached within 5-10 minutes. Note from figure 3.4 that the WBC maximum isn't reached until 30-60 minutes, thus a marker for tobacco clearly precedes the *in vivo* WBC response temporal pattern.

TABLE 03.21 ANALYSIS OF RBC, WBC, AND DIFFERENTIAL COUNT RESULTS IN TOBACCO SENSITIZED SMOKERS.

CASE	AGE	SEX	TIME HOUR	RBC 10^6mm^3	WBC 10^3mm^3	DIFFERENTIAL SEG	LYMPH	MONO	EOSIN	BASO
CAP	33	F	**0.00**	**5.04**	**8.25**	**55.0**	**41.0**	**2.0**	**2.0**	**0.0**
			0.25	5.05	8.85	55.0	45.0	0.0	2.0	0.0
			0.50	5.05	10.65	56.0	42.0	1.0	1.0	0.0
			0.75							
			1.00	5.05	10.10	55.0	43.0	0.0	2.0	0.0
			1.50	5.04	8.85	57.0	40.0	1.0	2.0	0.0
			2.00	5.05	8.50	53.0	44.0	0.0	3.0	0.0
			2.50	5.05	8.20	55.0	42.0	1.0	2.0	0.0
			3.00	5.02	8.15	53.0	45.0	2.0	0.0	0.0
============										
AVERAGE				**5.0**	**9.0**	**54.9**	**43.0**	**0.7**	**1.7**	**0.0**
SUM				35.3	63.3	384.0	301.0	5.0	12.0	0.0
SUM OF SQS				178.1	578.0	21078.0	12963.0	7.0	26.0	0.0
NUMBER (N)				7.0	7.0	7.0	7.0	7.0	7.0	7.0
SQRT OF N				2.6	2.6	2.6	2.6	2.6	2.6	2.6
MAXIMUM				5.1	10.7	57.0	45.0	2.0	3.0	0.0
MINIMUM				5.0	8.2	53.0	40.0	0.0	0.0	0.0
RANGE				0.0	2.5	4.0	5.0	2.0	3.0	0.0
STD ERROR(SEM)				0.0	0.4	0.6	0.7	0.3	0.4	0.0
SEM %				0.1	3.9	1.0	1.7	40.0	25.0	0.0
STD DEV (SD)				**0.0**	**0.9**	**1.5**	**1.9**	**0.8**	**1.1**	**0.0**
SD % COEF VAR				**0.2**	**10.4**	**2.8**	**4.4**	**105.8**	**66.1**	**0.0**

TABLE 03.22 ANALYSIS OF RBC, WBC, AND DIFFERENTIAL COUNT RESULTS IN TOBACCO SENSITIZED SMOKERS.

CASE	AGE	SEX	TIME HOUR	RBC 10^6mm^3	WBC 10^3mm^3	DIFFERENTIAL SEG	LYMPH	MONO	EOSIN	BASO
LEM	59	M	**0.00**	**5.23**	**7.70**	**73.0**	**26.0**	**1.0**	**0.0**	**0.0**
			0.25							
			0.50	5.23	8.90	69.0	29.0	2.0	0.0	0.0
			0.75							
			1.00	5.25	9.55	70.0	27.0	2.0	1.0	0.0
			1.50	5.29	8.65	71.0	25.0	2.0	1.0	1.0
			2.00	5.23	8.05	73.0	24.0	2.0	1.0	0.0
=============										
AVERAGE				**5.2**	**8.8**	**70.8**	**26.3**	**2.0**	**0.8**	**0.3**
SUM				21.0	35.2	283.0	105.0	8.0	3.0	1.0
SUM OF SQS				110.1	310.0	20031.0	2771.0	16.0	3.0	1.0
NUMBER (N)				4.0	4.0	4.0	4.0	4.0	4.0	4.0
SQRT OF N				2.0	2.0	2.0	2.0	2.0	2.0	2.0
MAXIMUM				5.3	9.6	73.0	29.0	2.0	1.0	1.0
MINIMUM				5.2	8.1	69.0	24.0	2.0	0.0	0.0
RANGE				0.1	1.5	4.0	5.0	0.0	1.0	1.0
STD ERROR(SEM)				0.0	0.4	1.0	1.3	0.0	0.3	0.3
SEM %				0.3	4.3	1.4	4.8	0.0	33.3	0.0
STD DEV (SD)				**0.0**	**0.8**	**2.0**	**2.5**	**0.0**	**0.5**	**0.5**
SD % COEF VAR				**0.6**	**8.5**	**2.8**	**9.5**	**0.0**	**66.7**	**0.0**

LEM (Table 03.22) also a regular smoker that was cytotoxic test positive for tobacco (sensitized) showed a significant WBC response with no comparable changes in the differential or RBC. As with PAD (Table 03.20) and CAP (Table 03.21) there was also a significant change in the WBC values (24.0% increase over the 0.0-time control at 1.0 hour) the 0.0-time control was relatively high (7.70 x10^3/mm^3) compared to other fasted subjects in these studies (see Tables 03.26 and 03.27) or in our studies (Chapter 2, Tables 02.5 and 02.6). These elevated initial or control WBC counts could reflect a response to the chronic exposure to tobacco or be unrelated to tobacco and possibly be a low-grade infection, common cold or a response to a seasonal airborne allergen. Given this elevated baseline the effects of tobacco still superimpose and permit us to measure significant increases over carefully controlled time frames.

TABLE 03.23 ANALYSIS OF RBC, WBC, AND DIFFERENTIAL COUNT RESULTS IN TOBACCO SENSITIZED SMOKERS.

						DIFFERENTIAL				
			TIME	RBC	WBC	**				
CASE	AGE	SEX	HOUR	10^6mm^3	10^3mm^3	SEG	LYMPH	MONO	EOSIN	BASO
BWV	26	M	**0.00**	**4.61**	**10.90**	**66.0**	**31.0**	**1.0**	**2.0**	**0.0**
			0.25	4.63	11.00	68.0	30.0	1.0	1.0	0.0
			0.50	4.62	10.95	65.0	33.0	1.0	1.0	0.0
			0.75	4.60	10.95	68.0	31.0	0.0	1.0	0.0
			1.00	4.62	11.00	67.0	29.0	2.0	2.0	0.0
			1.50	4.63	10.95	67.0	30.0	1.0	2.0	0.0
			2.00	4.61	10.95	68.0	30.0	1.0	1.0	0.0
===============										
AVERAGE				**4.6**	**11.0**	**67.2**	**30.5**	**1.0**	**1.3**	**0.0**
SUM				27.7	65.8	403.0	183.0	6.0	8.0	0.0
SUM OF SQS				127.8	721.6	27075.0	5591.0	8.0	12.0	0.0
NUMBER (N)				6.0	6.0	6.0	6.0	6.0	6.0	6.0
SQRT OF N				2.4	2.4	2.4	2.4	2.4	2.4	2.4
MAXIMUM				4.6	11.0	68.0	33.0	2.0	2.0	0.0
MINIMUM				4.6	11.0	65.0	29.0	0.0	1.0	0.0
RANGE				0.0	0.1	3.0	4.0	2.0	1.0	0.0
STD ERROR(SEM)				0.0	0.0	0.5	0.7	0.3	0.2	0.0
SEM %				0.1	0.1	0.7	2.2	33.3	12.5	0.0
STD DEV (SD)				**0.0**	**0.0**	**1.2**	**1.6**	**0.8**	**0.4**	**0.0**
SD % COEF VAR				**0.3**	**0.2**	**1.8**	**5.4**	**81.6**	**30.6**	**0.0**

By contrast RAG (Table 03.24) was an example of a regular smoker and cytotoxic test negative (nonsensitized) that showed no WBC response, and no perceptible changes were seen in the differential or RBC. Although there was no change over the intervals tested the average WBC value (8.4 x10^3/mm^3+/-0.4 S.D. (1.4% C.V.) ranged high compared to other fasted subjects in these studies. The RBC for this subject at 3.4 x10^6/mm^3 is well below the normal range (4.2 - 5.4) for females.

Another subject, SGP (Table 03.25) was a pipe smoker of ten years that was cytotoxic test negative for tobacco. There was a small WBC response (6 - 11% range at 0.5 - 1.5 hrs.) in response to smoking two filter king cigarettes while the differential and RBC parameters were stable over the intervals tested. The analysis of the differential and RBC as additional parallel hematological parameters strengthens the validity of the WBC response and provides established valid, safe, and efficacious tests as internal controls during the testing period, further supporting our contention that the WBC response is real and a specific indicator of potential immune-defense system involvement.

The results of the tobacco challenge WBC response studies where subjects were screened using *in vitro* Cytotoxic Testing then submitted to *in vivo* tobacco challenges where RBC, and WBC/differential parameters were followed in groups of tobacco nonsensitized smokers (Table 03.26), tobacco sensitized smokers (Table 03.27) and tobacco sensitized nonsmokers (Table 03.28) are of stunning clarity even within the small population that we have studied. As with coffee and the results of multiple foods WBC response challenges, tobacco corroborates our earlier findings that foods identified as sensitizing (allergenic) by *in vitro* Cytotoxic Testing give a positive WBC response. When the WBC averages for the four nonsensitized smokers (Table 03.26) were compared as controls (student t-Test) to eight sensitized smokers (Table 03.27) the differences seen at the 0.5 - 2.0 intervals were real and highly significant at the (95% confidence level or better). Similarly (Table 03.27) for sensitized smokers, when the 0.0-time averaged WBC controls were compared to the 0.0 - 3.0 averaged WBC test intervals the 0.5 - 1.0 tested as significantly (95% confidence level or better) different than the 0.0 hr. controls. The *in vivo* WBC Response was real within sensitized smokers and the differences between sensitized and nonsensitized smokers are specific and statistically significant at the 95% confidence level or better. The WBC response is real not just in smokers, but just in sensitized smokers! There are many publications in the literature, some cited earlier in this section, that had they had known about tobacco smokers nonsensitized/ tobacco smokers sensitized or even coffee drinker nonsensitized/coffee drinker sensitized would have sorted the data a few more times and refined their results with more definitive conclusions. Some old WBC questions reasked with a few new answers.

TABLE 03.24 ANALYSES OF RBC, WBC, AND DIFFERENTIAL COUNT RESULTS IN TOBACCO NONSENSITIZED SMOKERS.

						DIFFERENTIAL				
			TIME	RBC	WBC	******	******	******	******	******
CASE	AGE	SEX	HOUR	10^6mm^3	10^3mm^3	SEG	LYMPH	MONO	EOSIN	BASO
RAG	27	F	**0.00**	**3.43**	**8.30**	**65.0**	**32.0**	**2.0**	**1.0**	**0.0**
			0.25							
			0.50	3.46	8.45	68.0	28.0	2.0	2.0	0.0
			0.75							
			1.00	3.44	8.55	69.0	28.0	1.0	2.0	0.0
			1.50	3.43	8.45	66.0	30.0	1.0	3.0	0.0
			2.00	3.44	8.25	68.0	29.0	1.0	2.0	0.0
			2.50	3.45	8.25	65.0	30.0	2.0	3.0	0.0
			3.00	3.43	8.25	67.0	30.0	1.0	2.0	0.0
===============										
AVERAGE				**3.4**	**8.4**	**66.9**	**29.6**	**1.4**	**2.1**	**0.0**
SUM				24.1	58.5	468.0	207.0	10.0	15.0	0.0
SUM OF SQS				82.8	489.0	31304.0	6133.0	16.0	35.0	0.0
NUMBER (N)				7.0	7.0	7.0	7.0	7.0	7.0	7.0
SQRT OF N				2.6	2.6	2.6	2.6	2.6	2.6	2.6
MAXIMUM				3.5	8.6	69.0	32.0	2.0	3.0	0.0
MINIMUM				3.4	8.3	65.0	28.0	1.0	1.0	0.0
RANGE				0.0	0.3	4.0	4.0	1.0	2.0	0.0
STD ERROR (SEM)				0.0	0.0	0.6	0.6	0.1	0.3	0.0
SEM %				0.1	0.5	0.9	1.9	10.0	13.3	0.0
STD DEV (SD)				**0.0**	**0.1**	**1.5**	**1.5**	**0.4**	**0.8**	**0.0**
SD % COEF VAR				**0.3**	**1.4**	**2.3**	**5.1**	**26.5**	**35.3**	**0.0**

TABLE 03.25 ANALYSES OF RBC, WBC, AND DIFFERENTIAL COUNT IN TOBACCO NONSENSITIZED SMOKERS.

						DIFFERENTIAL				
			TIME	RBC	WBC	************	************	************	************	************
CASE	AGE	SEX	HOUR	10^6mm^3	10^3mm^3	SEG	LYMPH	MONO	EOSIN	BASO
SGP	35	M	**0.00**	**5.52**	**5.45**	**55.0**	**40.0**	**3.0**	**1.0**	**1.0**
			0.25							
			0.50	5.57	5.80	57.0	40.0	2.0	1.0	0.0
			0.75							
			1.00	5.55	5.90	55.0	41.0	2.0	2.0	0.0
			1.50	5.61	6.05	58.0	40.0	1.0	1.0	0.0
			2.00	5.54	5.40	59.0	38.0	2.0	1.0	0.0
			2.50	5.51	5.30	60.0	37.0	2.0	1.0	0.0
			3.00	5.51	5.25	57.0	40.0	2.0	1.0	0.0
			3.50	5.54	5.25	54.0	44.0	2.0	2.0	0.0
			4.00	5.54	5.30	55.0	42.0	1.0	1.0	0.0
===============										
AVERAGE				**5.5**	**5.5**	**56.7**	**40.2**	**1.9**	**1.2**	**0.1**
SUM				49.9	49.7	510.0	362.0	17.0	11.0	1.0
SUM OF SQS				276.3	275.2	28934.0	14594.0	35.0	15.0	1.0
NUMBER (N)				9.0	9.0	9.0	9.0	9.0	9.0	9.0
SQRT OF N				3.0	3.0	3.0	3.0	3.0	3.0	3.0
MAXIMUM				5.6	6.1	60.0	44.0	3.0	2.0	1.0
MINIMUM				5.5	5.3	54.0	37.0	1.0	1.0	0.0
RANGE				0.1	0.8	6.0	7.0	2.0	1.0	1.0
STD ERROR (SEM)				0.0	0.1	0.7	0.8	0.2	0.1	0.1
SEM %				0.2	1.6	1.2	1.9	11.8	9.1	0.0
STD DEV (SD)				**0.0**	**0.3**	**2.0**	**2.3**	**0.7**	**0.3**	**0.3**
SD % COEF VAR				**0.6**	**4.8**	**3.5**	**5.8**	**35.3**	**27.3**	**0.0**

In most of the literature cases cited above the WBC differences between smokers and nonsmokers was variable and small, and in some cases became significant only after hundreds even thousands of cases are analyzed statistically. Considering our finding (Ulett and Perry, 1975; 1974) specific sensitizations to tobacco and even coffee could be used to further sort out the data and establish controls, even in small samples, to eliminate the variability that may blur or bias the final results. For example, even those sensitized by both tobacco and coffee could be isolated and studied compared to those sensitized separately or not sensitized at all as statistical evidence suggesting the possibility of a cause and effect relationship between tobacco and/or coffee and some forms of cardiovascular disease has been reported (Lesmes et al, 1992; Freidman et al 1974; U.S. Dept of Health, 1973; Boston Collaborative Drug Program, 1972; Blue, 1970; Harkavy, 1968; Fontana, 1960). Based on our findings and separate from tobacco's pharmacological or toxicological effects it is speculated that the conclusions of these cardiovascular studies may be strengthened by considering the possibility that tobacco and/or coffee may be specific, allergenic, genetically controlled, sensitizers and that individuals sensitized by them may be more likely to manifest the signs of cardiovascular disease in unique ways, therefore treated differently.

TABLE 03.26 TOBACCO NONSENSITIZED SMOKERS WBC($X10^3$) COMPARISONS AT HOURS AFTER EXPOSURE TO TOBACCO.											
			CTL/WBC	WBC	WBC	WBC	WBC	WBC	WBC	WBC	WBC
CASE	AGE	SEX	**0.0**	0.25	0.5	0.75	1.0	1.5	2.0	2.5	3.0
CASE RAG	27	F	**8.30**		8.45		8.55	8.45	8.25	8.25	8.25
CASE SSA	24	F	**6.65**	6.60	6.65	6.70	6.65	6.55	6.70	6.65	6.70
CASE EJL	24	M	**7.30**	7.30	7.30	7.40	7.30	7.45	7.30	7.40	7.30
CASE SGP	35	M	**5.45**		5.80	7.40	5.90	6.05	5.40	5.30	5.25
=============	=====										
AVERAGE	27.5		**6.93**	**7.02**	**7.10**	**7.17**	**7.10**	**7.13**	**6.91**	**6.90**	**6.88**
WBC(SUM)			27.7	13.9	28.2	21.5	28.4	28.5	27.7	27.6	27.5
WBCSQ(SUM)			196.1	96.9	202.6	154.4	205.4	206.4	195.4	195.1	193.8
VS SENS 0.0-3.0			**P<0.10**	**P<0.10**	**P<0.01**	**P<0.025**	**P<0.05**	**P<0.05**	**P<0.025**	**P<0.25**	**P<0.30**
NUMBER (N)	4.0		4.0	2.0	4.0	3.0	4.0	4.0	4.0	4.0	4.0
SQRT OF N	2.0		2.0	1.4	2.0	1.7	2.0	2.0	2.0	2.0	2.0
MAXIMUM	35.0		8.3	8.5	8.5	7.4	8.6	8.5	8.3	8.3	8.3
MINIMUM	24.0		5.5	5.8	5.8	6.7	5.9	6.1	5.4	5.3	5.3
RANGE	11.0		2.9	2.7	2.7	0.7	2.7	2.4	2.9	3.0	3.0
STD ERROR(SEM)	2.8		0.7	1.3	0.7	0.2	0.7	0.6	0.7	0.7	0.8
SEM %	10.0		10.3	18.9	9.3	3.3	9.3	8.4	10.3	10.7	10.9
STD DEV (SD)	**5.5**		**1.4**	**1.9**	**1.3**	**0.4**	**1.3**	**1.2**	**1.4**	**1.5**	**1.5**
SD % COEF VAR	**20.0**		**20.6**	**26.7**	**18.7**	**5.6**	**18.7**	**16.8**	**20.6**	**21.4**	**21.8**
EXPECTED -			4.1	3.3	4.5	6.4	4.5	4.7	4.1	4.0	3.9
RANGE (2SD)			9.8	10.8	9.8	8.0	9.8	9.5	9.8	9.9	9.9

TABLE 03.27 TOBACCO SENSITIZED SMOKER'S WBC($X10^3$) COMPARISONS AT HOURS AFTER EXPOSURE TO TOBACCO.											
			CTL/WBC	WBC	WBC	WBC	WBC	WBC	WBC	WBC	WBC
CASE	AGE	SEX	**0.0**	0.25	0.5	0.75	1.0	1.5	2.0	2.5	3.0
CASE PAD	25	F	**8.45**	9.30	9.55	9.80	10.40	9.15	8.70	8.50	8.35
CASE MLB	52	F	**10.30**	12.95	15.20	14.30	12.75	10.40	10.20		
CASE DBK		F	**6.80**	7.95	9.70	9.25	8.45	7.50	6.70	6.60	6.65
CASE CAP	33	F	**8.25**	8.85	10.65		10.10	8.85	8.50	8.20	8.15
CASE RXW	33	F	**6.25**	7.45	9.25	8.65	8.10	7.50	6.95	6.50	6.15
CASE EXM	46	M	**7.05**		8.15		8.80	9.70	8.70	7.60	7.30
CASE LEM	59	M	**7.70**		8.90		9.55	8.65	8.05		
CASE BWV	26	M	**10.90**	11.00	10.95	10.95	11.00	10.95	10.95		
=============	=====										
AVERAGE	39.1		**8.21**	9.58	10.29	10.59	9.89	9.09	8.59	7.48	7.32
WBC(SUM)			65.7	57.5	82.4	53.0	79.2	72.7	68.8	37.4	36.6
WBCSQ(SUM)			553.7	572.2	874.2	580.8	792.2	680.6	610.8	283.1	271.5
0.0 CTL/0.5-3.0				**P<0.10**	**P<0.025**	**P<0.01**	**P<0.01**	**P<0.15**	**P<0.35**	**P<0.20**	**P<0.15**
NUMBER (N)	7.0		8.00	6.00	8.00	5.00	8.00	8.00	8.00	5.00	5.00
SQRT OF N	2.6		2.8	2.4	2.8	2.2	2.8	2.8	2.8	2.2	2.2
MAXIMUM	59.0		10.90	12.95	15.20	14.30	12.75	10.95	10.95	8.50	8.35
MINIMUM	25.0		6.25	7.45	8.15	8.65	8.10	7.50	6.70	6.50	6.15
RANGE	34.0		4.7	5.5	7.1	5.7	4.7	3.5	4.3	2.0	2.2
STD ERROR(SEM)	4.9		0.58	0.92	0.88	1.13	0.58	0.43	0.53	0.40	0.44
SEM %	12.4		7.08	9.57	8.56	10.67	5.87	4.75	6.18	5.35	6.01
STD DEV (SD)	**12.9**		**1.64**	**2.25**	**2.49**	**2.53**	**1.64**	**1.22**	**1.50**	**0.89**	**0.98**
SD % COEF VAR	**32.8**		**20.0**	**23.4**	**24.2**	**23.9**	**16.6**	**13.4**	**17.5**	**12.0**	**13.4**
EXPECTED -			4.9	5.1	5.3	5.5	6.6	6.6	5.6	5.7	5.4
RANGE (2SD)			11.5	14.1	15.3	15.6	13.2	11.5	11.6	9.3	9.3

The RBC results for these five subjects (Tables 03.24 - 03.26) were also analyzed carefully. The nonsensitized smokers (Table 03.26) showed no variations in either the RBC or the WBC/differentials. The tobacco sensitized smokers (Table 03.27) that showed significant changes in the WBC showed no comparable immediate variations of the RBC or differential. And lastly no variations of the RBCs were seen among the sensitized nonsmokers (Table 03.28) while significant WBC responses were being documented. The RBC and differential data provided excellent internal controls to help interpret the WBC response.

Five subjects sensitized to tobacco as determined by *in vitro* Cytotoxic Testing were confined in a 10' x 10' office space and passively exposed to tobacco smoke from two filter king-sized cigarettes for 10 minutes. The average WBC response (Table 03.28) when compared to the 0.0-time controls proved to be significantly (95% confidence level or better) different at the 0.5 - 1.5 test intervals. In this example three of the subjects showed an out-right WBC response two showed no change in their WBC response. One subject (JHS) came to our program knowing in her heart that passive smoke seriously affected her in small amounts and all occasions, but got little or no support from family, friends, or community when she expressed her concerns and became ill in front of them. We observed after only a few minutes of experimental passive exposure her reactions were immediate and of near anaphylactic proportion. These tests helped to document her fears and significantly reduce the psychosomatic component of her torment and from her tormentors. Two subjects tested with JHS (GAU and JAS) were medical doctors and showed moderate positive WBC responses. Two other subjects (JEM and IXA) tested on another occasion were cytotoxic test positive for tobacco and had complaints about passive exposure to smoke but showed no positive WBC response to passive exposure to tobacco smoke indicating this test protocol needs additional investigation.

TABLE 03.28 TOBACCO SENSITIZED NONSMOKERS WBC ($X10^3$) COMPARISONS AT HOURS AFTER PASSIVE EXPOSURE TO TOBACCO.

			CTL/WBC	WBC	WBC	WBC	WBC	WBC	WBC	WBC	WBC
CASE	AGE	SEX	**0.0**	0.25	0.5	0.75	1.0	1.5	2.0	2.5	3.0
CASE JHS		F	**7.10**	10.15	13.45	12.95	11.95	10.90	9.45	8.20	7.55
CASE GAU	55	M	**6.55**	7.90	8.00	7.20	6.80	6.50			
CASE JAS		M	**7.50**	8.50	8.55	8.10	7.90	7.80			
CASE JEM	32	F	**10.85**	10.75	11.15	11.30		11.05	10.75	10.70	10.75
CASE IXA		F	**6.35**	6.30	6.35	6.30	6.40	6.35	6.30	6.25	6.30
=============	=====										
AVERAGE	43.5		**7.67**	**8.72**	**9.50**	**9.17**	**8.26**	**8.52**	**8.83**	**8.38**	**8.20**
WBC(SUM)			38.4	43.6	47.5	45.9	33.1	42.6	26.5	25.2	24.6
WBCSQ(SUM)			307.6	392.9	482.7	452.5	292.4	384.3	244.6	220.8	212.3
0.0 CTL/0.5-3.0				P<0.15	P<0.05	P<0.10	P<0.025	P<0.05	P<0.15	P<0.475	P<0.45
NUMBER (N)	2.0		5.0	5.0	5.0	5.0	4.0	5.0	3.0	3.0	3.0
SQRT OF N	1.4		2.2	2.2	2.2	2.2	2.0	2.2	1.7	1.7	1.7
MAXIMUM	55.0		10.9	10.8	13.5	13.0	12.0	11.1	10.8	10.7	10.8
MINIMUM	32.0		6.4	6.3	6.4	6.3	6.4	6.4	6.3	6.3	6.3
RANGE	23.0		4.5	4.5	7.1	6.7	5.6	4.7	4.5	4.5	4.5
STD ERROR(SEM)	11.5		0.9	0.9	1.4	1.3	1.4	0.9	1.5	1.5	1.5
SEM %	26.4		11.7	10.2	14.9	14.5	16.8	11.0	16.8	17.7	18.1
STD DEV (SD)	**16.3**		**2.0**	**2.0**	**3.2**	**3.0**	**2.8**	**2.1**	**2.6**	**2.6**	**2.6**
SD % COEF VAR	**37.4**		**26.2**	**22.8**	**33.4**	**32.4**	**33.6**	**24.7**	**29.1**	**30.6**	**31.3**
EXPECTED -			3.6	4.7	3.1	3.2	2.7	4.3	3.7	3.2	3.1
RANGE (2SD)			11.7	12.7	15.9	15.1	13.8	12.7	14.0	13.5	13.3

TABLE 03.29 TOBACCO NONSENSITIZED SMOKER'S RBC(X10³) COMPARISONS HRS AFTER TOBACCO EXPOSURE.

			CTL/RBC	RBC	RBC	RBC	RBC	RBC	RBC	RBC	RBC
CASE	AGE	SEX	**0.0**	0.25	0.5	0.75	1.0	1.5	2.0	2.5	3.0
CASE RAG	27	F	**3.43**		3.46		3.44	3.43	3.44	3.45	3.43
CASE SSA	24	F	**3.66**	3.63	3.65	3.65	3.68	3.66	3.64	3.63	3.63
CASE EJL	24	M	**4.58**	4.58	4.56	4.59	4.56	4.57	4.57	4.58	4.60
CASE SGP	35	M	**5.52**		5.57		5.55	5.61	5.54	5.51	5.51
===============	======										
AVERAGE	27.5		**4.30**	**4.10**	**4.31**	**4.24**	**4.31**	**4.32**	**4.30**	**4.29**	**4.29**
WBC(SUM)			17.2	8.2	17.2	8.2	17.2	17.3	17.2	17.2	17.2
WBCSQ(SUM)			76.6	34.1	77.0	34.4	76.9	77.5	76.6	76.3	76.4
EXP CTL A/EXP B											
NUMBER (N)	4.0		4.0	2.0	4.0	2.0	4.0	4.0	4.0	4.0	4.0
SQRT OF N	2.0		2.0	1.4	2.0	1.4	2.0	2.0	2.0	2.0	2.0
MAXIMUM	35.0		5.5	4.6	5.6	4.6	5.5	5.6	5.5	5.5	5.5
MINIMUM	24.0		3.4	3.6	3.5	3.7	3.4	3.4	3.4	3.4	3.4
RANGE	11.0		2.1	0.9	2.1	0.9	2.1	2.2	2.1	2.1	2.1
STD ERROR (SEM)	2.8		0.5	0.5	0.5	0.5	0.5	0.5	0.5	0.5	0.5
SEM %	10.0		12.1	11.5	12.2	11.1	12.2	12.6	12.2	12.0	12.1
STD DEV (SD)	**5.5**		**1.0**	**0.7**	**1.1**	**0.7**	**1.1**	**1.1**	**1.1**	**1.0**	**1.0**
SD % COEF VAR	**20.0**		**24.3**	**16.3**	**24.5**	**15.7**	**24.4**	**25.2**	**24.4**	**24.1**	**24.2**
EXPECTED -			2.2	2.8	2.2	2.9	2.2	2.1	2.2	2.2	2.2
RANGE (2SD)			6.4	5.4	6.4	5.6	6.4	6.5	6.4	6.4	6.4

The Journal of the American Medical Association (1996) published Phase I. (1988 - 1991) of a longitudinal study documenting exposure of the U.S. population to environmental tobacco smoke as reflected by serum cotinine (nicotine metabolite). Tracking a marker or a direct action of passive tobacco smoke in the bodies of others is in its self a significant accomplishment, but to relate these involvement's to increased incidences of asthma (Barber et al, 1996), atopy and delayed maturation of T-cell competence (Hoit et al 1992; Seidman and Mashiach, 1991), low birth weight (Martin, 1986), or general long term clinical effects of active and passive smoking during pregnancy (Makin et al, 1986) brings these diagnostic accomplishments to a full circle of success when a patient or innocent subject can be treated or relieved of a threat to their health.

Epidemiological trend data (Goodwin, 2007) on cigarette use among adults parallels a three-fold increase in childhood asthma during the past 30 years. Environmental tobacco smoke is reported to have a major role promoting an epidemic of asthma in children. Tobacco smoke is not just nicotine and the multitude of offending compounds riche their havoc not only on the respiratory system, but on all susceptible surfaces, circulated sites, including the intima of blood vessels. For some this contact causes no detectible harm, for a few the slightest exposure can be significance. The contact may be an allergic versus a nonallergic in bronchial disease; however, tobacco has been reported (Armentia, 2007) as an allergen in bronchial disease in a study that included 180 patients. The study reported that tobacco appeared to be responsible for a specific IgE response. Studies like these will lead to detailed and specific explanations for some of the mechanisms that tobacco exposure triggers suggesting more effective care and management procedures.

TABLE 03.30 TOBACCO SENSITIZED SMOKER'S RBC(X10^6) COMPARISONS HRS AFTER TOBACCO EXPOSURE.

			CTL/RBC	RBC	RBC	RBC	RBC	RBC	RBC	RBC	RBC
CASE	AGE	SEX	**0.0**	0.25	0.5	0.75	1.0	1.5	2.0	2.5	3.0
CASE PAD	25	F	**5.25**	5.55	5.45	5.35	5.52	5.54	5.55	5.55	5.54
CASE MLB	52	F	**3.85**	3.87	3.86	3.86	3.87	3.86	3.85	3.86	3.88
CASE DBK		F	**4.33**	4.31	4.33	4.36	4.33	4.35	4.33	4.34	4.35
CASE CAP	33	F	**5.04**	5.05	5.03		5.05	5.04	5.05	5.05	5.02
CASE RXW	33	F	**3.98**	3.96	3.96	3.95	3.98	3.97	3.97	3.95	3.94
CASE EXM	46	M	**5.32**		5.34		5.35	5.36	5.39	5.30	5.28
CASE LEM	59	M	**5.23**		5.23		5.25	5.29	5.23		
===============	======										
AVERAGE	41.3		**4.71**	**4.55**	**4.74**	**4.38**	**4.76**	**4.77**	**4.77**	**4.67**	**4.67**
WBC(SUM)			33.0	22.7	33.2	17.5	33.3	33.4	33.4	28.0	28.0
WBCSQ(SUM)			157.9	105.5	160.0	78.1	161.6	162.3	161.9	133.7	133.2
CTL B/EXP B											
NUMBER (N)	6.0		7.00	5.00	7.00	4.00	7.00	7.00	7.00	6.00	6.00
SQRT OF N	2.4		2.6	2.2	2.6	2.0	2.6	2.6	2.6	2.4	2.4
MAXIMUM	59.0		5.32	5.55	5.45	5.35	5.52	5.54	5.55	5.55	5.54
MINIMUM	25.0		3.85	3.87	3.86	3.86	3.87	3.86	3.85	3.86	3.88
RANGE	34.0		1.5	1.7	1.6	1.5	1.7	1.7	1.7	1.7	1.7
STD ERROR (SEM)	5.7		0.21	0.34	0.23	0.37	0.24	0.24	0.24	0.28	0.28
SEM %	13.7		4.44	7.39	4.81	8.54	4.95	5.05	5.08	6.04	5.95
STD DEV (SD)	**13.9**		**0.55**	**0.75**	**0.60**	**0.75**	**0.62**	**0.64**	**0.64**	**0.69**	**0.68**
SD % COEF VAR	**33.6**		**11.8**	**16.5**	**12.7**	**17.1**	**13.1**	**13.3**	**13.4**	**14.8**	**14.6**
EXPECTED -			3.6	3.0	3.5	2.9	3.5	3.5	3.5	3.3	3.3
RANGE (2SD)			5.8	6.0	5.9	5.9	6.0	6.0	6.0	6.1	6.0

The effect of tobacco on our health has been (Lesmes et al, 1992; Blue, 1972; Friedman et al, 1972), and will continue to be, a very busy and highly complex occupation for a dedicated legion of healthcare workers. Although, one of the most preventable of causes of disease and death, tobacco use does it so many ways and by so many apparently different mechanisms that by its own complexity it has continued to escape its own demise for now. Why can some use tobacco and not be affected, while others appear to "get caught". Perhaps it's because they are Cytotoxic Test positive for tobacco and the immune system and target tissues are "awakened "as a target tissue.

NOTES:

TABLE 03.31 TOBACCO SENSITIVE NONSMOKER'S RBC ($x10^6$) COMPARISONS HRS AFTER PASSIVE SMOKE EXPOSURE.

			CTL/RBC	RBC	RBC	RBC	RBC	RBC	RBC	RBC	RBC
CASE	AGE	SEX	**0.0**	0.25	0.5	0.75	1.0	1.5	2.0	2.5	3.0
CASE JHS		F	**5.20**	5.17	5.20	5.19	5.21	5.17	5.21	5.18	5.20
CASE GAU	55	M									
CASE JAS		M									
CASE JEM	32	F	**3.47**	3.46	3.49	3.47		3.46	3.47	3.47	3.47
CASE IXA		F	**3.28**	3.28	3.30	3.30	3.27	3.26	3.25	3.25	3.37
===============	======										
AVERAGE	43.5		**3.98**	**3.97**	**3.99**	**3.99**	**4.24**	**3.96**	**3.98**	**3.97**	**4.01**
WBC(SUM)			11.9	11.9	12.0	12.0	8.5	11.9	11.9	11.9	12.0
WBCSQ(SUM)			49.8	49.4	50.0	49.8	37.8	49.3	49.7	49.4	50.4
EXP CTL A/EXP B											
NUMBER (N)	2.0		3.0	3.0	3.0	3.0	2.0	3.0	3.0	3.0	3.0
SQRT OF N	1.4		1.7	1.7	1.7	1.7	1.4	1.7	1.7	1.7	1.7
MAXIMUM	55.0		5.2	5.2	5.2	5.2	5.2	5.2	5.2	5.2	5.2
MINIMUM	32.0		3.3	3.3	3.3	3.3	3.3	3.3	3.3	3.3	3.4
RANGE	23.0		1.9	1.9	1.9	1.9	1.9	1.9	2.0	1.9	1.8
STD ERROR (SEM)	11.5		0.6	0.6	0.6	0.6	1.0	0.6	0.7	0.6	0.6
SEM %	26.4		16.0	15.9	15.9	15.9	22.9	16.1	16.4	16.2	15.3
STD DEV (SD)	**16.3**		**1.1**	**1.1**	**1.1**	**1.1**	**1.4**	**1.1**	**1.1**	**1.1**	**1.1**
SD % COEF VAR	**37.4**		**27.8**	**27.6**	**27.5**	**27.5**	**32.4**	**27.8**	**28.5**	**28.1**	**26.4**
EXPECTED -			1.8	1.8	1.8	1.8	1.5	1.8	1.7	1.7	1.9
RANGE (2SD)			6.2	6.2	6.2	6.2	7.0	6.2	6.2	6.2	6.1

Comments on single Sensitizing Foods: In this Chapter on Single Sensitizing Foods, we have described for you in great detail our findings that positive *in vitro* Cytotoxic Tested foods elicit a positive WBC response when this "food" challenge test was administered with the right controls and diagnostic procedures. The findings from our first analysis described in Chapter 1 were corroborated and extended. What does it all mean? We see the *in vivo* WBC Response as a common denominator of an allergenic immune defense system response to the allergens that we have also identified by Cytotoxic Testing that then affects a genetically susceptible target organ or tissue system (i.e., skin, CNS, gastrointestinal, cardiovascular intima, muscular etc.). Genetically each subject has been dealt foods and susceptible end organ(s) for which they are reactive. Within the first two years of life, they are exposed to most if not all of these foods. Now what happens?

Food allergens were placed in our trays and tested in alphabetic order. For purpose of analysis the data was organized by food groups. Historically investigators (Unger and Harris, 1974) discussed classifying food along botanical or genetic lines. The molecular biology of allergens (Bush, 1996), especially food allergens (Hefle, 1996), provide detailed and highly specific studies defining the nature and properties of allergens. Coffee probably exhibits multiple antigens that may or may not include caffeine. A five-ounce cup of instant coffee approximated at 160 mls is estimated to contain 66 mg of caffeine or 412.4 µgms/ml caffeine concentration. By comparison the caffeine content of percolated (693.8 µg /ml), dripolated (887.5 µg/ml) and brewed (937.5 µg/ml) forms of coffee have been approximated and further vary by brands and degree of freshness. By further comparison the caffeine content of a 12 oz cola is approximately 65 mgs, or (169.3 μg/ml), and a chocolate bar contains approximately 25 mgs of caffeine. The point here is that coffee does not equate to caffeine and that coffee more than likely contains one or more antigens that elicit the WBC response can be substantiated by analyzing the data presented here and in the case studies presented in Chapter 5 to see that those that react to chocolate and tea do not also react to coffee or some other combination of other caffeine containing foods. The same can be said for tobacco antigen(s) not equating to tobacco nicotine or other pharmacologically active tobacco molecules. Most food antigens primarily follow a gastrointestinal route of entry as opposed to antigens like tobacco or barley flour (Vidal and Gonzalez, 1995), or pollen that occasionally or uniquely enter through the oral-respiratory and respirator routes of entry. Tobacco has been widely chronicled as a multifaceted health hazard, however, the barley flour induced asthma scenario through the inhalation oral-respiratory route details a specific IgE mediated mechanism that showcases an allergic etiology. Interestingly, IgE and IgG appear to be regulated differently (Gerrard et al, 1980) when monitored in smokers and nonsmokers with IgG deficiency. Without IgG deficiency IgE levels ranged at 40.0 IU in obstructive lung disease patients and at 80.0 IU for patients with asthma associated disease. Even more to the point DNA sequencing techniques are being used that can readily determine the entire primary sequence of an allergen, while recombinant DNA procedures can be used to harvest quantities of allergens for studies. The basic concepts of molecular biology as reviewed by Horowitz (1996) are being applied to the field of clinical allergy with ever increasing fundamental discoveries such as the publication of the amino acid sequence analysis and cloning of the major peanut allergen Arah II (Stanley, et al, 1996).

Anderson (1994) in his review of adverse reactions to foods has stressed that true food allergies must meet rigid criteria before the claim can be made. He cited VanMetre's (1983) criteria for proving a new medical concept as valid, safe, and efficacious including using Double-Blind, Placebo-Controlled Food Challenge (DBPCFC) as the gold standard for defining true clinical allergy. He went on to report that most true allergies occur in children as documented using DBPCFC testing methods with 93% of the proven reactions attributed to just eight foods (egg, peanut, milk, soy, tree nuts, crustacean shellfish, fish, and wheat) in order of frequency as reported by Bock et.al. (1988). Proven adverse reactions to corn (2/481 patients) and chocolate (0/710 patients) were way below previous suspected adverse claims. Bock (1992) estimated from a two-year study of admissions to 73 Colorado emergency departments that 950 severe reactions to food occur annually. Yinginger et.al., (1989; 1988) report of an ultimate adverse reaction to peanut, crustacean shellfish, tree nuts and fish in order of frequency were most often associated with food anaphylaxis related deaths in older children and adults. Sampson, et.al. (1998; 1988) have also reported on fatal and near-fatal food induced anaphylaxis as part of their ongoing food allergy studies.

The cytotoxic tests and WBC response that we are reporting on in these studies do not represent evidence for proven allergies but do reopen some doors for identifying how allergens may be genetically selected for or activated and it does give us new places to look for diagnostic procedures or remedies. The WBC response represents an effect on a clinical parameter that has long been established as valid, safe, and efficacious as well as being a window into the workings of the immune-defense system. What is going on here? In these five subjects, as with other subjects tested, what we believe we are seeing is the front-end of a very complex sequence of events that in some cases manifests itself as an allergic reaction or food induced immune mediated disease. As formerly postulated (Ulett and Perry, 1975) we are looking at phase I of a possible three phase process where phase I is the genetic ability to recognize a food antigen(s), phase II initiates a molecular response(s), and phase III involves a genetically susceptible end-organ (i.e., skin, CNS, cardiovascular, gastrointestinal, etc.) that responds at the right time under the right conditions. This may explain why, in phase I, we see more positives than can be explained by current clinical paradigms. Our model may cover many "sub clinical" maladies of short- and long-term consequence that are hiding beneath a "not required to be fasted" WBC "normal range" and here-to-fore not dealt with unless they reached a "clinical" magnitude.

As we proposed earlier (Ulett and Perry, 1975) consider the following hypothetical situation, assuming an allergic etiology, as one basis for causing or aggravating cardiovascular disease: 1) specific antigen(s) in tobacco and/or coffee are involved, 2) antigen(s) initiate immunochemical events in the circulatory system, and 3) resultant immunochemically dependent changes affect a susceptible cardiovascular system target tissue. Thus, it would not be sufficient to merely smoke cigarettes or drink coffee alone, but this model suggests a rather specific requirement for sensitivity capable of initiating an allergic response, a capacity to carry an immunochemical response to some completion, and finally a target tissue capable of responding allergically in order to arrive at the final clinical picture. We went on to point out that these complex requirements may also explain why, even though the subjects in this study were sensitized to one out of three foods they were tested for, they do not have a comparable frequency of overt clinical reactions. Similarly, these models could account for why some can smoke and seemingly get away with it while others suffer many or all the traps of smoking.

The following models were used to conceptualize our interpretation of a positive cytotoxic test for:

$$\text{Coffee Ag(s)} + \text{Coffee Ab(s)} + \text{Target Cells} = \text{Positive Test} \qquad (3\text{-}1)$$

Any missing or insufficient component of this model would result in a negative cytotoxic test. Cytotoxic tests positive determined only at maximum leukocytosis require additional conceptualization. The amount of coffee antigen(s) was constant as were the conditions of the in vitro bioassay. The release of coffee Ab(s) would not be directly stimulated as coffee was not included among the challenge cytotoxic test positive foods used to elicit the WBC response and the leukocytosis maximum necessary for the special cytotoxic test. We also know that while the WBC differentials do show a comparable fluctuation during the total WBC increase there may be a mediator or at least mediating circumstances associated with this leukocytosis that may potentiate items that show as false negatives in prefasted cytotoxic tests. The first model is expanded to consider the apparent requirement for a mediator:

$$\text{Coffee Ag(s)} + \text{Coffee Ab(s)} + \text{Mediator} + \text{Target Cells} = \text{Positive Test} \qquad (3\text{-}2)$$

The mediator may be a small molecular weight chemical, or even an immune complex knowing today's possibilities. The mediator appears to be stimulated or produced by a number of sensitizing foods and apparently does not require a specific trigger. It may represent an intermediary which could be common to many or even all sensitizing foods reaching sufficient levels during leukocytosis to reveal these false negatives. Remember these false negatives can be validated with positive in vivo WBC responses.

The tobacco data (Tables 3.24 - 3.26) requires some further discussion. The results of the tobacco *in vivo* WBC response provides an alternative route of assimilation into the circulatory system that represents a way around having to consider the role the gastrointestinal tract may play in the assimilation of antigen(s). The following models were used to conceptualize our interpretation of a tobacco cytotoxic test positive for:

Tobacco Ag(s) (extract) + Tobacco Ab(s) + Target Cells = Positive Test (3-3)

The tobacco data (Table 3.25) had one subject (BWV) that was cytotoxic test positive for tobacco and failed to show a positive WBC after smoking two filter king-sized cigarettes within 10-minutes, and two subjects (JEM and IXA, Table 3.26) although tobacco cytotoxic test positive failed to give a positive WBC response when passively exposed to tobacco smoke. The following models were used to conceptualize our interpretation of a tobacco WBC response positive for:

Tobacco Ag(s) (volatile/ash) + Tobacco Ab(s) + Target Cells = Positive Test (3-4)

The way the in vitro cytotoxic test is set up the WBC fraction (buffy coat) comes into direct contact with the entire crude preparation of smoking tobacco as was the case with coffee and other foods as conceptualized in models (3.1) and (3.2), however, in the *in vivo* WBC response test only the volatilized antigen(s) components of the tobacco enter via the oral-respiratory route and the ash components are minimized or completely eliminated. The speculation at this point indicates that the antigen(s) that resulted in the *in vitro* cytotoxic test positive reaction got altered or otherwise failed to get into the circulatory system during the *in vivo* WBC response test. As indicated above the tobacco test protocols have a different set of variables that warrants further investigation. With these documented and qualified exceptions for tobacco noted the in vitro cytotoxic test and the in vivo WBC response identified potential allergens with excellent agreement between the two methods. The cytotoxic test was particularly useful as a screening method as it permits the testing of a large number of potential sensitizers in several subjects within a five-hour period. Although the *in vivo* WBC response could test only one item in several subjects within a five-hour period, the procedure is the more sensitive, the more valid, the more automatable, and the more accepted of the two methods.

REFERENCES

Anderson JA: Milestones marking the knowledge of adverse reactions to food in the decade of the 1980s. Annals allergy 72:143-54, 1994.

Armentia, A: Tobacco as an allergen in bronchial disease. Ann. Allergy, Asthma, & Immunology: 98:329-336, 2007.

Bannon GA: Peanut allergen Ara I. Identifying the clinically relevant epitomes: J Allergy Clin Immunol: 97:329, 1996.

Barber K, Mussin E, and Taylor DK: Passive maternal exposure to tobacco smoke during pregnancy is associated with increased incidence of asthma. Ann Allergy Asthma Immunol: 76:427-30, 1966.

Black AP: A New Diagnostic method in allergy disease. Pediat 17: 717-724, 1956.

Blue JA: Cigarette asthma and tobacco allergy. Annals Allergy: 28:110-15 1970.

Bock SA: The incidence of severe adverse reactions to food in Colorado. J Allergy Clin Immunol 83:900-4, 1992.

Bock SA, Atkins, FM: Patterns of food hypersensitivity during sixteen years of double-blind placebo-controlled food challenges. J Pediatr 117:561-7, 1990.

Bock SA, Sampson HA, Atkins FM, et.al: Double-Blind placebo-controlled food challenge (DBPCFC) as an office procedure: A Manual. J Allergy Clin Immunol 829:86-7, 1988.

Boston Collaborative Drug Program, 1972.

Bryan WTK and Bryan M: Cytotoxic Reactions in the diagnosis of food allergy. Otolarnyngol Clinics of N Am 4:523-533, 1971.

Bryan WTK and Bryan M: Cytotoxic Reactions in the diagnosis of food allergy. Laryngoscope 79: 1453-1472, 1969.

Bryan WTK and Bryan M: The application of in vitro cytotoxic reactions to clinical diagnosis of food allergy. Laryngoscope 70: 810-824, 1960.

Burks AW: IgE-binding epitopes on Arah I. a legume vicilin protein and a major peanut allergen. J Allergy Clin Immunol 97:329, 1996

Bush RK: Molecular Biology of Allergy and Immunology: Molecular Biology of Allergens. In: Immunology and Allergy Clinics of North America, pgs. 535-563, 1996.

Corre F, Lellouch J, and Schwartz, D: Smoking and leukocytes-counts: Results of an epidemiological survey. Lancet 2:632-4, 1971.

Fishman S, Hobbs K, and Borisk L: Molecular Biology of Allergy and Immunology: Molecular Biology of Cytokines on Allergic Disease and Asthma. In: Immunology and Allergy Clinics of North America, pgs. 613-42, 1996.

Fontana VJ: Cardiovascular effects of nicotine and smoking. Ann. N.Y. Acad Sci: 90:138, 1960.

Friedman GD, Klatsky, AL, Siegelaub, AB: Leukocyte count in smokers. N. Engl. J. Med.: 290:1275-8 1974.

Friedman GD, Seltzer CC, Siegelaub AB, Feldman R and Collen MF: Smoking among white, black, and yellow men and women. Kaiser-Permanente Multiphasic health examination data: 1964-1968. 96:23-35, 1972.

Goodwin, R.D: Environmental tobacco smoke and the epidemic of asthma in children: the role of cigarette use. Annals of Allergy, Asthma & Immunology, 97:447-454, 2007

Haus E: Chronobiology of circulating blood cells and platelets. In: Touitou Y, Haus E, Eds Biologic Rhythms in Clinical Laboratory Medicine. Berlin: Springer – Verlag; pgs. 504-526, 1992.

Hefle S: Molecular Biology of Allergy and Immunology: The Molecular Biology of Food Allergens. In: Immunology and Allergy clinics of North America, 1996.

Hoit PG, Clough JB, Holt BT, et al: Genetic "risk" for atopy is associated with delayed postnatal maturation of T-Cell competence. Clin Exp Allergy 22:1093-9, 1992.

Horowitz RJ, Molecular Biology of Allergy, and Immunology: Basic Concepts of Molecular Biology. In: Immunology and Allergy clinics of North America, pgs. 485-515, 1996.

Howell RW: Smoking habits and laboratory tests. Lancet: July 18, pg. 152, 1970.

Keating MU, Jones RT, Worley NJ, et.al: Immunoassay of peanut allergy in food processing materials and finished foods. J Allergy Clin Immunol 86:41-4, 1990.

Krupski WC: The peripheral vascular consequences of smoking. Ann Vasc Surg 5:291-304, 1992.

Lakier JB: Smoking and cardiovascular disease. The Am J of Med 93 {suppl 1a):8-12, 1992.

Lesmes GR; Editor: Proceedings of a symposium: The effects of cigarette smoking: A global perspective. 93 (1A):1-56, 1992.

Martin TR and Bracken MB: Association of low birth weight and passive smoke exposure on pregnancy. Am J Epidermiol 124:633-421, 1986.

Makin J, Fried PA, and Walkinson B: A comparison of active and passive smoking during pregnancy: long term effects. Neurotoxical Teratol 13:5-12, 1991.

Metcalfe DD, Sampson HA, Simon RA: Food allergy-adverse reactions to food and food additives. Cambridge, MA: Blackwell Scientific Publishers; pgs. 1-417, 1991.

Middleton E, Reed CE, Ellis EF, Adkinson NF, Yunginger JW and Busse, WW; Editors: Allergy Principles and Practice Vols I and II. Chapter 71: Unconventional theories and unproven methods in allergy. Mosley 1993.

Ostolani C and Czedik-Eysenberg; Quest Editors: Seventh International Symposium On Immunological, Chemical and Clinical Problems Of Food Allergy. Italy 4-7 October 1998.

Porerleau OF: Nicotine and the central nervous system: Biohavioral effects of cigarette smoking. The Am J of Med 93 (suppl 1a):2-7, 1992.

Sampson HA: Fatal food-induced anaphylaxis. Allergy 53 ☹ suppl 46):125-30, 1998.

Sampson HA: 1996.

Sampson HA: Mendelson L, Rosen JP Fatal and near-fatal anaphylactic reactions to foods in children and adolescents. N Engl J Med 327:380-4, 1992.

Sampson HA: Editorial: Peanut anaphylaxis. J Allergy Clin Immunol 86:1-3, 1990.

Sampson HA, Scanlon SM: Natural history of food hypersensitivity in children with atopic dermatitis, J Pediatrics 115:23-7, 1989a.

Sampson HA, Broadbent KR, Bernhisel-Broadbent J: Spontaneous release of histamine from basophils and histamine-releasing factor in patients with atopic dermatitis and food hypersensitivity. N Engl J Med 312:228-32, 1989b.

Sampson HA: The role of food allergy and mediator release in atopic dermatitis. J Allergy Clin Immunol 81:635-45, 1988.

Sampson HA: Jolie PA. Increased plasma histamine concentration, after food challenges, in children with atopic dermatitis. N Engl J Med 311:372-6, 1984.

Sampson HA: Role of immediate food hypersensitivity in the pathogenesis of atopic dermatitis. J. Allergy Clin Immunol 71:473-80, 1983.

Seidman DS and Mashiach s: Involuntary smoking and pregnancy. Eur J Obstet Gyn Reproductive Biol 41:105-16, 1991.

Stanley J, et al: Amino acid sequence analysis and cloning of the major peanut allergen Arah II. J Allergy Clin Immunol; 97:329, 1996.

Ulett GA and Perry SG: Cytotoxic testing and leukocyte increase as an index to food sensitivity. II. Coffee and Tobacco. Annals Allergy 34: 150-160, 1975.

Ulett GA and Perry SG: Cytotoxic testing and leukocyte increase as an index to food sensitivity. Annals Allergy 33: 23-32, 1974.

Unger L and Harris MC: Steppingstones in Allergy. Chapter V: More about Troublemakers in Allergy. Foods: Ann Allergy 33:299-311, 1974.

VanMetre T: Critique of controversial and unproven procedures for diagnostic and therapy of allergic disorders. Pediatr Clin N Am 30:807-13, 1983.

Vidal C and Gonzalez-Quintela A: Food induced and occupational asthma due to barley flour. Ann Allergy Asthma & Immunology, 75:121-24, 1995.

Winkle P and Statland BE: The acute effect of cigarette smoking on the concentrations of blood leukocyte types in healthy young women. Am. J. Clin. Pathol 75:781-785, 1981.

Winkle P, Statland BE, Saunders A, et al: Within-day physiological variation of the concentration of leukocyte types in healthy subjects as assayed by two automated leukocyte differential analyzers. Am. J Clin Pathol 75:693-700, 1981.

Yunginger JW, Sweeney KG, Sturner WQ, et.al: Fatal food-induced anaphylaxis. J Am Med Assoc 260:1450-2, 1988.

Yunginger JW, Squillace DL, Jones R, et al: Fatal anaphylactic reactions induced by peanuts. Allergy Proc, 100:249-253, 1989.

NOTES:

CHAPTER 4

Cerebral Allergy

THE CENTRAL NERVOUS system (CNS) of all the organ systems in the body may be the most sensitive and in tune with the body's many bio-environments and feedback circuits whether morphological or molecular. Rowe (1944) discussed a number of CNS allergic responses and related them to: 1) clinical conditions, 2) diagnostics and 3) potential remedies, as one of the early pioneers in the field of CNS allergies. A review of the literature (Davison, 1952) correlated many of the CNS's symptoms with signs attributable to allergic etiologies. Kitter and Baldwin (1972) studied the role of allergic factors in children with minimal brain dysfunction syndrome by following 20 children with severe learning problems and abnormal electroencephalographs (EEGs) that showed improvement with treatment for potential allergy(s). Neurological manifestations of allergic disease were the subject of a five-year neuropsychiatric private practice study (Campbell, 1973) that probed the etiology of puzzling neurologic syndromes where 654 consecutive EEG patients were analyzed.

The EEG described in animals for the first time by Canton (1875) and in humans by Berger (1929) as largely post synaptic extracellular potentials one tenth cortical cell potential strength characteristically localized in defined areas of the cortex, has become a common routine laboratory test used to establish baseline and/or abnormal function of CNS regions. The EEG International 10/20 Electrode Placement System (10% or 20% of the total distance nasion to inion divides the head into eight left and right regions; 1) frontal pole (FP), 2) frontal lobe (F), 3) central parietal (CP), 3) temporal (T), and 4) occipital (O)) was used (Figure 4.0).

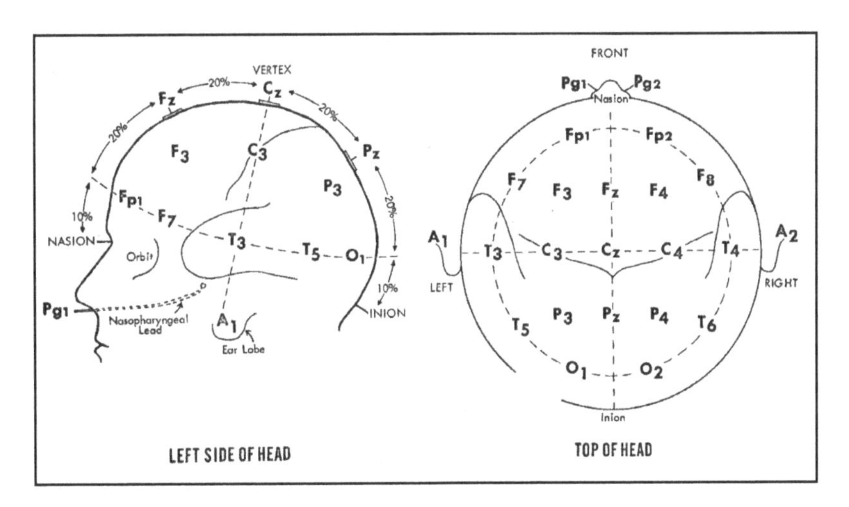

Figure 04.0

The EEG with computer analysis was one additional parameter measured during our initial studies (Ulett and Perry, 1974; 1975) detailed in Chapter 2. These early defining experiments (Tables 02.01, 02.02 and Figure 02.01) included classically recorded 0.0-time EEGs with EEGs recorded at 30-minute intervals during the allergenic and nonallergenic food challenges at the same time we were monitoring the dramatic WBC changes in response to these allergenic food challenges. We report in this chapter what might be one of our most startling findings, that allergenic and nonallergenic food challenges can be directly detected by computer analyzed EEGs. Specifically frontal lobe computer analyzed EEGs (Tables 04.01 – 04.20), showed distinct and significant activity, while parietal lobe computer analyzed EEGs (Tables 04.21 and 04.22) appear to reflect minimal if any involvement thus providing an internal control for the observed frontal lobe phenomena (Figure 04.1).

THE EEG AND THE IMMUNE DEFENSE SYSTEM: A research protocol was developed to test cerebral effects of test positive, potentially allergenic, foods using computer analyzed EEGs to detect and document Central Nervous System (CNS) areas of involvement as the possible target tissue. The clinical objective would analyze subjects having specific complaints regarding symptoms involving or affecting the CNS that were tested for food sensitivities according to the method described by Ulett and Perry (1974; 1975). On follow up the subjects would be monitored for various clinical parameters (EEG, EKG, blood pressure, GRS, pulse, and any attendant symptomatology over the time course of the leukocytosis response to selected test positive food(s). The protocol recommended that five normal males, five normal females, five Schizophrenics, and five Alcoholics be analyzed as a pilot study to establish the clinical feasibility for using Cerebral Allergy as an approach for the treatment of Neurological disorders. Reported here are the remarkable results of the first normal male studied. Unfortunately, the complete study was terminated! Figure 04.01 provides a summary of the primary wave and three representative component parietal and frontal (Tables 04.19 – 04.22) wave patterns expressed as the percent of the respective 0.0-time computer analyzed EEG controls for the allergenic and nonallergenic test periods detailed in Chapter 2 (Figure 02.01 and Table 02.02). At the top left of figure 04.01 is the comparison of the parietal results for the allergenic and nonallergenic challenge test-period primary waves. Both parietal (allergenic and nonallergenic) primary waves do not vary significantly from the 0.0-time controls over the 4.0-hour test period, as was the case with the three representative's parietal component waves at 7.5, 13.0 and 20.0 Hz. Note that the parietal component wave at 13.0 Hz varied characteristically and significantly ($P<0.05$ or greater) from the frontal component wave for 13.0 Hz. See Tables 04.01 – 04.18 detailing the differences between all of the frontal primary and component waves during the allergenic and nonallergenic test periods. As seen here in Figure 04.01 the control frontal primary wave for the allergenic test period does not differ greatly from its 0.0-time control, or the parietal allergenic and nonallergenic test period waves. The frontal primary wave nonallergenic test period, however, displayed very temporal, characteristic, and significantly ($P<0.05$) different results compared to the 0.0-time controls, the nonallergenic vs. allergenic comparison test period, or the parietal lobe results (internal control)!

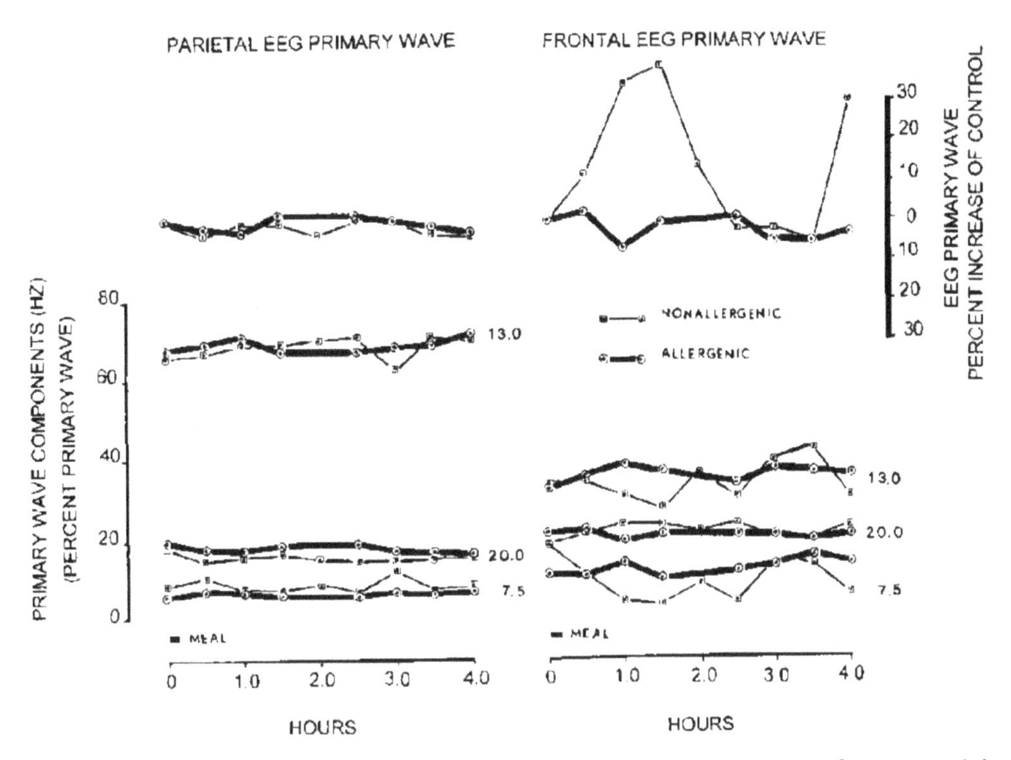

Figure 04.01 Comparison of primary and key component EEG waves for Frontal (Tables 04.19 and 04.20) and Parietal regions (Tables 04.20 and 04.21) during allergenic and nonallergenic food challenge tests.

We observed no significant global difference in the averaged recorded 0.0-time controls for the primary waves or component primary waves between the allergenic/nonallergenic sessions conducted six days apart (Tables 04.19 – 04.20). We did observe dramatic and highly significant changes in the primary waves, including shifts of the component primary waves, in the timed test series relative to the corresponding 0.0-time controls (Tables 04.1 – 04.18). These patterns of change were very different in the frontal lobe (Tables 04.19, 04.20) vs. the parietal lobe (Tables 04.21, 04.22) during the allergenic and nonallergenic food challenge sessions.

In our original experiments (Tables 02.01 and 02.02) we detailed a dramatic WBC response pattern (Figure 02.01) for allergenic and nonallergenic challenge test periods. In Figure 04.02 we plot the Table 02.0 WBC data and include the lymphocyte and segmented neutrophils (segs) results expressed as the percent of the respective 0.0-time controls. The differences between the allergenic and nonallergenic test-period patterns are unique and distinctive from each other. The lymphocyte pattern shows two very interesting relationships. In addition to the difference between allergenic and nonallergenic test periods note the relationship between the patterns for total WBC relative to the distinct pattern for the allergenic food challenge lymphocyte component of the differential. Initially lymphocytes are triggered to increase followed by a decline during the increase of the total WBC with a lymphocyte recovery to the control level during the total WBC decline to control levels potentially reflects an immune response to the allergenic food challenge not seen with nonallergenic food challenge. Note that the end of the allergenic challenge test period (3.5 – 4.0 hours) that has returned to the control level looks very much like the entire nonallergenic test period (0-4.0 hours). Here we have a window to look for immune, molecular, and clinical clues to this puzzle.

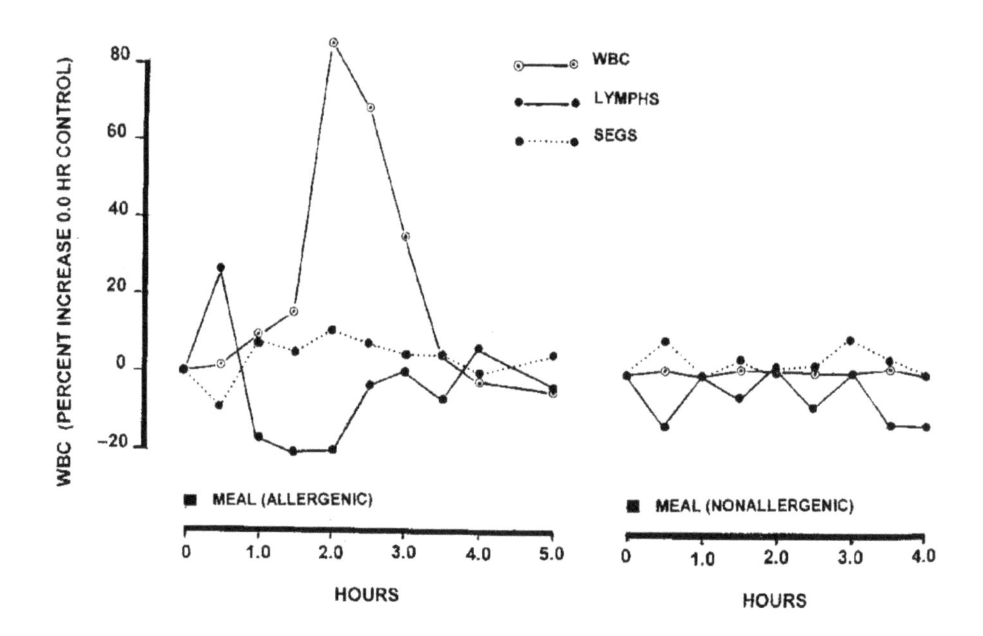

Figure 04.02 A comparison of temporal patterns of the WBC and two key differential components, segs (segmented neutrophils) and lymphs (lymphocytes) during allergic and nonallergic challenge sessions (see Table 02.02).

THE EEG AND THE IMMUNE RESPONSE: The frontal and parietal lobe primary wave patterns (FIGURE 04.01) were synchronized with temporal changes in the WBC detailed in Chapter 2 (figure 02.01 and Table 02.02). This observation in and of itself was exciting and when compared to the WBC results (figure 04.02) gave us an unbelievable correlation (see figure 04.03) and a reason to be ecstatic about the possibilities for further investigations!

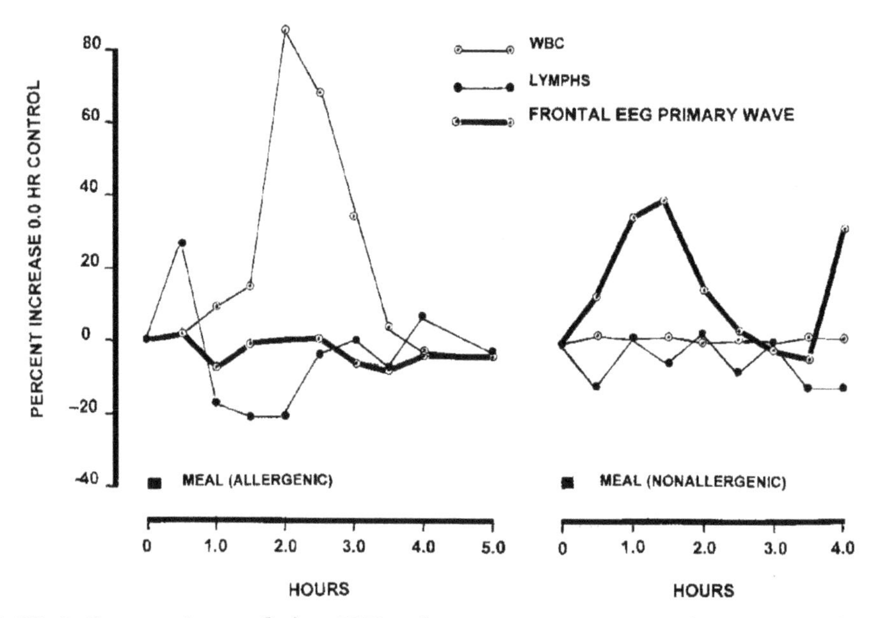

Figure 04.03 A Comparison of the EEG primary wave temporal pattern relative to the temporal patterns for the WBC and the differential Lymphocytes during allergic and nonallergic test sessions.

The average timed primary EEG wave patterns for allergenic and nonallergenic challenges (Figure 04.01) were very different in relation to the WBC and lymphocyte patterns (Figures 04.02 and 04.03). The component primary waves, specifically the allergenic challenge 7.5 and 13.0 components appear to temporally coordinate with the WBC and lymphocyte patterns (Figure 04.04). The parietal primary wave and component wave patterns (Figure 04.05) did not relate to the patterns seen in the frontal lobe and, in fact, revealed a characteristic, distinct and separate set of results that also served as an internal control for the observed frontal lobe patterns. The detailed studies (Tables 01 – 22) leading to these observations will be reviewed in great detail in the subsequent pages. There is a great deal of information that remains to be extracted from this set of data. As is generally the case one set of experiments suggest many more experiments and many new avenues to explore.

COMPUTER ANALYZED ELECTROENCEPHALOGRAPHS: The 0.0-time control (Table 04.01) provided the basis for comparison for a timed series of samples taken at 30-minute intervals after an allergenic food challenge (Tables 04.02 – 04.09). Using the classical International (10 – 20) Electrode Placement procedure for parietal, frontal and occipital EEGs, responses were recorded. EKGs and GSRs were also recorded. Tapes of the temporal and parietal EEG recordings were computer analyzed. The computer analyzed EEG sampling rate was 320/sec, the epoch size was 20 secs (50 microvolt = 7.0 mm/10 Hz = 14 mm). Wavelength frequencies, as also described by Ward (1958), in the alpha range were 8 – 13 Hz. In the beta range were greater than 14 Hz, in the theta range were 4 – 7 Hz, and in the delta range were less than 4 Hz.

TABLE 4.1 IN VIVO CYTOTOXIC TESTED FOOD (SENSITIZING) CHALLENGE WITH TEMPORAL ANALYSIS OF EEG RESULTS.

			FRONTAL (F3,F4 10/20) EEG PRIMARY WAVE (320/SEC SAMPLING/20 SEC EPOCHS)							
ANALYTE			3.5	7.5	13.0	20.0	26.6	40.0	90.0	
RANGE		AVG								
UNITS	TIME	PRIMARY	%	%	%	%	%	%	%	%
CASE/AGE/SEX	HOUR	WAVE								
SGP/35/M	0.0	13.2	34.1	11.7	20.2	13.1	7.6	6.9	4.3	0.8
"ALLERGENIC"		19.0	0.0	12.8	33.0	22.4	11.2	13.2	6.6	0.6
MEAL(10 MIN)		19.5	0.0	10.8	34.8	21.7	12.1	12.1	7.3	0.8
(CHICKEN)		18.9	0.0	12.4	36.2	22.4	10.2	11.0	6.6	0.8
(CHOCOLATE)		18.7	0.0	8.0	38.9	25.4	10.0	10.5	6.3	0.7
(PEANUT)		18.8	0.0	11.8	34.2	24.7	10.5	11.7	6.3	0.7
(TEA)		20.2	0.0	11.5	30.2	21.8	12.2	16.4	7.1	0.6
5/30/73		20.7	0.0	10.0	29.2	25.0	13.2	13.7	7.8	1.0
9:30AM		19.4	0.8	12.1	33.5	23.2	11.5	11.1	7.0	0.7
========										
AVERAGE		**18.7**	**3.9**	**11.2**	**32.2**	**22.2**	**10.9**	**11.8**	**6.6**	**0.7**
SUM		168.4	34.9	101.1	290.2	199.7	98.5	106.6	59.3	6.7
SUM OF SQS		3188.7	1163.5	1153.0	9588.3	4539.8	1099.2	1316.3	398.5	5.1
NUMBER (N)		9.0	9.0	9.0	9.0	9.0	9.0	9.0	9.0	9.0
SQRT OF N		3.0	3.0	3.0	3.0	3.0	3.0	3.0	3.0	3.0
MAXIMUM		20.7	34.1	12.8	38.9	25.4	13.2	16.4	7.8	1.0
MINIMUM		13.2	0.0	8.0	20.2	13.1	7.6	6.9	4.3	0.6
RANGE		7.5	34.1	4.8	18.7	12.3	5.6	9.5	3.5	0.4
STD ERROR (SEM)		0.8	3.8	0.5	2.1	1.4	0.6	1.1	0.4	0.0
SEM %		4.5	97.7	4.7	6.4	6.2	5.7	8.9	5.9	6.0
STD DEV (SD)		**2.5**	**11.4**	**1.6**	**6.2**	**4.1**	**1.9**	**3.2**	**1.2**	**0.1**
SD % COEF VAR		**13.4**	**293.1**	**14.2**	**19.3**	**18.5**	**17.1**	**26.7**	**17.7**	**17.9**
EXPECTED-		13.7	-18.9	8.0	19.8	14.0	7.2	5.5	4.3	0.5
RANGE (2SD)		23.7	26.6	14.4	44.7	30.4	14.7	18.2	8.9	1.0

ALLERGENIC TEST SERIES AND THE EEG: The average primary wave for the 0.0-time control (Table 4.1) was 18.7 +/-2.5 SD (13.4%CV). The percentage distribution of the primary wave into the component primary waves (3.5, 7.5, 13.0, 20.0, 26.6, 40.0 and 90.0) provides a highly specific and unique detailed expansion of the primary wave of the EEG. The averages for the component primary waves +/- the standard deviation (SD) and the percent coefficient of variation (CV) provide key statistical information about the data for each timed interval. The SD and CV are useful indicators of variation in the data sample. The expected range calculated at +/- 2SDs sets limits that one can reasonably expect for the primary or component primary waves to fall within for a given individual(s) for the timed intervals and conditions studied. The computer analyzed EEG primary wave and component primary waves provide diagnostic information useful to interpreting CNS status similar to the way the WBC and differential help to monitor and provide diagnostic information on the bodies systemic well-being as monitored from the peripheral circulation. These average data points for the 0.0-time control were analyzed in comparison to the average data points at the designated 30-minute test intervals.

At the 0.5-hour interval (Table 4.2) there were no significant primary waves or component primary wave differences as compared to the 0.0-time control as determined by application of the student T Test to corresponding sets of data? Significant differences were defined as achieving the 95% confidence level or better and are recorded in the tables for each timed interval. Human cortical electrical potentials range from 5 – 50 micro volts with 1 second down to 20 milliseconds duration. The most prominent component wave frequency, the alpha or Berger (1929) rhythm (8 – 13 cycles per second) is seen here averaging 35.7 +/-5.0 S.D. (13.9%C.V.) percent of the primary wave at the 0.5-hour mark. These alpha frequency waves are characteristic of subjects that have closed eyes and are physically and mentally relaxed. Sensory stimulation, physical or mental activities abolish the "at rest" pattern. These alpha patterns increase, with body temperature, with surprise, fear, or nervousness, and with change as a function of age. We will describe unique and distinctly different alpha wave patterns for the frontal and parietal regions during allergenic or nonallergenic test periods. Refer to Figs 4.4 and 4.5 to follow the component alpha wave as patterns unfold. Following the 0.5-hour interval the frontal 8 – 13 Hz alpha wave pattern is about to take a significant turn as will the component 4.0 – 7.5 Hz theta and >13 Hz beta frequency wave patterns during allergenic/nonallergenic challenges.

TABLE 4.2 IN VIVO CYTOTOXIC TESTED FOOD (SENSITIZING) CHALLENGE WITH TEMPORAL ANALYSIS OF EEG RESULTS.

			FRONTAL (F3,F4 10/20) EEG PRIMARY WAVE (320/SEC SAMPLING/20 SEC EPOCHS)							
ANALYTE		AVG	3.5	7.5	13.0	20.0	26.6	40.0	90.0	
RANGE	TIME	PRIMARY								
UNITS	HOUR	WAVE	%	%	%	%	%	%	%	%
CASE/AGE/SEX										
SGP/35/M	**0.5**	18.6	5.0	13.5	30.5	20.5	9.6	12.3	6.8	0.9
"ALLERGENIC"		19.3	1.3	13.9	32.4	21.3	10.2	13.2	6.4	1.0
MEAL(10 MIN)		19.1	0.0	14.4	34.7	20.8	9.3	12.9	7.2	0.5
(CHICKEN)		19.7	0.0	10.6	36.2	21.3	12.4	11.5	7.4	0.5
(CHOCOLATE)		18.1	0.0	9.3	41.8	21.1	9.7	11.5	5.9	0.5
(PEANUT)		20.4	0.0	11.2	32.0	23.5	11.0	13.8	7.5	1.0
(TEA)		20.1	0.0	11.6	32.9	22.9	11.4	12.3	7.7	0.9
5/30/73		17.7	0.0	9.2	46.2	20.1	8.5	9.6	5.4	0.8
9:30AM		18.6	0.8	10.2	35.7	25.2	11.9	9.1	6.1	0.9
		19.3	0.0	12.1	34.4	22.0	13.0	11.3	6.5	0.7
========										
AVERAGE		**19.1**	**0.7**	**11.6**	**35.7**	**21.9**	**10.7**	**11.8**	**6.7**	**0.8**
SUM		190.9	7.1	116.0	356.8	218.7	107.0	117.5	66.9	7.7
SUM OF SQS		3650.9	27.3	1376.8	12940.5	4805.2	1164.4	1400.8	452.8	6.3
0HR CTL/0.5 EXP					P<0.10					
NUMBER (N)		10.0	10.0	10.0	10.0	10.0	10.0	10.0	10.0	10.0
SQRT OF N		3.2	3.2	3.2	3.2	3.2	3.2	3.2	3.2	3.2
MAXIMUM		20.4	5.0	14.4	46.2	25.2	13.0	13.8	7.7	1.0
MINIMUM		17.7	0.0	9.2	30.5	20.1	8.5	9.1	5.4	0.5
RANGE		2.7	5.0	5.2	15.7	5.1	4.5	4.7	2.3	0.5
STD ERROR (SEM)		0.3	0.5	0.5	1.6	0.5	0.5	0.5	0.2	0.1
SEM %		1.4	70.4	4.5	4.4	2.3	4.2	4.0	3.4	6.5
STD DEV (SD)		**0.9**	**1.6**	**1.6**	**5.0**	**1.6**	**1.4**	**1.5**	**0.7**	**0.2**
SD % COEF VAR		**4.5**	**222.7**	**14.2**	**13.9**	**7.4**	**13.3**	**12.6**	**10.9**	**20.5**
EXPECTED-		17.4	-2.5	8.3	25.8	18.6	7.9	8.8	5.2	0.5
RANGE (2SD)		20.8	3.9	14.9	45.6	25.1	13.5	14.7	8.1	1.1

Following the 0.5-hour interval (Figure 4.4) the frontal lobe component alpha wave patterns (8 – 13 Hz) is about to take a significant up-turn as will the theta wave component (4.0 – 7.5 Hz). The beta component wave (>13 Hz frequencies wave patterns) is tracking below the zero control and very different than the non-allergenic challenge meal. The allergic/nonallergic challenges show distinct and separate experimental profiles.

At the 1.0-hour interval (Table 4.3) the frontal lobe primary wave decreased significantly (P<0.05) from the 0.0 hour 18.7+/-2.5 S.D.(13.4%CV) to 17.3+/-1.2 S.D.(7.0%CV) while the theta (7.5 Hz) and alpha (8 -13 Hz) components of the frontal lobe primary waves shifted up significantly (P<0.0025). The component waves at 20.0 and 26.6 Hz showed no significant change, and the component waves at 40.0 and 90.0 Hz shifted down significantly (P<0.0025) as compared to the 0.0-time control values. For now, these differences are left for further studies

Continuing to track the timed data sets for the components of the primary wave a very interesting pattern begins to unfold in the frontal and parietal lobes. In figure 4.1 we saw some surprising apparently opposite primary wave temporal lobe patterns unfold, beneath the surface within the net sum of the respective allergenic and non-allergenic component primary waves the real story of these studies begins to be revealed. In figures 4.2 and 4.3 the timed primary wave components for allergenic/nonallergenic were compared to total WBC and component lymphocyte differential taken simultaneously for each experimental session. A remarkable story emerges.

TABLE 4.3 IN VIVO CYTOTOXIC TESTED FOOD (SENSITIZING) CHALLENGE WITH TEMPORAL ANALYSIS OF EEG RESULTS.

			FRONTAL (F3,F4 10/20) EEG PRIMARY WAVE (320/SEC SAMPLING/20 SEC EPOCHS)							
			**							
ANALYTE		AVG	3.5	7.5	13.0	20.0	26.6	40.0	90.0	
RANGE	TIME	PRIMARY								
UNITS	HOUR	WAVE	%	%	%	%	%	%	%	%
CASE/AGE/SEX										
SGP/35/M	**1.0**	16.0	8.0	18.0	34.1	13.6	8.8	10.4	5.0	0.8
"ALLERGENIC"		17.9	0.7	11.9	38.7	23.1	9.1	10.2	5.5	0.6
MEAL(10 MIN)		17.2	0.0	17.2	36.2	21.1	11.4	8.2	5.2	0.6
(CHICKEN)		16.6	0.0	15.5	38.6	22.6	9.9	8.4	4.5	0.4
(CHOCOLATE)		15.3	0.0	18.7	43.5	18.0	8.3	6.4	4.1	0.6
(PEANUT)		19.3	0.8	11.2	35.2	21.8	10.8	12.6	6.5	0.9
(TEA)		17.4	0.0	16.8	38.3	19.5	9.6	9.5	5.7	0.5
5/30/73		18.5	0.0	14.9	35.8	19.9	10.8	11.4	6.5	0.5
9:30AM		17.7	0.7	12.0	40.8	18.8	11.0	10.7	5.2	0.6
		17.3	0.8	13.6	39.8	21.5	9.2	9.2	5.4	0.4
		16.6	0.7	11.9	45.2	19.7	9.5	7.9	4.4	0.6
========										
AVERAGE		**17.3**	**1.1**	**14.7**	**38.7**	**20.0**	**9.9**	**9.5**	**5.3**	**0.6**
SUM		189.8	11.7	161.7	426.2	219.6	108.4	104.9	58.0	6.5
SUM OF SQS		3287.5	66.8	2451.7	16632.4	4453.8	1078.4	1031.5	311.9	4.1
0HR CTL/1.0 EXP		P<0.05		P<0.0025	P<0.0025	P<0.10	P<0.10	P<0.025	P<0.0025	
NUMBER (N)		11.0	11.0	11.0	11.0	11.0	11.0	11.0	11.0	11.0
SQRT OF N		3.3	3.3	3.3	3.3	3.3	3.3	3.3	3.3	3.3
MAXIMUM		19.3	8.0	18.7	45.2	23.1	11.4	12.6	6.5	0.9
MINIMUM		15.3	0.0	11.2	34.1	13.6	8.3	6.4	4.1	0.4
RANGE		4.0	8.0	7.5	11.1	9.5	3.1	6.2	2.4	0.5
STD ERROR (SEM)		0.4	0.7	0.7	1.0	0.9	0.3	0.6	0.2	0.0
SEM %		2.1	68.4	4.6	2.6	4.3	2.9	5.9	4.1	7.7
STD DEV (SD)		**1.2**	**2.4**	**2.3**	**3.3**	**2.9**	**0.9**	**1.9**	**0.7**	**0.2**
SD % COEF VAR		**7.0**	**226.8**	**15.4**	**8.6**	**14.3**	**9.5**	**19.6**	**13.7**	**25.5**
EXPECTED-		14.8	-3.8	10.2	32.1	14.2	8.0	5.8	3.8	0.3
RANGE (2SD)		19.7	5.9	19.2	45.4	25.7	11.7	13.3	6.7	0.9

At the 1.0-hour interval we see the theta and alpha waves rise significantly above the 0.0-hour controls preceding the rise of the total WBC count and lagging behind the significant rise of the differential lymphocytes (Figure 4.4). The frontal lobe theta rhythms with origins in the hippocampus constitute 14.7+/-2.3 S.D. (15.4%CV) per cent of the primary wave, that at the 1.0-hour interval had significantly (P<0.0025) risen 31.3 percent above the 0.0-hour control level 0.5 hour after a lymphocyte spike and one hour in advance of the WBC maximum (Figure 4.4). **Lymphocytes sound an alarm, theta, beta waves**

alert the frontal lobe, marginal WBCs released from endothelial lining, a second alarm is sounded in the frontal lobe perhaps to remarginalize the WBCs or to recruit other tissues to respond immunologically or allergically, who knows, I love to speculate!

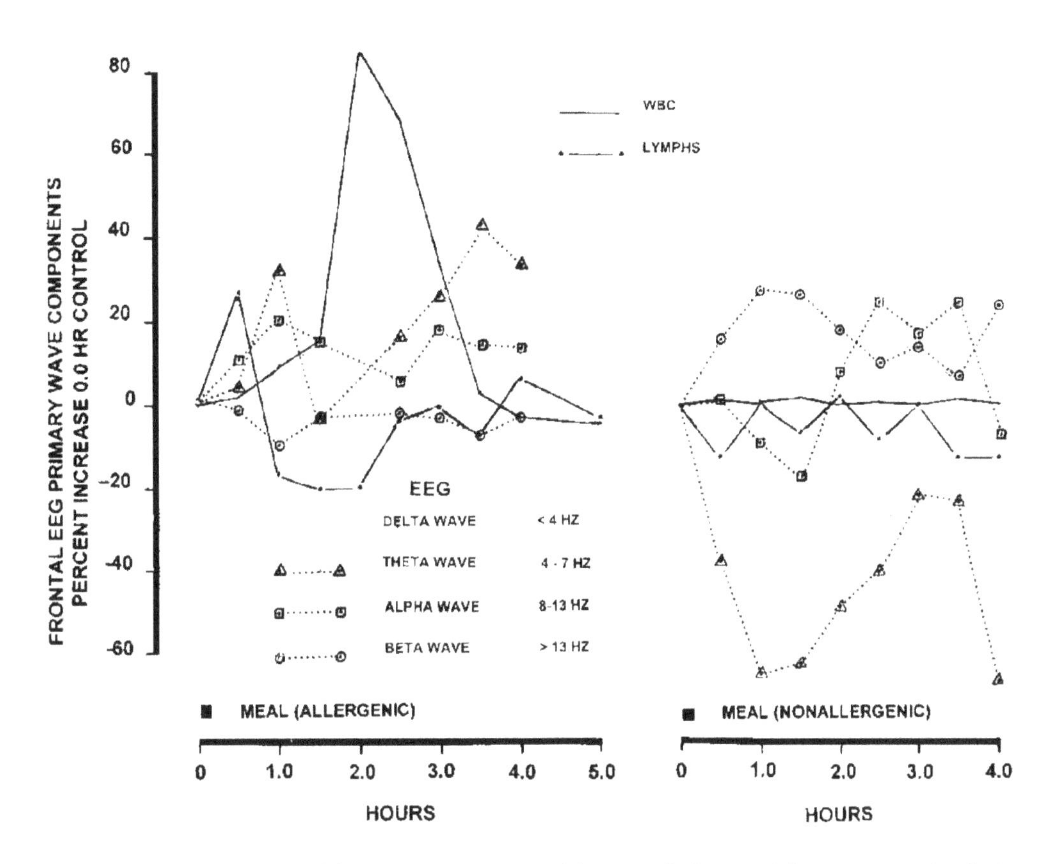

Figure 4.4 A Comparison of key components (theta, alpha and beta waves) of the EEG primary wave temporal pattern relative to the temporal patterns for the WBC and the differential Lymphocyte during allergic and nonallergic test sessions.

Hypothesis: Allergenic meal antigen(s) is detected by the immune defense system (lymphocytes) detected (immunosensory) in the frontal lobe (theta waves), and marginal cells (WBC maxima) are released via immune-visceral efferent messages to the vascular endothelial lining (intima):

Ag(s) + Ab(s) + Target Cells = Positive Test + Cortical Sensory Coding (4.1)

This advances the molecular models proposed in Chapter 3 into the realm of electrical sensory coding (Morrel, 1967; Adey, 1967). We will continue to track the primary wave and components of the primary wave bearing in mind that many molecular and genetic connections underlie these results.

TABLE 4.4 IN VIVO CYTOTOXIC TESTED FOOD (SENSITIZING) CHALLENGE WITH TEMPORAL ANALYSIS OF EEG RESULTS.

			FRONTAL (F3,F4 10/20) EEG PRIMARY WAVE (320/SEC SAMPLING/20 SEC EPOCHS)							
ANALYTE RANGE UNITS CASE/AGE/SEX	TIME HOUR	AVG PRIMARY WAVE	3.5 %	7.5 %	13.0 %	20.0 %	26.6 %	40.0 %	90.0 %	%
SGP/35/M	**1.5**	17.5	10.9	13.7	26.8	22.2	7.9	10.6	6.8	0.7
"ALLERGENIC"		19.9	0.7	9.8	37.6	19.6	10.7	12.6	7.9	0.9
MEAL(10 MIN)		18.1	0.0	10.4	42.2	21.9	8.7	9.5	6.4	0.6
(CHICKEN)		18.0	0.0	9.4	43.3	20.7	11.2	8.9	5.7	0.6
(CHOCOLATE)		19.1	0.7	11.6	35.3	21.1	11.4	12.0	6.8	0.8
(PEANUT)		18.4	0.0	11.7	39.6	20.9	9.8	10.5	6.7	0.6
(TEA)		18.0	0.7	10.0	39.6	22.8	10.4	10.1	5.8	0.4
5/30/73		20.4	0.0	9.5	31.5	24.5	11.8	14.5	7.1	0.8
9:30AM		17.9	0.0	11.1	38.9	23.0	10.9	10.5	4.9	0.6
========										
AVERAGE		**18.6**	**1.4**	**10.8**	**37.2**	**21.9**	**10.3**	**11.0**	**6.5**	**0.7**
SUM		167.3	13.0	97.2	334.8	196.7	92.8	99.2	58.1	6.0
SUM OF SQS		3117.8	120.3	1065.2	12675.6	4316.2	970.2	1117.3	381.3	4.2
0HR CTL/1.5 EXP					P<0.05					
NUMBER (N)		9.0	9.0	9.0	9.0	9.0	9.0	9.0	9.0	9.0
SQRT OF N		3.0	3.0	3.0	3.0	3.0	3.0	3.0	3.0	3.0
MAXIMUM		20.4	10.9	13.7	43.3	24.5	11.8	14.5	7.9	0.9
MINIMUM		17.5	0.0	9.4	26.8	19.6	7.9	8.9	4.9	0.4
RANGE		2.9	10.9	4.3	16.5	4.9	3.9	5.6	3.0	0.5
STD ERROR (SEM)		0.3	1.2	0.5	1.8	0.5	0.4	0.6	0.3	0.1
SEM %		1.7	83.8	4.4	4.9	2.5	4.2	5.6	5.2	8.3
STD DEV (SD)		**1.0**	**3.6**	**1.4**	**5.5**	**1.6**	**1.3**	**1.9**	**1.0**	**0.2**
SD % COEF VAR		**5.2**	**251.5**	**13.3**	**14.8**	**7.5**	**12.6**	**16.9**	**15.5**	**25.0**
EXPECTED-		16.7	-5.8	7.9	26.2	18.6	7.7	7.3	4.5	0.3
RANGE (2SD)		20.5	8.7	13.7	48.2	25.1	12.9	14.8	8.5	1.0

By the 1.5-hour interval (Table 4.4) only the 13.0 Hz alpha wave component constituting 37.2+/-5.5 SD(14.8%) percent of the primary wave remained a 15.5% significant (P<0.05) increase over the 32.2+/-6.2SD(19.3%CV) 0.0-time control. These are significant changes and shifts that are being temporally tracked following the ingestion of foods identify as allergenic/nonallergenic, however the temporal pattern keeps the full meaning of these results in perspective. At 1.5 hours the theta component wave falls back to control levels while the WBC begins a significant rise. The theta wave component lagged behind the lymphs that fell significantly below control levels (Figure 4.4) while the total WBC began the significant rise. At the 1.5-hour interval the proposed model (4.1) moves forward:

Allergen(s) + Immune System + Frontal CNS =>
End-organ Sensitivity/Positive Test + Cortical Sensory Coding (4.2)

We were unable to make recordings at the 2.0-hour interval for the sensitizing food response; therefore, the corresponding table 4.5 has been omitted. At the 2.5-hour interval (Table 4.6) there was no significant change in the average primary wave at 18.8+/-1.1 SD (5.8% CV) as compared to the 0.0-time control, nor were there any significant shifts in the percentage distributions of the primary wave components. Note the theta wave in figure 4.4 shows a rising trend at 12.9+/-3.6 SD(27.6%CV) per cent of the primary wave, however the 27.6% coefficient of variation in this data sample proves too variable (rule of thumb cut-off <20%CV) to test significant difference from 0.0-hour control. By comparison note the %CVs of the primary wave (5.8%CV) or the component waves at 13Hz (10.1%CV) and 20Hz (6.9%).

TABLE 4.6 IN VIVO CYTOTOXIC TESTED FOOD (SENSITIZING) CHALLENGE WITH TEMPORAL ANALYSIS OF EEG RESULTS.

ANALYTE RANGE UNITS CASE/AGE/SEX	TIME HOUR	AVG PRIMARY WAVE	FRONTAL (F3,F4 10/20) EEG PRIMARY WAVE (320/SEC SAMPLING/20 SEC EPOCHS) 3.5 %	 7.5 %	 13.0 %	 20.0 %	 26.6 %	 40.0 %	 90.0 %	 %
SGP/35/M	**2.5**	17.9	7.5	13.7	28.5	21.1	10.5	11.6	6.0	0.8
"ALLERGENIC"		17.5	0.8	13.3	38.8	21.6	9.7	9.3	5.6	0.6
MEAL(10 MIN)		20.8	0.7	8.9	30.1	24.0	11.5	16.0	7.8	0.8
(CHICKEN)		17.5	0.0	19.6	34.8	19.5	10.0	9.6	5.9	0.6
(CHOCOLATE)		19.6	0.7	12.4	35.2	21.7	9.6	12.1	7.3	0.8
(PEANUT)		19.5	0.0	10.7	33.8	21.8	13.8	12.4	6.5	0.8
(TEA)		19.2	0.0	12.1	33.8	22.0	11.8	12.0	7.2	0.6
5/30/73		19.1	0.0	13.3	34.5	21.3	11.6	12.6	5.8	0.7
9:30AM		18.5	0.0	12.5	37.4	23.7	8.8	10.5	6.3	0.7
========										
AVERAGE		**18.8**	**1.1**	**12.9**	**34.1**	**21.9**	**10.8**	**11.8**	**6.5**	**0.7**
SUM		169.6	9.7	116.5	306.9	196.7	97.3	106.1	58.4	6.4
SUM OF SQS		3205.7	57.9	1575.8	10547.7	4313.5	1070.4	1282.4	383.7	4.6
0HR CTL/2.5 EXP										
NUMBER (N)		9.0	9.0	9.0	9.0	9.0	9.0	9.0	9.0	9.0
SQRT OF N		3.0	3.0	3.0	3.0	3.0	3.0	3.0	3.0	3.0
MAXIMUM		20.8	7.5	19.6	38.8	24.0	13.8	16.0	7.8	0.8
MINIMUM		17.5	0.0	8.9	28.5	19.5	8.8	9.3	5.6	0.6
RANGE		3.3	7.5	10.7	10.3	4.5	5.0	6.7	2.2	0.2
STD ERROR (SEM)		0.4	0.8	1.2	1.1	0.5	0.6	0.7	0.2	0.0
SEM %		1.9	77.3	9.2	3.4	2.3	5.1	6.3	3.8	3.1
STD DEV (SD)		**1.1**	**2.5**	**3.6**	**3.4**	**1.5**	**1.7**	**2.2**	**0.7**	**0.1**
SD % COEF VAR		**5.8**	**232.0**	**27.6**	**10.1**	**6.9**	**15.4**	**18.9**	**11.3**	**9.4**
EXPECTED-		16.6	-3.9	5.8	27.2	18.9	7.5	7.3	5.0	0.6
RANGE (2SD)		21.0	6.1	20.1	41.0	24.9	14.1	16.3	8.0	0.8

At the 2.5-hour interval there was no significant change in the primary wave at 18.8+/-1.1 SD (5.8% CV) as compared to the 0.0-hour control, nor were there any significant shifts in the percentage distributions of the primary wave components. Note the theta wave in Fig 4.4 theta wave plot shows a rising trend at 12.9+/-3.6 SD (27.6% CV) per cent of the primary wave, however the 27.6% coefficient of variation in this small data sample proves too variable to test significantly different from the 0.0-hour control. What happens if the low and high values are dropped from the theta wave data sample? By comparison note the %CVs of the primary wave (5.8%) or the component waves at 13 Hz (10.1%) and 20 Hz (6.9%).

At the 3.0-hour interval (Table 4.7) the average primary wave at 17.7+/-1.1 SD (6.2 %CV) did not differ significantly from the 0.0-time control values continuing to show a low variation and a narrow-expected range within the samples taken during the challenge period. The same trend of shifts in the average component waves seen at 1.0-hour interval was repeated at 3.0 hours. The 7.5 Hz and 13.0 Hz components increased significantly, the 20.0 Hz and 26.6 Hz showing no change, and the 40.0 Hz and 90.0 Hz showing significant decreases as compared to their 0.0-time controls.

Note the high variation of the 3.0- and 3.5-hour theta wave components relative to the companion component waves for these time periods. Is this chaos among order or are we walking a delicate tightrope of preciously small samples sizes that add up to these observations. Although interesting, the snapshots taken at thirty-min intervals only tell part of the story. The parallel tracking of multiple parameters with time tells the more relevant aspect of these findings (Figure 4.4).

TABLE 4.7 IN VIVO CYTOTOXIC TESTED FOOD (SENSITIZING) CHALLENGE WITH TEMPORAL ANALYSIS OF EEG RESULTS.

			FRONTAL (F3,F4 10/20) EEG PRIMARY WAVE (320/SEC SAMPLING/20 SEC EPOCHS)							
			**							
ANALYTE		AVG	3.5	7.5	13.0	20.0	26.6	40.0	90.0	
RANGE	TIME	PRIMARY								
UNITS	HOUR	WAVE	%	%	%	%	%	%	%	%
CASE/AGE/SEX										
SGP/35/M	**3.0**	17.5	0.0	17.5	36.5	21.2	9.3	9.1	5.3	0.9
"ALLERGENIC"		16.8	0.0	13.1	41.8	21.6	9.5	8.3	5.0	0.4
MEAL(10 MIN)		17.0	0.7	16.5	37.0	21.2	9.6	9.5	4.8	0.6
(CHICKEN)		17.7	0.0	12.5	39.8	21.1	10.7	9.4	5.9	0.5
(CHOCOLATE)		16.7	0.7	20.8	34.3	19.4	9.9	9.5	4.8	0.4
(PEANUT)		20.0	0.0	10.0	33.4	24.5	10.9	12.4	7.8	0.8
(TEA)		18.2	0.0	8.6	43.7	20.7	10.0	10.9	5.5	0.5
5/30/73		18.2	0.7	15.9	33.5	24.1	8.7	10.3	5.6	0.8
9:30AM		17.1	0.7	12.3	42.4	19.9	10.6	8.6	5.0	0.5
========										
AVERAGE		**17.7**	**0.3**	**14.1**	**38.0**	**21.5**	**9.9**	**9.8**	**5.5**	**0.6**
SUM		159.2	2.8	127.2	342.4	193.7	89.2	88.0	49.7	5.4
SUM OF SQS		2824.6	2.0	1917.1	13154.3	4192.6	888.3	873.2	281.4	3.5
0HR CTL/3.0 EXP		P<0.15		P<0.05	P<0.01	P<0.30	P<0.10	P<0.05	P<0.025	
NUMBER (N)		9.0	9.0	9.0	9.0	9.0	9.0	9.0	9.0	9.0
SQRT OF N		3.0	3.0	3.0	3.0	3.0	3.0	3.0	3.0	3.0
MAXIMUM		20.0	0.7	20.8	43.7	24.5	10.9	12.4	7.8	0.9
MINIMUM		16.7	0.0	8.6	33.4	19.4	8.7	8.3	4.8	0.4
RANGE		3.3	0.7	12.2	10.3	5.1	2.2	4.1	3.0	0.5
STD ERROR (SEM)		0.4	0.1	1.4	1.1	0.6	0.2	0.5	0.3	0.1
SEM %		2.1	25.0	9.6	3.0	2.6	2.5	4.7	6.0	9.3
STD DEV (SD)		**1.1**	**0.2**	**4.1**	**3.4**	**1.7**	**0.7**	**1.4**	**1.0**	**0.2**
SD % COEF VAR		**6.2**	**75.0**	**28.8**	**9.0**	**7.9**	**7.4**	**14.0**	**18.1**	**27.8**
EXPECTED-		15.5	-0.2	6.0	31.2	18.1	8.4	7.0	3.5	0.3
RANGE (2SD)		19.9	0.8	22.3	44.9	24.9	11.4	12.5	7.5	0.9

By the 3.5-hour interval (Table 4.8) the average primary wave at 17.6+/-1.5 SD (8.4 %CV) still did not differ significantly from the 0.0-time control values. The 7.5 Hz and 13.0 Hz components increased significantly (the largest and most significant seen for 7.5 Hz), only the 20.0 Hz showed no change, with the 26.6, 40.0 and 90.0 Hz components showing significant decreases as compared to the 0.0-time controls. Continuing to track the frontal lobe allergenic test series primary wave component waves, the theta wave (Table 4.4) tracked to a 3.5-hour interval 18% high (P<0.01) above the 0.0-time control. This included the 1.0-hour theta spike that fell back to control levels at the 1.5-hour interval before capping a steady 2-hour upward trend to the 3.5-hour interval. Meanwhile the lymphs that peaked at the 0.5-hour interval fell below the 0.0-hour control levels at the 1.0-, 1.5-, and 2.0-hour intervals before tracking back to control levels for the 2.5 – 4.0-hour intervals. During the 1.0-hour frontal lobe theta wave spike the initial WBC was underway and as the initial theta wave fell back to the 0.0-hour control levels the WBC rose sharply to a high at the 2.0-hour mark. While the frontal lobe theta wave rose in the 2.5 – 3.5-hour time frame while the WBC fell sharply back to the 0.0-hour control levels. The primary wave alpha component wave with the theta component–at the 0.5-hour interval, then tracked less dramatically above 0.0-hour control through the 4.0-hour mark of this test series. Only the beta wave 2.5-hour interval did not test significantly above the 0.0-hour control level.

This experimental model was based on a ten-minute exposure to the test meals, with what appears to be a recovery after about four hours. But, then its lunch time, dinner time and here we go again. Some of the foods we expose ourselves are dose dependent. How about snacks, coffee, or tobacco? Do you have "Food Fever" symptoms to go along with "Hay Fever" symptoms that give you that run down feeling? If positive for coffee or tobacco are you predisposed to cardiovascular disease or cancer? Many studies in the literature have not screened their patients and therefore have failed with thousands of patients and years of research to answer this question.

TABLE 4.8 IN VIVO CYTOTOXIC TESTED FOOD (SENSITIZING) CHALLENGE WITH TEMPORAL ANALYSIS OF EEG RESULTS.

			FRONTAL (F3,F4 10/20) EEG PRIMARY WAVE (320/SEC SAMPLING/20 SEC EPOCHS)							
ANALYTE RANGE UNITS CASE/AGE/SEX	TIME HOUR	AVG PRIMARY WAVE	3.5 %	7.5 %	13.0 %	20.0 %	26.6 %	40.0 %	90.0 %	%
SGP/35/M	**3.5**	14.8	8.4	24.2	30.5	16.2	6.0	8.2	4.7	0.5
"ALLERGENIC"		17.8	0.0	14.0	40.3	20.5	8.0	9.6	6.7	0.6
MEAL(10 MIN)		18.9	0.8	15.0	34.2	21.3	9.0	11.7	6.9	0.7
(CHICKEN)		17.4	0.0	13.3	40.2	22.6	8.3	8.7	5.9	0.7
(CHOCOLATE)		16.7	0.0	18.8	38.5	20.8	8.9	6.9	5.1	0.6
(PEANUT)		17.7	0.7	17.9	36.0	17.6	11.2	10.0	5.8	0.6
(TEA)		17.6	0.0	17.5	35.6	21.2	10.2	9.1	5.6	0.6
5/30/73		17.8	0.0	14.4	37.0	23.0	8.8	10.9	5.4	0.6
9:30AM		17.7	0.7	10.7	43.5	19.7	9.8	9.5	5.2	0.7
		19.5	0.0	14.3	31.8	23.3	10.1	12.5	7.1	0.7
========										
AVERAGE		**17.6**	**1.1**	**16.0**	**36.8**	**20.6**	**9.0**	**9.7**	**5.8**	**0.6**
SUM		175.9	10.6	160.1	367.6	206.2	90.3	97.1	58.4	6.3
SUM OF SQS		3108.2	72.2	2690.0	13658.1	4299.0	834.1	967.7	347.0	4.0
0HR CTL/3.5 EXP		P<0.10		P<0.0025	P<0.025	P<0.15	P<0.01	P<0.025	P<0.05	
NUMBER (N)		10.0	10.0	10.0	10.0	10.0	10.0	10.0	10.0	10.0
SQRT OF N		3.2	3.2	3.2	3.2	3.2	3.2	3.2	3.2	3.2
MAXIMUM		19.5	8.4	24.2	43.5	23.3	11.2	12.5	7.1	0.7
MINIMUM		14.8	0.0	10.7	30.5	16.2	6.0	6.9	4.7	0.5
RANGE		4.7	8.4	13.5	13.0	7.1	5.2	5.6	2.4	0.2
STD ERROR (SEM)		0.5	0.8	1.4	1.3	0.7	0.5	0.6	0.2	0.0
SEM %		2.7	79.2	8.4	3.5	3.4	5.8	5.8	4.1	3.2
STD DEV (SD)		**1.5**	**2.7**	**4.3**	**4.1**	**2.2**	**1.6**	**1.8**	**0.8**	**0.1**
SD % COEF VAR		**8.4**	**250.6**	**26.7**	**11.2**	**10.9**	**18.2**	**18.2**	**13.0**	**10.0**
EXPECTED-		14.6	-4.3	7.5	28.5	16.1	5.7	6.2	4.3	0.5
RANGE (2SD)		20.6	6.4	24.5	45.0	25.1	12.3	13.3	7.4	0.8

At 4.0 hours (Table 4.9), the last timed interval tested in this allergenic challenge meal series; the average primary waves at 18.0+/-1.0 SD (5.6 %CV) did not vary significantly from the 0.0-time control. The primary wave series consistently exhibited narrow variation in each data sample as reflected by the SDs and CVs. The shifts among the various average component waves showed the 7.5 Hz and 13.0 Hz significantly increased, the 20.0 Hz was unchanged as it had remained throughout this entire series, the 26.6 Hz was significantly decreased, while the 40.0 Hz and 90.0 Hz were not significantly changed, all as compared to the 0.0-time control.

These frontal lobe average primary and component waves for the time intervals of the May 30th allergenic challenge meal results are summarized in table 4.19 and figure 4.4. Comparisons to corresponding results for the nonallergenic challenge meal (Table 4.20) and for both challenge meals as to their effects on the parietal lobes (Tables 4.21 and 4.22) have yet to be reviewed. The parietal lobe data provide the real time internal controls that backs up the zero-time controls just reviewed. To be presented is another real time internal control that reveals some very interesting, albeit contrasting data for the nonallergenic challenge data from the parietal lobe EEG recordings six days later on June 3rd. We have presented a few of the highlights buried in all this data, that you have seen, now what needs to be done?

TABLE 4.9 IN VIVO CYTOTOXIC TESTED FOOD (SENSITIZING) CHALLENGE WITH TEMPORAL ANALYSIS OF EEG RESULTS.

			FRONTAL (F3,F4 10/20) EEG PRIMARY WAVE (320/SEC SAMPLING/20 SEC EPOCHS)							
			**							
ANALYTE		AVG	3.5	7.5	13.0	20.0	26.6	40.0	90.0	
RANGE	TIME	PRIMARY								
UNITS	HOUR	WAVE	%	%	%	%	%	%	%	%
CASE/AGE/SEX										
SGP/35/M	**4.0**	18.6	1.9	15.4	32.3	22.1	9.8	10.3	7.2	0.8
"ALLERGENIC"		16.3	0.0	18.0	40.6	18.3	8.5	9.4	4.7	0.4
MEAL(10 MIN)		17.0	0.8	18.4	35.8	19.1	9.2	11.0	5.2	0.4
(CHICKEN)		17.7	0.0	15.2	35.0	24.6	8.8	10.1	5.5	0.6
(CHOCOLATE)		18.0	0.0	14.8	34.5	24.3	10.3	10.0	5.2	0.6
(PEANUT)		19.5	0.0	14.2	33.6	21.2	10.2	12.8	7.2	0.8
(TEA)		16.4	0.0	14.9	43.2	19.5	8.8	8.2	4.7	0.5
5/30/73		18.6	0.0	11.6	40.3	20.2	9.9	11.0	6.1	0.8
9:30AM		19.0	0.7	13.5	33.2	24.4	9.1	11.1	7.1	0.9
		18.9	0.0	13.4	37.7	21.7	7.7	11.8	7.0	0.8
========										
AVERAGE		**18.0**	**0.3**	**14.9**	**36.6**	**21.5**	**9.2**	**10.6**	**6.0**	**0.7**
SUM		180.0	3.4	149.4	366.2	215.4	92.3	105.7	59.9	6.6
SUM OF SQS		3251.3	4.7	2269.8	13531.4	4687.7	858.1	1132.0	368.8	4.7
0HR CTL/4.0 EXP		P<0.15		P<0.0025	P<0.025	P<0.35	P<0.005	P<0.15	P<0.15	
NUMBER (N)		10.0	10.0	10.0	10.0	10.0	10.0	10.0	10.0	10.0
SQRT OF N		3.2	3.2	3.2	3.2	3.2	3.2	3.2	3.2	3.2
MAXIMUM		19.5	1.9	18.4	43.2	24.6	10.3	12.8	7.2	0.9
MINIMUM		16.3	0.0	11.6	32.3	18.3	7.7	8.2	4.7	0.4
RANGE		3.2	1.9	6.8	10.9	6.3	2.6	4.6	2.5	0.5
STD ERROR (SEM)		0.3	0.2	0.7	1.1	0.6	0.3	0.5	0.3	0.1
SEM %		1.8	55.9	4.6	3.0	2.9	2.8	4.4	4.2	7.6
STD DEV (SD)		**1.0**	**0.6**	**2.2**	**3.4**	**2.0**	**0.8**	**1.5**	**0.8**	**0.2**
SD % COEF VAR		**5.6**	**176.7**	**14.4**	**9.4**	**9.2**	**8.9**	**13.8**	**13.2**	**24.0**
EXPECTED-		16.0	-0.9	10.6	29.7	17.6	7.6	7.7	4.4	0.3
RANGE (2SD)		20.0	1.5	19.2	43.5	25.5	10.9	13.5	7.6	1.0

Nonallergic Test Series and the EEG: The 0.0-hour nonallergenic challenge meal control (Table 4.10) provided the beginning for comparison of the frontal lobe computer analyzed EEG series (Tables 4.11 – 4.18) recorded at 30-minute test intervals set up the same way the tests were for the allergenic series just analyzed (Tables 4.1 – 4.9) above. The average primary wave for the 0.0-time nonallergenic control was 17.0 +/-0.8SD (4.5 %CV) as compared to a more significantly (P<0.025) different with higher variability 18.7 +/-2.5SD (13.4 %CV) for the allergenic 0.0-time control measured six days earlier. Therefore, at the beginning of this test period the base lines differed significantly, but as we will detail below the temporal pattern of the nonallergenic primary waves (Tables 4.10 – 4.18) were dramatically different from the temporal pattern displayed by the allergenic primary waves (Tables 4.1 – 4.9). That is, the relationships of primary wave at each 30-minute test interval relative to the 0.0-time control were very different as were the corresponding relationships for the component primary waves.

In Fig 4.3 the difference between the frontal lobe EEG primary wave during the allergenic test period as compared to the frontal lobe EEG primary wave during the nonallergenic test period was dramatic and surprising in the backdrop of the WBC and lymphocyte result (Ulett and Perry; 1974, 1975). The allergenic meal test period primary wave differs significantly below its zero-time control only at the one-hour mark. We have revealed that the net distribution for the component waves of the primary wave tell a very different story that ties in very interestingly with the WBC and lymphocyte data. In Fig 4.1 the parietal lobe data for both the allergenic and nonallergenic test periods reveal a pattern that varies little from the zero-time control and indeed appears to be a real time internal control for the frontal lobe data, remains to be reviewed in detail. First we will look into the details for what we originally thought would be the real real-time nonallergenic test period control.

TABLE 4.10 IN VIVO CYTOTOXIC TESTED FOOD (NONSENSITIZING) CHALLENGE WITH TEMPORAL ANALYSIS OF EEG RESULTS.

ANALYTE RANGE UNITS CASE/AGE/SEX	TIME HOUR	AVG PRIMARY WAVE	FRONTAL (F3,F4 10/20) EEG PRIMARY WAVE (320/SEC SAMPLING/20 SEC EPOCHS) 3.5 %	7.5 %	13.0 %	20.0 %	26.6 %	40.0 %	90.0 %	%
SGP/35/M	**0.0**	15.6	5.1	23.9	33.9	14.5	8.8	7.7	5.0	0.6
"NONALLERGENIC"		16.3	3.2	18.2	37.5	18.4	8.6	7.9	5.6	0.5
MEAL(10 MIN)		17.2	3.5	21.5	27.5	21.2	9.4	10.3	6.2	0.4
(CHERRY)		17.9	1.4	19.5	32.5	18.3	11.2	9.9	6.5	0.5
(EGG)		16.6	1.7	20.0	35.7	17.8	9.2	9.5	5.6	0.5
(ORANGE)		17.7	0.8	16.7	35.3	20.7	9.7	10.2	5.9	0.6
(PORK)		16.3	0.0	20.7	36.4	19.8	9.4	8.2	4.9	0.5
6/5/73		17.9	1.6	18.9	34.1	18.9	8.4	10.7	6.4	0.8
9:00AM		17.9	2.4	13.8	36.9	21.7	8.2	10.1	5.8	1.0
========										
AVERAGE		**17.0**	**2.2**	**19.2**	**34.4**	**19.0**	**9.2**	**9.4**	**5.8**	**0.6**
SUM		153.4	19.7	173.2	309.8	171.3	82.9	84.5	51.9	5.4
SUM OF SQS		2620.7	62.3	3400.0	10737.9	3298.6	770.1	803.8	301.8	3.5
NUMBER (N)		9.0	9.0	9.0	9.0	9.0	9.0	9.0	9.0	9.0
SQRT OF N		3.0	3.0	3.0	3.0	3.0	3.0	3.0	3.0	3.0
MAXIMUM		17.9	5.1	23.9	37.5	21.7	11.2	10.7	6.5	1.0
MINIMUM		15.6	0.0	13.8	27.5	14.5	8.2	7.7	4.9	0.4
RANGE		2.3	5.1	10.1	10.0	7.2	3.0	3.0	1.6	0.6
STD ERROR (SEM)		0.3	0.6	1.1	1.1	0.8	0.3	0.3	0.2	0.1
SEM %		1.5	25.9	5.8	3.2	4.2	3.6	3.6	3.1	11.1
STD DEV (SD)		**0.8**	**1.7**	**3.4**	**3.3**	**2.4**	**1.0**	**1.0**	**0.5**	**0.2**
SD % COEF VAR		**4.5**	**77.7**	**17.5**	**9.7**	**12.6**	**10.9**	**10.7**	**9.2**	**33.3**
EXPECTED-		15.5	-1.2	12.5	27.8	14.2	7.2	7.4	4.7	0.2
RANGE (2SD)		18.6	5.6	26.0	41.1	23.8	11.2	11.4	6.8	1.0

The average primary wave for the frontal lobe allergenic 0.0-time control (Table 4.1) was 18.7 +/-2.5 SD (13.4%CV). This compared significantly to the frontal lobe nonallergenic 0.0-time control (Table 4.10) of 17.0 +/-0.8 SD (4.5% CV). The 30 min intervals for primary and component waves were compared to these 0.0 min values to calculate the percent increase or decrease from the control values during the four experimental periods. The patterns for the nonallergic challenge were very different than the allergic series (Figures 4.3, 4.4 and 4.5).

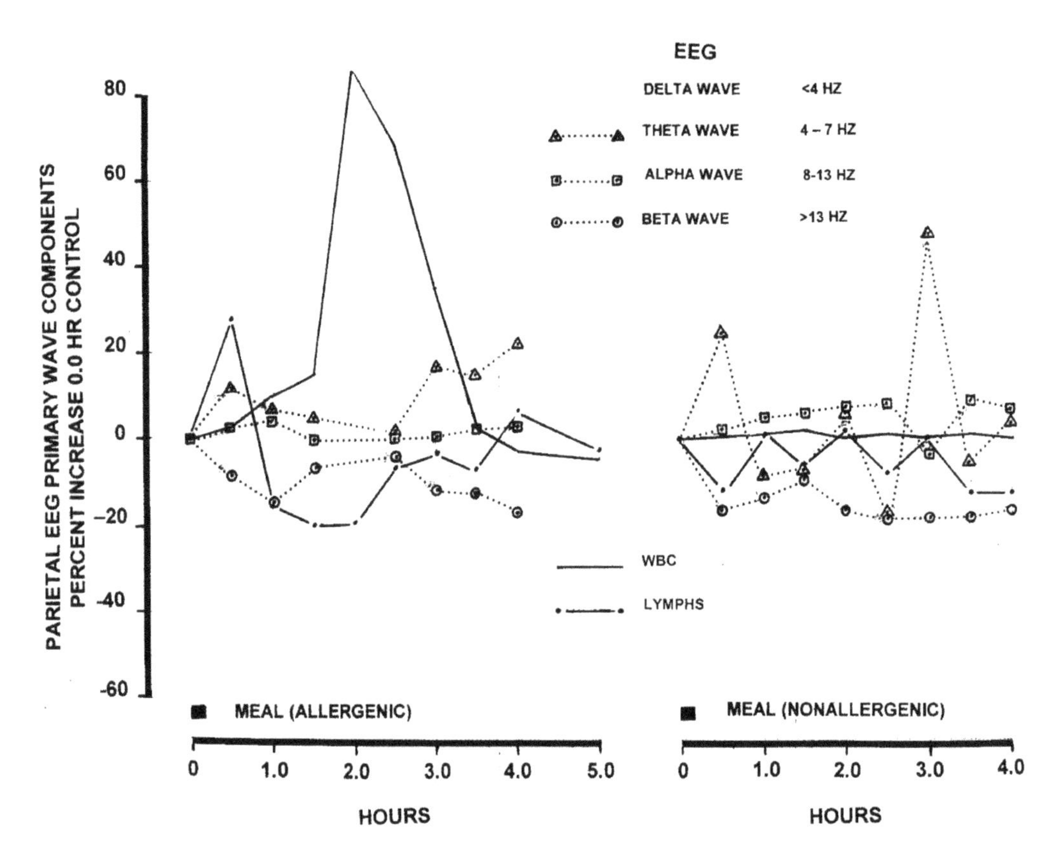

Figure 4.5A Comparison of key components (theta, alpha and beta waves) of the EEG primary wave parietal lobe pattern relative to the temporal patterns for the WBC and the differential Lymphocyte during allergic and nonallergic test sessions.

With the allergenic challenge there were no significant changes at the 0.5 hour (Figure 4.3)for the primary wave, but with the nonallergenic challenge the average primary wave had increased significantly to 19.1 +/-0.8SD (4.4 %CV). The component primary waves (Figure 4.4) were significantly (P<0.0005) down at 4.0-7.5 Hz (theta wave) where we saw only increases in the allergenic challenge, unchanged at 8-13 Hz (alpha wave), significantly (P<0.01) increased at 20 Hz where we did not see any change for the entire allergenic test period, unchanged at 26.6 Hz, and significantly increased at 40.0 Hz (P<0.0025) and 90.0 Hz (P<0.0005).

This is already a markedly different pattern than seen with the allergenic challenge. EEG data collected from the parietal lobe concurrently with the frontal lobe show (Figure 4.5) a different pattern. This was somewhat surprising considering the parietal lobe as being an afferent sensory center it might be receiving visceral afferent signals or even target organ mixed somatic and visceral sensory signals via the hippocampus. Instead, the frontal lobe (worry lobe) not the parietal lobe reflects the activity.

At 0.5 hour the average primary wave is significantly (P<0.0005) increased to 19.1 +/-0.8 (4.4 %CV) over the 0.0-time control. The average primary component waves are still decreased at 7.5 Hz (P<0.0005) while the 13.0 Hz (P<0.40) and 26.6 Hz (P<0.40) components were indistinguishable from the zero-time control. While the 20.0, 40.0 and 90.0 Hz components are all increased significantly by no less than P<0.01 over the 0.0-time nonallergenic challenge control

TABLE 4.11 IN VIVO CYTOTOXIC TESTED FOOD (NONSENSITIZING) CHALLENGE WITH TEMPORAL ANALYSIS OF EEG RESULTS.

			FRONTAL (F3,F4 10/20) EEG PRIMARY WAVE (320/SEC SAMPLING/20 SEC EPOCHS)							
ANALYTE RANGE UNITS CASE/AGE/SEX	TIME HOUR	AVG PRIMARY WAVE	3.5 %	7.5 %	13.0 %	20.0 %	26.6 %	40.0 %	90.0 %	%
SGP/35/M	**0.5**	19.9	9.1	7.6	27.0	21.4	10.4	14.4	8.4	0.8
"NONALLERGENIC"		19.8	0.0	13.0	29.4	25.7	11.5	12.3	7.5	0.6
MEAL(10 MIN)		18.2	0.0	10.3	42.5	21.4	7.3	10.5	6.9	0.6
(CHERRY)		20.6	1.7	10.1	30.9	23.8	10.1	14.0	8.2	1.0
(EGG)		19.2	0.0	13.4	36.2	19.7	10.2	12.2	7.1	0.9
(ORANGE)		18.5	0.0	10.4	39.5	22.6	9.2	10.9	6.7	0.6
(PORK)		18.1	0.7	14.1	40.1	17.9	8.8	10.4	6.8	0.7
6/5/73		19.0	0.0	13.3	35.0	22.2	9.1	12.2	7.4	0.7
9:00AM		18.9	0.0	15.9	34.5	22.5	8.2	10.5	7.4	0.9
========										
AVERAGE		**19.1**	**1.3**	**12.0**	**35.0**	**21.9**	**9.4**	**11.9**	**7.4**	**0.8**
SUM		172.2	11.5	108.1	315.1	197.2	84.8	107.4	66.4	6.8
SUM OF SQS		3300.4	86.2	1351.1	11248.4	4361.2	811.9	1299.8	492.7	5.3
0HR CTL/0.5 EXP		P<0.0005		P<0.0005	P<0.40	P<0.01	P<0.40	P<0.0025	P<0.0005	
NUMBER (N)		9.0	9.0	9.0	9.0	9.0	9.0	9.0	9.0	9.0
SQRT OF N		3.0	3.0	3.0	3.0	3.0	3.0	3.0	3.0	3.0
MAXIMUM		20.6	9.1	15.9	42.5	25.7	11.5	14.4	8.4	1.0
MINIMUM		18.1	0.0	7.6	27.0	17.9	7.3	10.4	6.7	0.6
RANGE		2.5	9.1	8.3	15.5	7.8	4.2	4.0	1.7	0.4
STD ERROR (SEM)		0.3	1.0	0.9	1.7	0.9	0.5	0.4	0.2	0.0
SEM %		1.5	79.1	7.7	4.9	4.0	5.0	3.7	2.6	5.9
STD DEV (SD)		**0.8**	**3.0**	**2.8**	**5.2**	**2.6**	**1.4**	**1.3**	**0.6**	**0.1**
SD % COEF VAR		**4.4**	**237.4**	**23.0**	**14.8**	**11.9**	**14.9**	**11.2**	**7.7**	**17.6**
EXPECTED-		17.5	-4.8	6.5	24.7	16.7	6.6	9.3	6.2	0.5
RANGE (2SD)		20.8	7.3	17.5	45.3	27.1	12.2	14.6	8.5	1.0

At 1.0 hour the average primary wave is significantly (P<0.0005) increased to 22.9 +/-1.0 (4.5 %CV) over the 0.0-time control. The average primary component waves are still decreased at 7.5 (P<0.0005) and 13.0 (P<0.0125), while the 20.0, 26.6, 40.0 and 90.0 are all increased significantly (P<0.0005) over the 0.0-time nonallergenic challenge control. The same pattern of relationships held at the 1.5-hour test interval

NOTES:

TABLE 4.12 IN VIVO CYTOTOXIC TESTED FOOD (NONSENITIZING) CHALLENGE WITH TEMPORAL ANALYSIS OF EEG RESULTS.

ANALYTE RANGE UNITS CASE/AGE/SEX	TIME HOUR	AVG PRIMARY WAVE	FRONTAL (F3,F4 10/20) EEG PRIMARY WAVE (320/SEC SAMPLING/20 SEC EPOCHS) 3.5 %	7.5 %	13.0 %	20.0 %	26.6 %	40.0 %	90.0 %	 %
SGP/35/M	**1.0**	23.8	0.0	4.9	28.5	24.0	10.6	20.0	11.1	0.9
"NONALLERGENIC"		22.1	0.0	7.2	29.6	26.0	11.8	13.9	10.3	1.0
MEAL(10 MIN)		23.6	0.0	2.6	33.1	21.6	11.5	19.7	10.3	1.1
(CHERRY)		22.8	0.0	4.4	32.9	23.1	10.4	18.9	9.1	1.0
(EGG)		24.1	0.0	2.9	26.9	27.9	12.0	17.9	10.9	1.3
(ORANGE)		22.3	0.0	5.3	30.4	25.5	12.6	16.0	9.2	0.8
(PORK)		23.0	0.0	2.3	33.5	23.8	11.5	17.1	10.6	0.9
6/5/73		23.2	0.0	5.6	30.5	22.6	11.5	18.4	10.1	1.1
9:00AM		21.0	0.0	7.5	34.4	23.3	10.3	14.3	9.1	0.7
========										
AVERAGE		**22.9**	**0.0**	**4.7**	**31.1**	**24.2**	**11.4**	**17.4**	**10.1**	**1.0**
SUM		205.9	0.0	42.7	279.8	217.8	102.2	156.2	90.7	8.8
SUM OF SQS		4718.0	0.0	231.4	8750.1	5300.9	1165.4	2750.4	918.8	8.9
0HR CTL/1.0 EXP		P<0.0005		P<0.0005	P<0.0125	P<0.0005	P<0.0005	P<0.0005	P<0.0005	
NUMBER (N)		9.0	9.0	9.0	9.0	9.0	9.0	9.0	9.0	9.0
SQRT OF N		3.0	3.0	3.0	3.0	3.0	3.0	3.0	3.0	3.0
MAXIMUM		24.1	0.0	7.5	34.4	27.9	12.6	20.0	11.1	1.3
MINIMUM		21.0	0.0	2.3	26.9	21.6	10.3	13.9	9.1	0.7
RANGE		3.1	0.0	5.2	7.5	6.3	2.3	6.1	2.0	0.6
STD ERROR (SEM)		0.3	0.0	0.6	0.8	0.7	0.3	0.7	0.2	0.1
SEM %		1.5		12.2	2.7	2.9	2.3	3.9	2.2	6.8
STD DEV (SD)		**1.0**	**0.0**	**1.7**	**2.5**	**2.1**	**0.8**	**2.0**	**0.7**	**0.2**
SD % COEF VAR		**4.5**		**36.5**	**8.0**	**8.7**	**6.8**	**11.7**	**6.6**	**20.5**
EXPECTED-		20.8	0.0	1.3	26.1	20.0	9.8	13.3	8.7	0.6
RANGE (2SD)		24.9	0.0	8.2	36.1	28.4	12.9	21.4	11.4	1.4

The same pattern of relationships held at the 1.0-hour and the 1.5-hour test intervals.

NOTES:

TABLE 4.13 IN VIVO CYTOTOXIC TESTED FOOD (NONSENSITIZING) CHALLENGE WITH TEMPORAL ANALYSIS OF EEG RESULTS.

			FRONTAL (F3,F4 10/20) EEG PRIMARY WAVE (320/SEC SAMPLING/20 SEC EPOCHS)							
ANALYTE		AVG	3.5	7.5	13.0	20.0	26.6	40.0	90.0	
RANGE	TIME	PRIMARY								
UNITS	HOUR	WAVE	%	%	%	%	%	%	%	%
CASE/AGE/SEX										
SGP/35/M	**1.5**	24.9	0.0	4.1	24.1	23.4	12.6	23.1	11.3	1.2
"NONALLERGENIC"		26.9	0.0	1.6	21.9	22.9	15.0	23.1	14.0	1.3
MEAL(10 MIN)		26.1	0.0	1.6	25:2	24.8	12.0	22.0	12.7	1.5
(CHERRY)		26.0	0.0	1.0	25.4	23.5	14.2	21.9	12.5	1.3
(EGG)		25.3	0.0	2.4	22.6	23.9	16.8	21.9	11.4	1.0
(ORANGE)		21.7	0.0	9.6	30.7	22.7	10.3	16.9	8.9	0.9
(PORK)		21.5	0.0	7.8	31.3	24.3	11.2	15.2	9.1	1.0
6/5/73		19.5	0.0	5.5	40.8	25.9	7.9	11.5	7.2	0.8
9:00AM		21.4	0.0	5.6	33.8	24.7	10.7	15.3	9.0	0.7
========										
AVERAGE		**23.7**	**0.0**	**4.4**	**28.4**	**24.0**	**12.3**	**19.0**	**10.7**	**1.1**
SUM		213.3	0.0	39.2	255.8	216.1	110.7	170.9	96.1	9.7
SUM OF SQS		5112.3	0.0	243.3	7580.6	5197.2	1420.1	3393.4	1066.1	11.0
0HR CTL/1.5 EXP		P<0.0005		P<0.0005	P<0.01	P<0.0005	P<0.0025	P<0.0005	P<0.0005	
NUMBER (N)		9.0	9.0	9.0	9.0	9.0	9.0	9.0	9.0	9.0
SQRT OF N		3.0	3.0	3.0	3.0	3.0	3.0	3.0	3.0	3.0
MAXIMUM		26.9	0.0	9.6	40.8	25.9	16.8	23.1	14.0	1.5
MINIMUM		19.5	0.0	1.0	21.9	22.7	7.9	11.5	7.2	0.7
RANGE		7.4	0.0	8.6	18.9	3.2	8.9	11.6	6.8	0.8
STD ERROR (SEM)		0.8	0.0	1.0	2.1	0.4	1.0	1.3	0.8	0.1
SEM %		3.5		21.9	7.4	1.5	8.0	6.8	7.1	8.2
STD DEV (SD)		**2.5**	**0.0**	**2.9**	**6.3**	**1.1**	**3.0**	**3.9**	**2.3**	**0.3**
SD % COEF VAR		**10.4**		**65.8**	**22.2**	**4.4**	**24.1**	**20.4**	**21.2**	**24.7**
EXPECTED-		18.8	0.0	-1.4	15.8	21.9	6.4	11.3	6.1	0.5
RANGE (2SD)		28.6	0.0	10.1	41.0	26.1	18.2	26.7	15.2	1.6

NOTES:

TABLE 4.14 IN VIVO CYTOTOXIC TESTED FOOD (NONSENSITIZING) CHALLENGE WITH TEMPORAL ANALYSIS OF EEG RESULTS.

			FRONTAL (F3,F4 10/20) EEG PRIMARY WAVE (320/SEC SAMPLING/20 SEC EPOCHS)							
ANALYTE			3.5	7.5	13.0	20.0	26.6	40.0	90.0	
RANGE		AVG								
UNITS	TIME	PRIMARY	%	%	%	%	%	%	%	%
CASE/AGE/SEX	HOUR	WAVE								
SGP/35/M	**2.0**	19.5	0.8	12.8	34.6	21.3	9.8	12.9	7.1	0.6
"NONALLERGENIC"		20.8	0.0	9.5	31.9	22.6	12.2	14.2	8.6	0.7
MEAL(10 MIN)		18.2	0.8	10.7	44.3	18.2	7.6	10.9	6.4	0.9
(CHERRY)		20.5	0.0	10.8	31.6	23.7	10.7	14.3	8.1	0.8
(EGG)		19.6	0.0	10.8	33.8	24.3	12.3	10.9	7.1	0.7
(ORANGE)		19.6	0.0	6.8	36.3	25.4	11.9	12.4	6.3	0.7
(PORK)		18.3	0.0	7.2	44.6	20.4	10.0	11.3	5.8	0.5
6/5/73		19.0	0.0	11.2	37.0	22.9	9.5	11.1	7.4	0.8
9:00AM		19.1	0.0	10.1	40.6	20.4	9.7	11.2	6.8	1.0
		19.5	0.0	8.5	37.1	24.6	9.4	12.5	6.9	0.7
========										
AVERAGE		**19.4**	**0.2**	**9.8**	**37.2**	**22.4**	**10.3**	**12.2**	**7.1**	**0.7**
SUM		194.1	1.6	98.4	371.8	223.8	103.1	121.7	70.5	7.4
SUM OF SQS		3773.7	1.3	999.6	14018.9	5054.9	1082.7	1496.5	503.3	5.7
0HR CTL/2.0 EXP		P<0.0005		P<0.0005	P<0.10	P<0.0025				
NUMBER (N)		10.0	10.0	10.0	10.0	10.0	10.0	10.0	10.0	10.0
SQRT OF N		3.2	3.2	3.2	3.2	3.2	3.2	3.2	3.2	3.2
MAXIMUM		20.8	0.8	12.8	44.6	25.4	12.3	14.3	8.6	1.0
MINIMUM		18.2	0.0	6.8	31.6	18.2	7.6	10.9	5.8	0.5
RANGE		2.6	0.8	6.0	13.0	7.2	4.7	3.4	2.8	0.5
STD ERROR (SEM)		0.3	0.1	0.6	1.3	0.7	0.5	0.3	0.3	0.1
SEM %		1.3	50.0	6.1	3.5	3.2	4.6	2.8	4.0	6.8
STD DEV (SD)		**0.8**	**0.3**	**1.9**	**4.1**	**2.3**	**1.5**	**1.1**	**0.9**	**0.2**
SD % COEF VAR		**4.2**	**158.1**	**19.3**	**11.1**	**10.2**	**14.4**	**8.8**	**12.6**	**21.4**
EXPECTED-		17.8	-0.3	6.0	29.0	17.8	7.3	10.0	5.3	0.4
RANGE (2SD)		21.1	0.7	13.6	45.4	26.9	13.3	14.3	8.8	1.1

NOTES:

____

____

____

____

____

At 2.0 hours the average primary wave at 19.4 +/-0.8SD (4.2 %CV) is still significantly (P<0.0005) increased over yet falling back toward the 0.0-time control average value. A distinct and very different pattern for the component primary waves has unfolded. The 7.5 component trended toward yet significantly (P<0.0005) decreased from the 0.0 control values. The following component primary waves 13.0, 26.6, 40.0, 90.0 have shifted back to being indistinguishable from the control values, the 20.0 component remained significantly (P<0.0025) increased over the nonallergenic control. These components are exhibiting a pattern completely different than the allergenic test series.

TABLE 4.15 IN VIVO CYTOTOXIC TESTED FOOD (NONSENSITIZING) CHALLENGE WITH TEMPORAL ANALYSIS OF EEG RESULTS.

			FRONTAL (F3,F4 10/20) EEG PRIMARY WAVE (320/SEC SAMPLING/20 SEC EPOCHS)							
ANALYTE		AVG	3.5	7.5	13.0	20.0	26.6	40.0	90.0	
RANGE	TIME	PRIMARY								
UNITS	HOUR	WAVE	%	%	%	%	%	%	%	%
CASE/AGE/SEX										
SGP/35/M	**2.5**	24.3	0.0	6.2	30.4	21.9	9.6	17.5	13.2	1.2
"NONALLERGENIC"		19.1	0.0	11.6	41.1	18.5	8.2	11.7	7.7	0.8
MEAL(10 MIN)		18.3	0.9	11.7	39.2	19.9	10.8	9.9	6.5	0.7
(CHERRY)		16.6	0.0	8.8	47.2	22.9	8.8	6.8	4.8	0.6
(EGG)		16.2	1.3	13.4	44.1	19.4	8.1	8.0	4.7	0.4
(ORANGE)		15.6	0.0	13.6	48.4	19.7	7.5	6.2	3.9	0.5
(PORK)		16.8	0.0	9.0	46.2	21.7	10.0	8.6	3.6	0.7
6/5/73		14.9	0.0	12.1	49.9	21.1	6.4	7.0	3.0	0.4
9:00AM		15.8	0.7	15.3	41.5	20.7	10.2	7.0	3.7	0.4
		17.1	0.0	13.4	41.2	23.1	8.4	7.9	5.4	0.5
========										
AVERAGE		**17.5**	**0.3**	**11.5**	**42.9**	**20.9**	**8.8**	**9.1**	**5.7**	**0.6**
SUM		174.7	2.9	115.1	429.2	208.9	88.0	90.6	56.5	6.2
SUM OF SQS		3117.9	2.5	1158.8	16987.1	3956.4	686.9	875.2	387.2	4.2
0HR CTL/2.5 EXP				P<0.0005	P<0.05					
NUMBER (N)		10.0	10.0	10.0	10.0	10.0	10.0	10.0	10.0	10.0
SQRT OF N		3.2	3.2	3.2	3.2	3.2	3.2	3.2	3.2	3.2
MAXIMUM		24.3	1.3	15.3	49.9	23.1	10.8	17.5	13.2	1.2
MINIMUM		14.9	0.0	6.2	30.4	18.5	6.4	6.2	3.0	0.4
RANGE		9.4	1.3	9.1	19.5	4.6	4.4	11.3	10.2	0.8
STD ERROR (SEM)		0.9	0.1	0.9	2.0	0.5	0.4	1.1	1.0	0.1
SEM %		5.4	44.8	7.9	4.5	2.2	5.0	12.5	18.1	12.9
STD DEV (SD)		**3.0**	**0.4**	**2.9**	**6.2**	**1.5**	**1.4**	**3.6**	**3.2**	**0.3**
SD % COEF VAR		**17.0**	**141.8**	**25.0**	**14.4**	**7.0**	**15.8**	**39.4**	**57.1**	**40.8**
EXPECTED-		11.5	-0.5	5.8	30.6	18.0	6.0	1.9	-0.8	0.1
RANGE (2SD)		23.4	1.1	17.3	55.3	23.8	11.6	16.2	12.1	1.1

At 2.5 hours the average primary wave of 17.5 +/-3.0 (17 %CV) was at the control level. The component primary wave at 7.5 remained significantly (P<0.0005) decreased relative to the 0.0-time control, while the 13.0 component was significantly (P<0.05) increased, and all other components were indistinguishable from the nonallergenic control.

At 3.0 hours the average primary wave had decreased significantly (P<0.05) relative to the 0.0-time control. The 7.5 primary wave component was still decreased significantly (P<0.005). The 13.0 component shift to significantly (P<0.0005) increased was holding strong. The 20.0 component was still trending as significantly (P<0.01) increased, 26.6 was the same as the control. The components for 40.0 (P<0.01) and 90 (P<0.0005) were significantly decreased.

NOTES:

TABLE 4.16 IN VIVO CYTOTOXIC TESTED FOOD (NONSENITIZING) CHALLENGE WITH TEMPORAL ANALYSIS OF EEG RESULTS.

			FRONTAL (F3,F4 10/20) EEG PRIMARY WAVE (320/SEC SAMPLING/20 SEC EPOCHS)							
			**							
ANALYTE		AVG	3.5	7.5	13.0	20.0	26.6	40.0	90.0	
RANGE	TIME	PRIMARY								
UNITS	HOUR	WAVE	%	%	%	%	%	%	%	%
CASE/AGE/SEX										
SGP/35/M	**3.0**	16.0	0.8	18.3	40.3	19.6	7.7	8.4	4.2	0.5
"NONALLERGENIC"		16.2	1.0	16.2	40.2	20.1	9.0	8.3	4.4	0.6
MEAL(10 MIN)		15.8	0.0	15.1	43.8	21.4	8.0	7.5	3.8	0.3
(CHERRY)		16.4	0.0	17.5	37.3	23.5	9.5	7.1	4.3	0.6
(EGG)		16.4	0.0	16.6	40.2	20.6	10.0	6.7	5.2	0.5
(ORANGE)		17.7	0.0	9.1	40.8	24.1	11.6	8.8	5.1	0.5
(PORK)		17.0	0.0	16.8	41.9	18.1	8.1	8.6	5.6	0.6
6/5/73		16.9	1.7	13.0	37.9	22.7	10.2	9.2	4.2	0.7
9:00AM		16.9	0.0	11.2	42.5	24.0	8.0	8.7	4.8	0.5
		17.5	0.0	16.0	35.7	21.5	10.0	10.9	4.7	0.7
========										
AVERAGE		**16.7**	**0.4**	**15.0**	**40.1**	**21.6**	**9.2**	**8.4**	**4.6**	**0.6**
SUM		166.8	3.5	149.8	400.6	215.6	92.1	84.2	46.3	5.5
SUM OF SQS		2785.8	4.5	2322.6	16103.3	4684.7	862.8	721.5	217.1	3.2
0HR CTL/3.0 EXP				P<0.005	P<0.0005	P<0.01		P<0.05	P>0.0005	
NUMBER (N)		10.0	10.0	10.0	10.0	10.0	10.0	10.0	10.0	10.0
SQRT OF N		3.2	3.2	3.2	3.2	3.2	3.2	3.2	3.2	3.2
MAXIMUM		17.7	1.7	18.3	43.8	24.1	11.6	10.9	5.6	0.7
MINIMUM		15.8	0.0	9.1	35.7	18.1	7.7	6.7	3.8	0.3
RANGE		1.9	1.7	9.2	8.1	6.0	3.9	4.2	1.8	0.4
STD ERROR (SEM)		0.2	0.2	0.9	0.8	0.6	0.4	0.4	0.2	0.0
SEM %		1.1		6.1	2.0	2.8	4.2	5.0	3.9	7.3
STD DEV (SD)		**0.6**	**0.5**	**2.9**	**2.6**	**1.9**	**1.2**	**1.3**	**0.6**	**0.1**
SD % COEF VAR		**3.6**		**19.4**	**6.4**	**8.8**	**13.4**	**15.8**	**12.3**	**23.0**
EXPECTED-		15.5	-0.7	9.2	34.9	17.8	6.7	5.8	3.5	0.3
RANGE (2SD)		17.9	1.4	20.8	45.2	25.4	11.7	11.1	5.8	0.8

NOTES:

TABLE 4.17 IN VIVO CYTOTOXIC TESTED FOOD (NONSENSITIZING) CHALLENGE WITH TEMPORAL ANALYSIS OF EEG RESULTS.

			FRONTAL (F3,F4 10/20) EEG PRIMARY WAVE (320/SEC SAMPLING/20 SEC EPOCHS)							
ANALYTE		AVG	3.5	7.5	13.0	20.0	26.6	40.0	90.0	
RANGE	TIME	PRIMARY								
UNITS	HOUR	WAVE	%	%	%	%	%	%	%	%
CASE/AGE/SEX										
SGP/35/M	**3.5**	15.4	0.7	19.2	40.0	19.6	9.8	6.0	4.0	0.4
"NONALLERGENIC"		16.0	0.7	14.5	45.0	20.1	6.3	7.8	5.0	0.4
MEAL(10 MIN)		15.2	0.0	13.4	48.6	19.1	8.2	6.2	3.7	0.3
(CHERRY)		17.0	0.0	12.2	42.7	21.7	9.4	8.3	4.5	0.6
(EGG)		16.2	0.8	11.4	45.9	19.1	8.8	9.0	4.4	0.3
(ORANGE)		16.0	1.5	16.3	41.0	19.0	9.1	7.7	4.8	0.4
(PORK)		15.9	0.0	14.3	44.3	20.7	8.7	7.4	3.9	0.4
6/5/73		17.6	0.0	16.1	34.8	23.8	9.4	9.9	5.1	0.7
9:00AM										
========										
AVERAGE		**16.2**	**0.5**	**14.7**	**42.8**	**20.4**	**8.7**	**7.8**	**4.4**	**0.4**
SUM		129.3	3.7	117.4	342.3	163.1	69.7	62.3	35.4	3.5
SUM OF SQS		2094.2	3.9	1766.6	14771.6	3344.6	615.6	497.2	158.6	1.7
0HR CTL/3.5 EXP		P<0.05		P<0.0025	P<0.0005			P<0.01	P<0.0005	
NUMBER (N)		8.0	8.0	8.0	8.0	8.0	8.0	8.0	8.0	8.0
SQRT OF N		2.8	2.8	2.8	2.8	2.8	2.8	2.8	2.8	2.8
MAXIMUM		17.6	1.5	19.2	48.6	23.8	9.8	9.9	5.1	0.7
MINIMUM		15.2	0.0	11.4	34.8	19.0	6.3	6.0	3.7	0.3
RANGE		2.4	1.5	7.8	13.8	4.8	3.5	3.9	1.4	0.4
STD ERROR (SEM)		0.3	0.2	1.0	1.7	0.6	0.4	0.5	0.2	0.1
SEM %		1.9		6.6	4.0	2.9	5.0	6.3	4.0	11.4
STD DEV (SD)		**0.8**	**0.5**	**2.8**	**4.9**	**1.7**	**1.2**	**1.4**	**0.5**	**0.1**
SD % COEF VAR		**5.2**		**18.8**	**11.4**	**8.3**	**14.2**	**17.7**	**11.2**	**32.3**
EXPECTED-		14.5	-0.6	9.2	33.0	17.0	6.2	5.0	3.4	0.2
RANGE (2SD)		17.9	1.5	20.2	52.5	23.8	11.2	10.5	5.4	0.7

At 3.5 hours the average primary wave at 16.2 +/-0.8 (5.2 %CV) was still significantly (P<0.05) decreased from the nonallergenic control. The 7.5 primary wave component was still significantly (P<0.0025) decreased, the 13.0 component was significantly increased, the 20.0 and 26.6 component were not distinguishable from the control, the 40.0 (P<0.01) and 90.0 (P<0.0005) components were significantly decreased.

At 4.0 hours the average primary wave at 22.3 +/-1.5SD (6.9 %CV) increased significantly (P<0.005) shifting dramatically to the positive side of the nonallergenic control. The 7.5 component wave decreased significantly to its lowest value for the test series, the 13.0 component did not differ significantly from the control, the 20.0, 26.6, 40.0 and 90.0 were all decreased significantly (P<0.0005) from the nonallergenic control.

NOTES:

TABLE 4.18 IN VIVO CYTOTOXIC TESTED FOOD (NONSENITIZING) CHALLENGE WITH TEMPORAL ANALYSIS OF EEG RESULTS.

			FRONTAL (F3,F4 10/20) EEG PRIMARY WAVE (320/SEC SAMPLING/20 SEC EPOCHS)							
ANALYTE RANGE UNITS CASE/AGE/SEX	TIME HOUR	AVG PRIMARY WAVE	3.5 %	7.5 %	13.0 %	20.0 %	26.6 %	40.0 %	90.0 %	%
SGP/35/M	**4.0**	21.4	0.0	6.6	35.2	23.5	10.6	14.6	8.4	1.0
"NONALLERGENIC"		21.1	0.0	8.0	36.3	20.1	11.0	14.2	9.3	0.8
MEAL(10 MIN)		20.6	0.0	8.4	35.2	21.7	11.3	13.9	8.1	1.0
(CHERRY)		20.0	0.0	7.1	40.2	21.4	10.5	12.3	7.6	0.8
(EGG)		20.9	0.0	8.3	33.7	22.3	11.8	14.9	8.0	0.7
(ORANGE)		23.7	0.0	5.3	30.2	22.3	11.5	17.5	12.0	1.0
(PORK)		24.4	0.0	3.4	27.6	26.2	9.8	19.9	11.8	1.2
6/5/73		23.9	0.0	6.2	22.0	28.4	12.8	17.9	10.8	1.2
9:00AM		24.6	0.0	4.2	25.6	25.6	11.6	20.5	11.6	0.9
========										
AVERAGE		**22.3**	**0.0**	**6.4**	**31.8**	**23.5**	**11.2**	**16.2**	**9.7**	**1.0**
SUM		200.6	0.0	57.5	286.0	211.5	100.9	145.7	87.6	8.6
SUM OF SQS		4497.8	0.0	393.2	9360.7	5028.1	1137.2	2424.2	878.9	8.5
OHR CTL/4.0 EXP		P<0.005		P<0.0005		P<0.0005	P<0.0005	P<0.0005	P<0.0005	
NUMBER (N)		9.0	9.0	9.0	9.0	9.0	9.0	9.0	9.0	9.0
SQRT OF N		3.0	3.0	3.0	3.0	3.0	3.0	3.0	3.0	3.0
MAXIMUM		24.6	0.0	8.4	40.2	28.4	12.8	20.5	12.0	1.2
MINIMUM		20.0	0.0	3.4	22.0	20.1	9.8	12.3	7.6	0.7
RANGE		4.6	0.0	5.0	18.2	8.3	3.0	8.2	4.4	0.5
STD ERROR (SEM)		0.5	0.0	0.6	2.0	0.9	0.3	0.9	0.5	0.1
SEM %		2.3		8.7	6.4	3.9	3.0	5.6	5.0	5.8
STD DEV (SD)		**1.5**	**0.0**	**1.7**	**6.1**	**2.8**	**1.0**	**2.7**	**1.5**	**0.2**
SD % COEF VAR		**6.9**		**26.1**	**19.1**	**11.8**	**8.9**	**16.9**	**15.1**	**17.4**
EXPECTED-		19.2	0.0	3.1	19.6	18.0	9.2	10.7	6.8	0.6
RANGE (2SD)		25.4	0.0	9.7	43.9	29.0	13.2	21.7	12.7	1.3

FRONTAL LOBE ALLERGENIC AND NONLLERGENIC COMPARATIVE EEGs We have now carefully analyzed the origin and details for the primary and component computer analyzed EEG waves recorded for the timed intervals of the frontal lobe allergenic (Tables 4.1 – 4.9) and nonallergenic (Tables 4.10 – 4.18) test series (Fig 4.4). The significant shifts of the component primary waves playing "beneath the surface" of an average frontal lobe primary wave (Fig 4.1) that itself shows little or no significant change during the test period was an exciting finding. This is very intriguing and may bear an interpretative parallel with constant white counts that show significant shifts in corresponding differential white counts. To give expert clinicians studying specific clinical conditions these kinds of differences can be very meaningful. As you have seen the frontal lobe recording for a nonallergenic challenge the pattern of change was very different and maybe even opposite to what might be expected. This appears to be reduced in activity during the allergenic meal compared to the nonallergenic meal, but we have seen that before with drugs like Ritalin in the CNS.

NOTES:

TABLE 4.19 IN VIVO CYTOTOXIC TEST FOOD CHALLENGE COMPARISONS WITH TEMPORAL ANALYSIS OF EEG RESULTS.

			FRONTAL (F3,F4 10/20) EEG PRIMARY WAVE (320/SEC/20 SEC EPOCHS)							
ANALYTE			3.5	7.5	13.0	20.0	26.6	40.0	90.0	
RANGE		AVG								
UNITS	TIME	PRIMARY	%	%	%	%	%	%	%	%
CASE/AGE/SEX	HOUR	WAVE								
SGP/35/M	**0.0**	**18.7**	**3.9**	**11.2**	**32.2**	**22.2**	**10.9**	**11.8**	**6.6**	**0.7**
"ALLERGENIC"	0.5	19.1	0.7	11.6	35.7	21.9	10.7	11.8	6.7	0.8
MEAL(10 MIN)	1.0	17.3	1.1	14.7	38.7	20.0	9.9	9.5	5.3	0.6
(CHICKEN)	1.5	18.6	1.4	10.8	37.2	21.9	10.3	11.0	6.5	0.7
(CHOCOLATE)	2.0									
(PEANUT)	2.5	18.8	1.1	12.9	34.1	21.9	10.8	11.8	6.5	0.7
(TEA)	3.0	17.7	0.3	14.1	38.0	21.5	9.9	9.8	5.5	0.6
5/30/73	3.5	17.6	1.1	16.0	36.8	20.6	9.0	9.7	5.8	0.6
9:30AM	4.0	18.0	0.3	14.9	36.6	21.5	9.2	10.6	6.0	0.7
========										
AVERAGE		**18.2**	**0.9**	**13.6**	**36.7**	**21.3**	**10.0**	**10.6**	**6.0**	**0.7**
SUM		145.8	9.9	106.2	289.3	171.5	80.7	86.0	48.9	5.4
SUM OF SQS		2660.2	21.5	1436.0	10493.5	3680.5	817.7	931.5	300.9	3.7
NUMBER (N)		8.0	8.0	8.0	8.0	8.0	8.0	8.0	8.0	8.0
SQRT OF N		2.8	2.8	2.8	2.8	2.8	2.8	2.8	2.8	2.8
MAXIMUM		19.1	3.9	16.0	38.7	22.2	10.9	11.8	6.7	0.8
MINIMUM		17.3	0.3	10.8	32.2	20.0	9.0	9.5	5.3	0.6
RANGE		1.8	3.6	5.2	6.5	2.2	1.9	2.3	1.4	0.2
STD ERROR (SEM)		0.2	0.5	0.7	0.8	0.3	0.2	0.3	0.2	0.0
SEM %		1.2	52.5	4.8	2.2	1.3	2.4	2.7	2.9	3.7
STD DEV (SD)		**0.6**	**1.3**	**1.8**	**2.3**	**0.8**	**0.7**	**0.8**	**0.5**	**0.1**
SD % COEF VAR		**3.5**	**148.5**	**13.5**	**6.3**	**3.6**	**6.7**	**7.7**	**8.2**	**10.5**
EXPECTED-		16.9	-1.7	9.9	32.1	19.8	8.6	9.0	5.1	0.5
RANGE (2SD)		19.4	3.4	17.2	41.3	22.9	11.3	12.2	7.0	0.8

The frontal computer analyzed EEG primary wave and component primary wave recordings for the allergenic challenge (Table 4.19) were not significantly different from the nonallergenic challenge (Table 4.20). As we have seen (Tables 4.1 – 4.18) under the surface there are some very significant activities that indicate an order of magnitude more sensitive changes that suggest some very powerful relationships between diet, quality of diet and involvement of the frontal, frontal-limbic system of the CNS.

Although the values for the nonallergenic challenge primary wave of 18.2 +/-0.6 (3.5 %CV) did not differ significantly (95% confidence level or better) from the comparable values for the allergenic challenge primary wave of 19.4 +/-2.5 (12.9 %CV). The greater variation in the nonallergenic values as reflected in the higher coefficient of variation and broader expected range signals something under the surface.

As with Ritalin administration where an opposite outcome demanded a new way to look at the operating mechanism for this drug so here we have an allergic challenge sounding an alarm in the frontal lobe (I ate, I'm not OK, ...disrupted thinking and planning), where as a nonallergenic challenge is processed by the frontal lobe in a different way prompting recognition of a challenge then resumption of normal activity (I ate, I'm OK, ...thinking and planning as usual).

TABLE 4.20 IN VIVO CYTOTOXIC TEST FOOD CHALLENGE COMPARISONS WITH TEMPORAL ANALYSIS OF EEG RESULTS

			FRONTAL (F3,F4 10/20) EEG PRIMARY WAVE (320/SEC/20 SEC EPOCHS)							
			**							
ANALYTE		AVG	3.5	7.5	13.0	20.0	26.6	40.0	90.0	
RANGE	TIME	PRIMARY								
UNITS	HOUR	WAVE	%	%	%	%	%	%	%	%
CASE/AGE/SEX										
SGP/35/M	**0.0**	**17.0**	**2.2**	**19.2**	**34.4**	**19.0**	**9.2**	**9.4**	**5.8**	**0.6**
"NONALLERGENIC"	0.5	19.1	1.3	12.0	35.0	21.9	9.4	11.9	7.4	0.8
MEAL(10 MIN)	1.0	22.9	0.0	4.7	31.1	24.2	11.4	17.4	10.1	1.0
(CHERRY)	1.5	23.7	0.0	4.4	28.4	24.0	12.3	19.0	10.7	1.1
(EGG)	2.0	19.4	0.2	9.8	37.2	22.4	10.3	12.2	7.1	0.7
(ORANGE)	2.5	17.5	0.3	11.5	42.9	20.9	8.8	9.1	5.7	0.6
(PORK)	3.0	16.7	0.4	15.0	40.1	21.6	9.2	8.4	4.6	0.6
6/5/73	3.5	16.2	0.5	14.7	42.8	20.4	8.7	7.8	4.4	0.4
9:00AM	4.0	22.3	0.0	6.4	31.8	23.5	11.2	16.2	9.7	1.0
========										
AVERAGE		**19.7**	**0.3**	**9.8**	**36.2**	**22.4**	**10.2**	**12.8**	**7.5**	**0.8**
SUM		174.8	4.9	97.7	323.7	197.9	90.5	111.4	65.5	6.8
SUM OF SQS		3461.1	7.1	1264.4	11857.5	4375.8	923.6	1519.2	522.4	5.6
OHR CTL/0.5 EXP					P<0.10					
NUMBER (N)		9.0	9.0	9.0	9.0	9.0	9.0	9.0	9.0	9.0
SQRT OF N		3.0	3.0	3.0	3.0	3.0	3.0	3.0	3.0	3.0
MAXIMUM		23.7	2.2	19.2	42.9	24.2	12.3	19.0	10.7	1.1
MINIMUM		16.2	0.0	4.4	28.4	19.0	8.7	7.8	4.4	0.4
RANGE		7.5	2.2	14.8	14.5	5.2	3.6	11.2	6.3	0.7
STD ERROR (SEM)		0.8	0.2	1.6	1.6	0.6	0.4	1.2	0.7	0.1
SEM %		4.2	72.4	16.8	4.5	2.6	3.9	9.8	9.4	10.0
STD DEV (SD)		**2.5**	**0.7**	**4.9**	**4.8**	**1.7**	**1.2**	**3.7**	**2.1**	**0.2**
SD % COEF VAR		**12.7**	**217.3**	**50.3**	**13.4**	**7.8**	**11.8**	**29.3**	**28.1**	**30.1**
EXPECTED-		14.7	-1.1	-0.1	26.5	18.9	7.8	5.3	3.3	0.3
RANGE (2SD)		24.7	1.8	19.7	45.8	25.8	12.6	20.2	11.7	1.2

NOTES:

TABLE 4.21 IN VIVO CYTOTOXIC TEST FOOD CHALLENGE COMPARISONS WITH TEMPORAL ANALYSIS OF EEG RESULTS.

ANALYTE RANGE UNITS CASE/AGE/SEX	TIME HOUR	AVG PRIMARY WAVE	PARIETAL (P3,P4 10/20) EEG PRIMARY WAVE (320/SEC/20 SEC EPOCHS) 3.5 %	7.5 %	13.0 %	20.0 %	26.6 %	40.0 %	90.0 %	%
SGP/35/M	**0.0**	**12.5**	**0.0**	**6.1**	**67.7**	**19.9**	**3.1**	**2.0**	**0.9**	**0.1**
"ALLERGENIC"	0.5	12.2	0.0	6.8	69.4	18.1	2.7	1.8	0.8	0.1
MEAL(10 MIN)	1.0	12.1	0.0	6.5	70.8	16.9	3.1	1.6	0.8	0.1
(CHICKEN)	1.5	12.7	0.0	6.4	67.6	18.6	3.5	2.5	1.2	0.1
(CHOCOLATE)	2.0									
(PEANUT)	2.5	12.7	0.1	6.2	67.8	19.0	3.2	2.1	1.2	0.2
(TEA)	3.0	12.5	0.0	7.1	68.3	17.5	3.3	2.2	1.1	0.1
5/30/73	3.5	12.3	0.0	7.0	69.4	17.3	3.0	2.0	1.1	0.1
9:30AM	4.0	12.1	0.1	7.4	70.2	16.4	2.9	1.7	0.9	0.1
========										
AVERAGE		**12.4**	**0.0**	**6.8**	**69.1**	**17.7**	**3.1**	**2.0**	**1.0**	**0.1**
SUM		99.1	0.2	53.5	551.2	143.7	24.8	15.9	8.0	0.9
SUM OF SQS		1228.0	0.0	359.3	37988.2	2590.7	77.3	32.2	8.2	0.1
NUMBER (N)		8.0	8.0	8.0	8.0	8.0	8.0	8.0	8.0	8.0
SQRT OF N		2.8	2.8	2.8	2.8	2.8	2.8	2.8	2.8	2.8
MAXIMUM		12.7	0.1	7.4	70.8	19.9	3.5	2.5	1.2	0.2
MINIMUM		12.1	0.0	6.1	67.6	16.4	2.7	1.6	0.8	0.1
RANGE		0.6	0.1	1.3	3.2	3.5	0.8	0.9	0.4	0.1
STD ERROR (SEM)		0.1	0.0	0.2	0.4	0.4	0.1	0.1	0.1	0.0
SEM %		0.6	43.8	2.4	0.6	2.5	3.2	5.7	4.9	10.9
STD DEV (SD)		**0.2**	**0.0**	**0.5**	**1.1**	**1.2**	**0.3**	**0.3**	**0.1**	**0.0**
SD % COEF VAR		**1.7**	**123.7**	**6.8**	**1.6**	**7.0**	**9.1**	**16.0**	**13.9**	**30.9**
EXPECTED-		11.9	0.0	5.9	66.8	15.2	2.5	1.3	0.7	0.0
RANGE (2SD)		12.8	0.1	7.7	71.3	20.2	3.7	2.6	1.3	0.2

NOTES:

TABLE 4.22 IN VIVO CYTOTOXIC TEST FOOD CHALLENGE COMPARISONS WITH TEMPORAL ANALYSIS OF EEG RESULTS

ANALYTE			PARIETAL (P3,P4 10/20) EEG PRIMARY WAVE (320/SEC/20 SEC EPOCHS)							
ANALYTE		AVG	3.5	7.5	13.0	20.0	26.6	40.0	90.0	
RANGE	TIME	PRIMARY								
UNITS	HOUR	WAVE	%	%	%	%	%	%	%	%
CASE/AGE/SEX										
SGP/35/M	**0.0**	**12.3**	**0.4**	**8.6**	**65.9**	**18.6**	**3.2**	**2.0**	**1.0**	**0.1**
"NONALLERGENIC"	0.5	11.9	0.6	10.7	66.9	15.5	2.8	2.3	1.0	0.1
MEAL(10 MIN)	1.0	12.1	0.7	7.9	68.9	16.0	3.1	2.1	0.9	0.1
(CHERRY)	1.5	12.1	0.1	8.0	69.2	16.8	2.8	1.9	0.9	0.1
(EGG)	2.0	11.8	0.1	9.1	70.2	15.5	2.4	1.6	0.8	0.1
(ORANGE)	2.5	12.3	0.1	7.1	70.9	15.1	2.9	2.2	1.2	0.1
(PORK)	3.0	12.3	0.6	12.7	63.1	15.1	3.7	2.7	1.4	0.1
6/5/73	3.5	11.9	0.1	8.1	71.4	15.1	2.8	1.4	0.9	0.1
9:00AM	4.0	11.8	0.0	8.9	70.3	15.4	2.6	1.6	0.9	0.1
========										
AVERAGE		**12.0**	**0.3**	**9.1**	**68.9**	**15.6**	**2.9**	**2.0**	**1.0**	**0.1**
SUM		108.5	2.7	81.1	616.8	143.1	26.3	17.8	9.0	0.9
SUM OF SQS		1308.4	1.4	754.2	42330.8	2285.9	78.0	36.5	9.3	0.1
OHR CTL/0.5 EXP					P<0.10					
NUMBER (N)		9.0	9.0	9.0	9.0	9.0	9.0	9.0	9.0	9.0
SQRT OF N		3.0	3.0	3.0	3.0	3.0	3.0	3.0	3.0	3.0
MAXIMUM		12.3	0.7	12.7	71.4	18.6	3.7	2.7	1.4	0.1
MINIMUM		11.8	0.0	7.1	63.1	15.1	2.4	1.4	0.8	0.1
RANGE		0.5	0.7	5.6	8.3	3.5	1.3	1.3	0.6	0.0
STD ERROR (SEM)		0.1	0.1	0.6	0.9	0.4	0.1	0.1	0.1	0.0
SEM %		0.5	27.1	6.9	1.3	2.5	5.0	7.3	6.7	0.0
STD DEV (SD)		**0.2**	**0.2**	**1.9**	**2.8**	**1.2**	**0.4**	**0.4**	**0.2**	**0.0**
SD % COEF VAR		**1.4**	**81.2**	**20.6**	**4.0**	**7.5**	**15.0**	**21.9**	**20.0**	**0.0**
EXPECTED-		11.7	-0.2	5.3	63.3	13.2	2.0	1.1	0.6	0.1
RANGE (2SD)		12.4	0.8	12.8	74.4	17.9	3.8	2.8	1.4	0.1

REFERENCRES

Adey, W: Intrinsic organization of cerebral tissue in alerting, Orienting and discrimination responses. The Neurosciences. Eds. Quarton, G; Melnechuk, T and Schmitt, F. The Rockefeller University Press N.Y. pgs. 615-633, 1967.

Berger, H: Uber das electrenkephalogramm der menchen. Arch Psychiat Nervenkrankh, 87:527-570, 1929.

Campbell, MB: Neurological manifestations of allergic disease. Annals of Allergy 31:485-497, 1973.

Canton, R: The electric currents of the brain. Br Med J 2:279, 1875.

Davison, AM: Allergy of the nervous system. Quart Rev Allerg 6:157, 1952.

Kitter, FJ and Baldwin DC: The role of allergic factors in the child with minimal brain dysfunction syndrome. Ann Allerg 28:203, 1970.

Morrell, F: Electrical signs of sensory coding. The Neurosciences. Eds. Quarton, G; Melnechuk, T and Schmitt, F. The Rockefeller University Press N.Y. pgs. 452-469, 1967.

Rowe, AH: Clinical allergy in the nervous system. J. of Nervous and Mental Diseases. 99:834-41, 1944.

Ward, AA, Jr: A Symposium on Dendrites, Electroencephalog and Clin Neurophysiol., suppl. 10:7-60, 1958.

NOTES:

CHAPTER 5

Familial Relationships

TO FOLLOW ARE detailed case studies that include Cytotoxic Test screening and corresponding food sensitivity profile results for principal offspring, parents and in some cases sibling(s) and grandparents. These data open a whole new arena of study as they relate to the genetic origins and distributions of manifested food sensitivities as we have identified them using the in-vitro Cytotoxic Testing and in-vivo Challenge WBC response testing procedures (Ulett and Perry, 1974, 1975). Additional detailed data supporting these procedures have been discussed in Chapters One through Four. This data will also lay the foundation for a population study to follow in Chapter 6.0.

Case Study 05.01.00.1-7 SGP X CAL-P (BPP & Family): This extended family case study (Tables 05.01.MEMBERS.1 - 7) involves BPP age 3, parents in their mid-thirties, with two sibling brothers ages six and newborn. Both parental grandmothers were also included in the study. The paternal grandfathers were deceased. A sister of the father (SGP) and her six offspring, including two sets of twins are also covered in a separate case study (Tables 05.01.BPP.1-7).

BPP's history indicated that "allergies" were a contributing factor to frequent complaints of discomfort. He was branded a "colicky kid" due to his inability; to sleep through the night, constant runny nose, frequent earaches, diarrhea, and general fussy behavior. These complaints were first dealt with through his Pediatrician. BPP's diet was changed from milk to various substitutes and eventually apple juice. On advice of his Pediatrician, his diet was randomly tested, by rotating his exposure to yellow then green vegetables, rotating various grains or food combinations. Not achieving much success BPP was referred to Dr Jack Strominger, a Pediatric Allergist associated with Washington University Medical School in St, Louis, MO.

What follows is a detailed analysis of the whole family using our standard clinical procedures and then current testing schemes (Ulett and Perry; 1974, 1975). This clinical approach had been used by Dr Ulett and his Pediatrician wife at The Missouri Institute of Psychiatry to treat children with a whole host of neurological and psychiatric complaints with significant success in controlling their clinical complaints and in some cases achieving side benefits that included weight loss and clearing of acne or rash problems. Our first attempt to provide quantitative support for a family genetic model of distribution of food sensitivities based on Cytotoxic Test results is detailed in Tables 05.01.MEMBERS.1 - 7. The Cytotoxic Test results provided a basis for designing diets avoiding Cytotoxic Test positive foods and using only foods that were confirmed Cytotoxic Test negative. This approach showed immediate improvement for BPP's list of clinical complaints. We will begin with analyzing the results of the cytotoxic testing in family members in an attempt to determine the correlation of given food sensitivities and the genetic origins from the parents. Understanding the allergic etiology prompts new clinical treatments.

NOTES:

TABLE 05.01.MEMBERS.01 SGPxCAL-P FAMILIAL CYTOTOXIC TEST POSITIVE SENSITIVITIES TO FRUITS.

SUBJECT	G-FATHER RFL	G-MOTHER LCP-L	G-FATHER RRP	G-MOTHER DMS-P	FATHER SGP	MOTHER CAL-P	SON RRP	SON BPP	SON SSP
AGE TESTED	NO TESTS	62	NO TESTS	59	35	34	6	3	NEWBORN
APPLE		1		1	1	1	1	1	1
BANANA		1		0	1	1	0	1	0
CANTALOUPE		0		0	0	1	0	1	0
CHERRY		0		0	0	1	0	0	0
HONEYDEW		0		0	0	1	0	0	0
ORANGE		1		0	0	1	1	1	1
PEACH		1		0	0	1	0	1	0
PEAR		0		0	0	1	0	1	0
PINEAPPLE		1		1	1	1	1	1	0
STRAWBERRY		1		0	0	1	0	0	0
TOMATO		0		0	0	0	1	1	1
VANILLA		1		0	1	1	0	1	1
WATERMELON		1		0	0	0	0	0	0
===============									
POSITIVES	0	8	0	2	4	11	4	9	4
NUMBER (N)	13	13	13	13	13	13	13	13	13
% POSITIVE	0.0	61.5	0.0	15.4	30.8	84.6	30.8	69.2	30.8

Tables 05.01.MEMBERS.01-7 detail the results of cytotoxic test screening for various food groupings and shows the comparisons among family members. For example, in Table 05.01.MEMBERS.01 (Fruits), apple was positive for all family members tested. Looking at the results for each offspring one can compare the result for each item tested to the result for the parent(s). This isolates the possible genetic origin for expressing a positive or negative test result. Also note that by comparing the vertical column of results one can assess the frequency of positive responses for an individual and this, expressed as a percentage (bottom of the table), permits a comparison of the over-all sensitivity of the individual within the family. In Table 05.01.MEMBERS.01 the three-year-old (BPP/69.2%) as compared to the mother (CAL-P/84.6%) and the maternal grandmother (LCP-L/61.5%) all showed a high level of sensitivity (i.e., greater than 60% positive) to the fruit category of foods. This table alone is enough to distinguish this family from other families and characteristically identify them as a unique genetic unit. Furthermore, individuals have their own unique genetic profiles! Those highly sensitized are not sensitized by the same foods! To prove this to yourself make a variety of such comparisons among the other case studies for families and family members presented below in this chapter. In Table 05.01.MEMBERS.02 (Grains) three family

members (RRP/57.1%, BPP/71.4% and LCP/57.1%) all exceed 50% sensitivity to various grains where no one was sensitivity to barley and all but one, a newborn showed a positive sensitivity to corn. Again, these results provide unique comparative differences among family members as well as distinguishing specific characteristics between fruit and grain food categories.

NOTES:

TABLE 05.01.MEMBERS.02 SGPxCAL-P FAMILIAL CYTOTOXIC TEST POSITIVE SENSITIVITIES TO GRAINS.

	G-FATHER	G-MOTHER	G-FATHER	G-MOTHER	FATHER	MOTHER	SON	SON	SON
SUBJECT	RFL	LCP-L	RRP	DMS-P	SGP	CAL-P	RRP	BPP	SSP
AGE TESTED	NO TESTS	62	NO TESTS	59	35	34	6	3	NEWBORN
BARLEY		0		0	0	0	0	0	0
CORN		1		1	1	1	1	1	0
MALT		0		0	0	0	1	0	0
OAT		1		0	1	1	1	1	0
RICE		1		0	0	0	1	1	0
RYE		0		0	0	0	0	1	0
WHEAT		1		0	0	1	0	1	0
===============									
POSITIVES	0	4	0	1	2	3	4	5	0
NUMBER (N)	7	7	7	7	7	7	7	7	7
% POSITIVE	0.0	57.1	0.0	14.3	28.6	42.9	57.1	71.4	0.0

TABLE 05.01.MEMBERS.03 SGPxCAL-P FAMILIAL CYTOTOXIC TEST POSITIVE SENSITIVITIES TO VEGETABLES.

	G-FATHER	G-MOTHER	G-FATHER	G-MOTHER	FATHER	MOTHER	SON	SON	SON
SUBJECT	RFL	LCP-L	RRP	DMS-P	SGP	CAL-P	RRP	BPP	SSP
AGE TESTED	NO TESTS	62	NO TESTS	59	35	34	6	3	NEWBORN
BROCCOLI		0		0	0	1	0	0	0
CABBAGE		1		0	0	0	0	0	0
CARROT		1		1	0	1	1	1	0
CUCUMBER		0		0	1	0	0	0	0
HOPS		1		0	0	0	0	1	0
HORSERADISH		0		0	0	0	0	0	0
LETTUCE		1		0	0	0	0	0	0
ONION		0		1	0	0	0	0	0
PEA		0		0	0	0	0	1	0
PEANUT		0		1	1	1	1	1	0
POTATO		1		0	1	1	0	1	0
RADISH		0		0	0	0	0	0	1
SOYBEAN		0		1	1	1	1	1	1
SPINACH		0		0	0	0	0	1	0
STRINGBEAN		0		0	1	0	0	0	0
SUGAR, BEET		0		0	0	0	0	0	0
SUGAR, CANE		0		0	0	0	0	1	0
===============									
POSITIVES	0	5	0	4	5	5	3	8	2
NUMBER (N)	17	17	17	17	17	17	17	17	17
% POSITIVE	0.0	29.4	0.0	23.5	29.4	29.4	17.6	47.1	11.8

Table 05.01.MEMBERS.03 (Vegetables) the overall pattern of sensitivity is significantly different than that observed among family members for the above food categories (Tables 05.01.MEMBERS.01 and 05.01. MEMBERS.02). Note that BPP age three has showed a very high percentage of sensitivity in all categories analyzed to this point in this case study. The fish and seafood category (Table 05.01.MEMBERS.04) showed very little positive sensitivity for any family member. In Table 05.01.MEMBERS.05 (Meat, Fowl, Egg, and Milk) BPP again shows a very high sensitivity (63.6%) with other family members sharing some while some were unique to BPP or shared with just a few. Note that for beef, lamb, pork, and veal the only positive appeared in the mother (CAL-P) while chicken appeared positive in all family members and milk appeared in all family members except for the paternal grandmother (DMS-P). The BPP profile was distinct and perhaps quite diagnostic regards poultry and milk products. Milk was already singled out by the Pediatric Allergist as a major contributing food item both suspected as a causative factor and the target of remedies pursued for BPP.

TABLE 05.01.MEMBERS.04 SGPxCAL-P FAMILIAL CYTOTOXIC TEST POSITIVE SENSITIVITIES TO FISH & SEAFOOD.

	G-FATHER	G-MOTHER	G-FATHER	G-MOTHER	FATHER	MOTHER	SON	SON	SON
SUBJECT	RFL	LCP-L	RRP	DMS-P	SGP	CAL-P	RRP	BPP	SSP
AGE TESTED	NO TESTS	62	NO TESTS	59	35	34	6	3	NEWBORN
CARP		0		0	0	0	0	0	0
CRAB		0		0	0	0	0	0	0
LOBSTER		0		0	1	0	0	0	0
SHRIMP		0		0	0	0	0	0	0
SWORDFISH		0		0	0	0	0	0	0
TROUT		0		0	0	0	0	0	0
TUNA		0		1	1	1	0	1	0
===============									
POSITIVES	0	0	0	1	2	1	0	1	0
NUMBER (N)	7	7	7	7	7	7	7	7	7
% POSITIVE	0.0	0.0	0.0	14.3	28.6	14.3	0.0	14.3	0.0

TABLE 05.01.MEMBERS.05 SGPxCAL-P FAMILY CYTOTOXIC POSITIVE SENSITIVITIES TO MEAT, FOWL, EGG & MILK.

	G-FATHER	G-MOTHER	G-FATHER	G-MOTHER	FATHER	MOTHER	SON	SON	SON
SUBJECT	RFL	62	RRP	DMS-P	SGP	CAL-P	RRP	BPP	SSP
AGE TESTED	NO TESTS	62.0	NO TESTS	59	35	34	6	3	NEWBORN
BEEF		0		0	0	0	0	0	0
LAMB		0		0	0	1	0	0	0
PORK		0		0	0	0	0	0	0
VEAL		0		0	0	0	0	0	0
CHICKEN		1		1	1	1	1	1	1
TURKEY, Lt.		0		0	0	1	0	1	0
EGG		1		0	1	0	1	1	1
MILK		1		0	1	1	1	1	1
CHEESE/AMER.		0		0	0	0	1	1	1
CHEESE/ROQUE.		0		0	0	0	0	1	0
CHEESE/SWISS		1		0	0	0	0	1	0
===============									
POSITIVES	0	4	0	1	3	4	4	7	4
NUMBER (N)	11	11	11	11	11	11	11	11	11
% POSITIVE	0.0	36.4	0.0	9.1	27.3	36.4	36.4	63.6	36.4

Table 5.6 (Miscellaneous items) shows immediate family members as not distinct and separate in their overall sensitivities, however, note chocolate, coffee, and tea where both parents and all off-spring show positive sensitivities. It is also interesting that tobacco showed positive in the mother, maternal grandmother and all three male off-spring while showing negative in the father and paternal grandmother. The "tobacco trait" illustrates a potential specific genetic manifestation as detailed in Chapter 4.0. Tobacco, secondary exposure, was singled out by both Pediatrician and Pediatric Allergist in their treatment regimens for BPP.

TABLE 05.01.MEMBERS.06 SGPxCAL-P FAMILIAL CYTOTOXIC POSITIVE SENSITIVITIES TO MISCELLANEOUS ITEMS.

	G-FATHER	G-MOTHER	G-FATHER	G-MOTHER	FATHER	MOTHER	SON	SON	SON
SUBJECT	RFL	LCP-L	RRP	DMS-P	SGP	CAL-P	RRP	BPP	SSP
AGE TESTED	NO TESTS	62	NO TESTS	59	35	34	6	3	NEWBORN
CHOCOLATE		0		1	1	1	1	1	1
COFFEE		1		1	1	1	1	1	1
COTTONSEED		0		0	0	1	0	0	1
GARLIC		1		0	1	0	0	0	1
GASOLINE		0		0	0	0	0	0	0
MUSHROOM		0		0	0	0	0	0	0
MUSTARD		0		0	0	0	0	1	0
PEPPERMINT		0		0	1	1	1	1	0
TEA		1		0	1	1	1	1	1
TOBACCO		1		0	0	1	1	1	1
TURPENTINE		0		0	1	0	0	0	0
YEAST, BAKER'S		0		0	0	1	1	1	0
YEAST, BREWER'S		0		0	0	0	0	0	0
ALCOHOL, 1%		0		0	0	0	0	0	0
ALCOHOL, 2%		0		0	0	0	0	0	0
===============									
POSITIVES	0	4	0	2	6	7	6	7	6
NUMBER (N)	15	15	15	15	15	15	15	15	15
% POSITIVE	0.0	26.7	0.0	13.3	40.0	46.7	40.0	46.7	40.0

TABLE 05.01.MEMBERS.07 SUMMARY: SGPXCAL-P FAMILIAL CYTOTOXIC TEST POSITIVE FOOD SENSITIVITIES.

	G-FATHER	G-MOTHER	G-FATHER	G-MOTHER	FATHER	MOTHER	SON	SON	SON
SUBJECT	RFL	LCP-L	RRP	DMS-P	SGP	CAL-P	RRP	BPP	SSP
AGE TESTED	NO TESTS	62	NO TESTS	59	35	34	6	3	NEWBORN
POSITIVES	0	25	0	11	22	31	21	37	16
NUMBER (N)	70	70	70	70	70	70	70	70	70
% POSITIVE	0.0	35.7	0.0	15.7	31.4	44.3	30.0	52.9	22.9

Table 05.01.MEMBERS.01-7 provides summary totals of the data gathered from the SGP-CAL-P case study family. There is a huge amount of information to sort through and begin to make a case for a genetic model explaining the distribution of the capacity to monitor manifested traits in response to specific food antigen(s). BPP was positive for about every other item he was tested for. Consider then the possible combinations that could affect him when small amounts of a single item might not reach a significant level. Or consider that tobacco (Table 05.01.MEMBERS.06), beyond nicotine, contains >4000 known individual components (Lakier, 1992; Krupski, 1991) many of which could be antigens or haptens (i.e., caffeine bound to protein capable of initiating antigen allergic responses. Are those who are positive, including BPP' grandmother, mother and three brothers, more susceptible to ill effects than his father and those who are negative to tobacco? Many tobacco or smoking studies in the literature require thousands of participants over many years that result in statistical squeakers that may easily be resolved quicker, less expensively and more decisively by knowing who's positive and who's not. Furthermore, consider coffee as not just caffeine and potentially in combination with tobacco with the potential ramifications that could result under the radar of current clinical practice and medical research.

Those of you interested in genetic analysis have already arrived at some conclusions. The genetic aspects of allergy are not discussed very much. Isolating genes traits to specific chromosomes are discussed even less. The genetic code has been reduced to a sequence of four billion bases occupying our 23 pairs of chromosomes as specific genes and operating segments. Our proposal here is that the immune system white cells hold the code for go or no go to initiate an allergic response. Further genetic coding may be required in end organs to complete the allergic response. Essentially we are all walking periodic tables, a molecular miracle, separate and unique from each other. For those of you willing to endure at a snail's pace we proceed to analyze this data looking at the parents and at each offspring one at a time for the SGP x CAL-P case study.

Case Study 05.01.BPPb.01-7. SGP X CAL-P (BPP): Three-year-old BPP and his parents agreed, either positive or negative, on 38 of 70 (54.3%) of the items tested for. The father and mother agreed on 16 of 70 (22.9%) positives. Of the16 positive items that both parents agreed on 16 of 16 (100.0%) were passed as positives to BPP! Of BPP's 37 positives, 17 (45.9%) matched his father's positives, 25 (67.6%) matched his mom's positives and 16 (43.2%) matched both parents. The father exclusively matched zero of 37 of BPP's positives. The mother exclusively matched on nine (cantaloupe, orange, peach, pear, wheat, carrot, turkey, tobacco, and baker's yeast) of 37 (24.3%) of BPP's positives. BPP had 11 (29.7%) of 37 test positive items (tomato, rice, rye, hops, pea, spinach, cane sugar, American cheese, Roquefort cheese, Swiss cheese, and mustard) that did not appear in either parent.

TABLE 05.01.BPP.01 SGPxCAL-P (BPP) FAMILIAL CYTOTOXIC TEST POSITIVE SENSITIVITIES TO FRUITS.

	FATHER	MOTHER	SON	%POSITIVE	%POSITIVE	%POSITIVE	%NEGATIVE	PER CENT OF POSITIVE FOODS			
SUBJECT	SGP	CAL-P	BPP	SGP/CAP	SGP	CAP	SGP/CAP		SGP/CAP	SGP	CAP
AGE TESTED	35	34	3	BPP	BPP	BPP	BPP	BPP	BPP	BPP	BPP
APPLE	1	1	1	1					1		
BANANA	1	1	1	1					1		
CANTALOUPE	0	1	1			1					1
CHERRY	0	1	0								
HONEYDEW	0	1	0								
ORANGE	0	1	1			1					1
PEACH	0	1	1			1					1
PEAR	0	1	1			1					1
PINEAPPLE	1	1	1	1					1		
STRAWBERRY	0	1	0								
TOMATO	0	0	1					1			
VANILLA	1	1	1	1					1		
WATERMELON	0	0	0				0				
===============											
POSITIVES	4	11	9	4	0	4	1	1	4	0	4
NUMBER (N)	13	13	13	13	13	13	13	9	9	9	9
% POSITIVE	30.8	84.6	69.2	30.8	0.0	30.8	7.7	11.1	44.4	0.0	44.4

Collectively, the father and son matched on 17 (24.3%) of 70 items tested for, while the mother and son matched on 25 (35.7%) of 70 items. In the mother to father 25 to 17 ratio (1.471) shows a significant "donor" difference between the father and mother. For every gene trait the father has "donated" the mother has "donated" almost 50% more (47.1%) genes. Individually, the father and son agreed on one (1.4%) of 70 items tested for, whereas the mother and son agreed on nine (12.9%) of 70 items. In the nine to one ratio (9.0) there is a huge "donor" advantage for the mother; for every gene trait the father has "donated" the mother has "donated" as many as nine gene traits. If you factor in what positive food items each parent has to "offer" there is a mother to father 31 to 22 ratios (1.409) favoring the mother? For every gene the father has to "offer", the mother statistically has about 40% more genes to "offer", advantage mother. Collectively did the father even contribute.

Sensitivities to various fruits (Table 05.01.BPP.01) appeared positive in the father (SGP) in four of 13 (30.8%), the mother (CAL-P) 11 of 13 (84.6%) and the son (BBP) nine of 13 (69.2%). The son and both parents were positive for apple, banana, pineapple, and vanilla while all family members were negative for watermelon. Sensitivities to fruits occurred in the son and both parents with a frequency of four in nine (44.4%) of BPP'S positives. Of the son's (BPP) positives four of nine (44.4%) matched up with only the mother (CAL-P). None of BPP's nine positives matched up with only the father (SGP). The mother matched with BPP's positives in eight of nine (88.9%) occurrences!

Apple and pineapple appeared in the son and both parents. Most simplistically, and perhaps in the eyes of Gregor Mendel apple and pineapple may be passed to each offspring as follows:

AA x AA		
	A	A
A	AA	AA
A	AA	AA

5-1

where homozygous parents (5-1) with all dominant and no recessive genes are passed as possibilities to the 100% of the offspring expressing the trait (sensitivity). Or heterozygous parents (5-2) generate offspring capable of expressing the sensitivity (trait) in 75% of the offspring as follows:

Aa x Aa		
	A	a
A	AA	Aa
a	Aa	aa

5-2

Orange appeared only in the son(s) and the mother. In this case (Table 5.3) the father (SGP) did not show the trait (oo), while the mother (CAL-P) and all male offspring did show the trait (Oo, OO). At the very least, the situation with orange may be expressed as a paternal homozygous recessive (oo) parent and a maternal heterozygous (Oo) parent generated the possibility of a trait positive male offspring as follows:

XoYo x XOXo		
	Xo	Yo
XO	XOXo	XOYo
Xo	XoXo	XoYo

5-3

TABLE 05.01.BPP.02 SGPxCAL-P (BPP) FAMILIAL CYTOTOXIC TEST POSITIVE SENSITIVITIES TO GRAINS.

	FATHER	MOTHER	SON	%POSITIVE	%POSITIVE	%POSITIVE	%NEGATIVE	PER CENT OF POSITIVE FOODS			
SUBJECT	SGP	CAL-P	BPP	SGP/CAP	SGP	CAP	SGP/CAP		SGP/CAP	SGP	CAP
AGE TESTED	35	34	3	BPP	BPP	BPP	BPP	BPP	BPP	BPP	BPP
BARLEY	0	0	0				0				
CORN	1	1	1	1					1		
MALT	0	0	0				0				
OAT	1	1	1	1					1		
RICE	0	0	1					1			
RYE	0	0	1					1			
WHEAT	0	1	1			1					1
===============											
POSITIVES	2	3	5	2	0	1	2	2	2		1
NUMBER (N)	7	7	7	7	7	7	7	5	5	5	5
% POSITIVE	28.6	42.9	72.4	28.6	0.0	14.3	28.6	40.0	40.0	0.0	20.0

Tomato presents an example where both parents SGP and CAL-P did not test positive for tomato, nor did the paternal and maternal grandmothers. The grandfathers were not tested. Here we see the possibility where an X-linked recessive trait (i.e., hemophilia) is carried and not expressed in the heterozygous female but is expressed in a hemizygous (XtY) male offspring lacking the trait for linkage on the male Y chromosome. There are several examples throughout this study where the food sensitivity traits do not show up in the immediate parents yet are expressed in their offspring and there are probably several explanations (i.e., sex linked multiple alleles in calico cats or autosomal multiple alleles in chestnut horses) as to why this trait was not manifest in the immediate parents. This data is further analyzed (Tables 05.01.BPP.01-7) by looking at the offspring (BPP) pattern of sensitivity in relationship to the immediate parents. This is done for several categories of foods (Tables 05.01.BPP.01-6) with a summary (Table 05.01.BPP.07) for all items tested giving an index for overall sensitivity. Not only does this give the individual's overall sensitivity but it provides even greater detail for comparison to individual sibling and parent. This sets-up for more detailed comparison among family members as well as to other families and their individual members. We have singled out some specific examples selected from tables above. Here we have isolated in on BPP and his parents, analyzing how they compared to all foods tested in each category, then how they compared to the foods that BPP was positive or negative for so that we could uncover what sensitivities BPP had and how they matched up with his parents. By analyzing the results in each table, you can see that familial relationships change from food category to food category. Although CAL-P was a significant contributor for various fruits both parents were stronger than any individual parent with 43.2% of BPP's total positives, even though the mother (CAL-P) was also a strong total exclusive individual contributor with 24.3% of BPP's total positives, when analyzing the summary data in table 05.01.BPP.07. In this example CAL-P collectively matches-up with BPP's positives in 25 (16 +9) of 37 (67.6%) of the items tested for and matches up with 47 of 70 (67.1%) positive (25) and negative (22) total foods identified by the cytotoxic test procedure indicating a strong similar pattern testing between mother and son.

NOTES:

____

____

____

____

____

TABLE 05.01.BPP.03 SGPxCAL-P (BPP) FAMILIAL CYTOTOXIC TEST POSITIVE SENSITIVITIES TO VEGETABLES.

	FATHER	MOTHER	SON	%POSITIVE	%POSITIVE	%POSITIVE	%NEGATIVE	PER CENT OF POSITIVE FOODS			
SUBJECT	SGP	CAL-P	BPP	SGP/CAP	SGP	CAP	SGP/CAP		SGP/CAP	SGP	CAP
AGE TESTED	35	34	3	BPP	BPP	BPP	BPP	BPP	BPP	BPP	BPP
BROCCOLI	0	1	0								
CABBAGE	0	0	0				0				
CARROT	0	1	1			1					1
CUCUMBER	1	0	0								
HOPS	0	0	1					1			
HORSERADISH	0	0	0				0				
LETTUCE	0	0	0				0				
ONION	0	0	0				0				
PEA	0	0	1					1			
PEANUT	1	1	1	1					1		
POTATO	1	1	1	1					1		
RADISH	0	0	0				0				
SOYBEAN	1	1	1	1					1		
SPINACH	0	0	1					1			
STRINGBEAN	1	0	0								
SUGAR, BEET	0	0	0				0				
SUGAR, CANE	0	0	1					1			
===============											
POSITIVES	5	5	8	3	0	1	6	4	3	0	1
NUMBER (N)	17	17	17	17	17	17	17	8	8	8	8
% POSITIVE	29.4	29.4	47.1	17.6	0.0	5.9	35.3	50.0	37.5	0.0	12.5

TABLE 05.01.BPP.04 SGPxCAL-P (BPP) FAMILIAL CYTOTOXIC TEST POSITIVE SENSITIVITIES TO FISH & SEAFOOD.

	FATHER	MOTHER	SON	%POSITIVE	%POSITIVE	%POSITIVE	%NEGATIVE	PER CENT OF POSITIVE FOODS			
SUBJECT	SGP	CAL-P	BPP	SGP/CAP	SGP	CAP	SGP/CAP		SGP/CAP	SGP	CAP
AGE TESTED	35	34	3	BPP	BPP	BPP	BPP	BPP	BPP	BPP	BPP
CARP	0	0	0				0				
CRAB	0	0	0				0				
LOBSTER	1	0	0								
SHRIMP	0	0	0				0				
SWORDFISH	0	0	0				0				
TROUT	0	0	0				0				
TUNA	1	1	1	1					1		
===============											
POSITIVES	2	1	1	1	0	0	5	0	1	0	0
NUMBER (N)	7	7	7	7	7	7	7	1	1	1	1
% POSITIVE	28.6	14.3	14.3	14.3	0.0	0.0	71.4	0.0	100.0	0.0	0.0

TABLE 05.01.BPP.05 SGPxCAL-P (BPP) FAMILIAL CYTOTOXIC TEST POSITIVE SENSITIVITIES TO MEAT, FOWL, EGG & MILK.

	FATHER	MOTHER	SON	%POSITIVE	%POSITIVE	%POSITIVE	%NEGATIVE	PER CENT OF POSITIVE FOODS			
SUBJECT	SGP	CAL-P	BPP	SGP/CAP	SGP	CAP	SGP/CAP		SGP/CAP	SGP	CAP
AGE TESTED	35	34	3	BPP	BPP	BPP	BPP	BPP	BPP	BPP	BPP
BEEF	0	0	0				0				
LAMB	0	1	0								
PORK	0	0	0				0				
VEAL	0	0	0				0				
CHICKEN	1	1	1	1					1		
TURKEY, Lt.	0	1	1			1					1
EGG	1	0	1		1					1	
MILK	1	1	1	1					1		
CHEESE/AMER.	0	0	1					1			
CHEESE/ROQUE.	0	0	1					1			
CHEESE/SWISS	0	0	1					1			
===============											
POSITIVES	3	4	7	2	1	1	3	3	2	1	1
NUMBER (N)	11	11	11	11	11	11	11	7	7	7	7
% POSITIVE	27.3	36.4	63.6	18.2	9.1	9.1	27.3	42.9	28.6	14.3	14.3

TABLE 05.01.BPP.06 SGPxCAL-P (BPP) FAMILIAL CYTOTOXIC TEST POSITIVE SENSITIVITIES TO MISCELLANEOUS ITEMS.

	FATHER	MOTHER	SON	%POSITIVE	%POSITIVE	%POSITIVE	%NEGATIVE	PER CENT OF POSITIVE FOODS			
SUBJECT	SGP	CAL-P	BPP	SGP/CAP	SGP	CAP	SGP/CAP		SGP/CAP	SGP	CAP
AGE	35	34	3.0	BPP	BPP	BPP	BPP	BPP	BPP	BPP	BPP
CHOCOLATE	1	1	1	1					1		
COFFEE	1	1	1	1					1		
COTTONSEED	0	1	0								
GARLIC	1	0	0								
GASOLINE	0	0	0				0				
MUSHROOM	0	0	0				0				
MUSTARD	0	0	1					1			
PEPPERMINT	1	1	1	1					1		
TEA	1	1	1	1					1		
TOBACCO	0	1	1			1					1
TURPENTINE	1	0	0								
YEAST, BAKER'S	0	1	1			1					1
YEAST, BREWER'S	0	0	0				0				
ALCOHOL, 1%	0	0	0				0				
ALCOHOL, 2%	0	0	0				0				
===============											
POSITIVES	6	7	7	4	0	2	5	1	4	0	2
NUMBER (N)	15	15	15	15	15	15	15	7	7	7	7
% POSITIVE	40.0	46.7	46.7	26.7	0.0	13.3	33.3	14.3	57.1	0.0	28.6

TABLE 05.01.BPP.07 SUMMARY: SGPxCAL-P (BPP) FAMILIAL CYTOTOXIC TEST POSITIVE FOOD SENSITIVITIES.

	FATHER	MOTHER	SON	%POSITIVE	%POSITIVE	%POSITIVE	%NEGATIVE	PER CENT OF POSITIVE FOODS			
SUBJECT	SGP	CAL-P	BPP	SGP/CAP	SGP	CAP	SGP/CAP		SGP/CAP	SGP	CAP
AGE	35	34	3	BPP	BPP	BPP	BPP	BPP	BPP	BPP	BPP
POSITIVES	22	31	37	16	1	9	22	11	16	1	9
NUMBER (N)	70	70	70	70	70	70	70	37	37	37	37
% POSITIVE	31.4	44.3	52.9	22.9	1.4	12.9	31.4	29.7	43.2	2.7	24.3

An overall assessment of food Cytotoxic Test food sensitivity is presented in table 05.01.BPP.07. BPP immune system recognizes 52.9% (37 Of 70) of the allergy antigens tested for. Of the 37 positives 43.2% were also positive in both parents. The mother matched with an additional 24.3% while the father matched only one of the 37 positives. BPP had 11 positives that did not appear positive in either of his immediate parents. We will pay close attention to these relationships among siblings, relatives, and other family case studies. We strongly feel that the immune system plasma white cells are the first line of defense for a systemic response to an invading allergen. Within the immune system cells is the genetic information that dictates how a given individual can respond. In a second phase immunochemical events initiate a response. In a third phase an end organ(s) can be affected. End organs, also dictated by genetic and/or environmental factors, may be skin, cardiovascular, respiratory, gastrointestinal, or central nervous systems. A negative skin test may well be, and often is, a false negative. BPP first exhibited significant symptoms at 6 months. Cytotoxic test was performed at 6 months and later at 47 months with consistent results. At 6 months BPP's pediatrician wrestled with the clinical problems with minimal success, then recommended that a pediatric allergist be included in the struggle to find relief for BPP. A clinical evaluation of this combined approach is contained in the letter (Exhibit 5.0) from the Pediatric Allergist to the Pediatrician and a summary of some of the clinic visits with the Pediatric Allergist over a four-year period.

- **06/13/1972:** At 6 months the "colicky kid" BPP with earaches, persistent diarrhea, chronic runny noses, and sleep problems received his initial Cytotoxic Tests for food sensitivities. These test results were provided to his pediatrician and used to improve his diet management.

- **08/1972:** Pediatric Allergist referral initiated

- **08/16/1972:** At about 12 months had been off milk with diet management directed away from Cytotoxic Test positive food, emphasizing Cytotoxic Test negative foods, and avoiding foods that were not tested. Desensitization shots were administered weekly. Environmental management included the cessation of smoking in the house and minimizing direct exposure to the forced air HVAC system. Clinical improvements for BPP noted.

- **01/22/1973:** At 17 months desensitizing shots cut back to twice monthly. Additional clinical improvements noted. Earaches have gone away during the past few weeks. Some milk reintroduced to his diet. Happier baby. Mom smokes less in the house.

- **07/16/1973**: At 23 months desensitizing shots continue at twice monthly. The case history indicates BPP is doing very well and has had no earaches. He continues to take some milk. Mom still smokes in the house.

- **01/14/1974:** At 29 months desensitizing shots reduced to monthly. BPP doing very well with no earaches. BPP still taking milk Mom still smokes.

- **07/29/1974:** At 47 months it's noted that BPP had no colds last winter and no ear infections. A challenge Cytotoxic Test is done on follow up to the initial test done at 6 months. The results are consistent with the initial test. Skin tests were generally negative with only Outside Mold showing slight and House Dust a +1. Desensitizing shots continue monthly.

Exhibit 5.0

August 16, 1972

Dr. Steven Plax
8025 Dale Avenue
St. Louis, Missouri 63117

RE: Brant Perry

Dear Steve:

Thank you very much for asking me to see one year old Brant Perry, who has had problems with a serous otitis and diarrhea. He had been breast fed for a very short period of time and then had to be put on formula. Diarrhea started at about six months of age and from that time on, Brant did gain weight, but not as well as his sibling. Head congestion started at about the same time, with frequent colds etc. The ears started also sometime about then and have been continuous. Cytotoxic studies were done and many food problems were elaborated. At that time he was taken off of multiple foods and his diarrhea seemed better.

The family history is positive for recurrent bronchitis, hay fever, hives, eczema, food allergies and drug allergies. The environmental history reveals a forced air heating duct blowing through the bedroom. I have asked that in the wintertime the heating duct be shut. I have asked that the pad underneath the rug in the bedroom be checked, to make sure it is not animal hair. Mother smokes in the house. I have asked that there be no smoking in the apartment or car. Brant is off milk and I feel that he should stay off milk, but have not taken any other foods away from him. I have asked that the bottle be discontinued.

My physical examination did reveal some fluid on the right. Immune globulins, milk precipitins, secretory IgA and nasal smear are pending. His sweat test was normal. Skin tests showed a very nice reaction to histamine, but only minimal reactions to odd and assorted antigens. I think that this represents primarily a bacterial allergy, possibly with mold and dust as excitants and so will make up an extract to contain dusts and molds with a respiratory vaccine. This will be sent on to you with instructions. I would like Brant to receive allergy injections weekly, until I see him again in five months.

With kindest personal regards,

DBS/pr
Dictated but not read.

Donald B. Strominger, M. D.

With the best efforts of all involved, all involved are not sure that BPP didn't just outgrow his problems. There is much more clinical information and laboratory information about this case than has been reported here. The desensitation shots were monitored by tracking the total IgE and IgG levels. No specific tests identified the specific antigens that may have triggered the adverse clinical symptoms nor identified the specific IgE or IgG components of the immune system that were generated during the immune response. This information could be critical and lurks under the radar in the subtle details of the immune system, perhaps hidden in the nonfasting vs. fasting WBC count differences that were reported here in Chapters 2 and 3.

CASE STUDY 05.01.RRPa.1-7 SGP X CAL-P (RRP): Six-year-old RRP and his parents (Tables 05.01.RRP.1-7) agreed, either positive or negative, on 39 of 70 (55.7%) of the items tested for. The father and mother agreed on 16 of 70 (22.9%) positives. Of the16 positive items that both parents agreed on 12 of 16 (75.0%) were passed as positives to RRP. Of the son's 21 positives, 13 (61.9%) matched his father's positives and 16 (76.2%) matched his mom's positives, and both parents matched on 12 of his 21 (57.1%) items. The father exclusively matched on only one (egg) of 21 (4.8%) of RRP's positives. The mother exclusively matched on four (orange, carrot, tobacco, and baker's yeast) of 21 (19.0%) of RRP's positives. RRP had four (19.0%) of 21 test positive items (tomato, malt, rice, and American cheese) that did not appear in either parent. Collectively, the father and son matched on 13 (18.6%) of 70 items tested for, while the mother and son matched on 16 (22.9%) of 70 items. In the mother to father 16 to 13 ratio (1.231) shows a "donor" advantage where the mother statistically appears to have "donated" about 20% more gene traits. Individually, the mother to father four to one ratio (4.000) shows a "donor" advantage where the mother statistically appears to have "donated" four gene traits for every paternal gene trait "donated". If you factor in what positive food items each parent has to "offer" there is a mother to father 22 to 31 ratios (1.409) favoring the mother. For every gene the father has to "offer", the mother statistically has about 40% more genes to "offer", advantage mother. This compares closely with his sibling brother BPP reported in the previous case.

TABLE 05.01.RRP.01 SGPxCAL-P (RRP) FAMILIAL DISTRIBUTION OF CYTOTOXIC TEST POSITIVE SENSITIVITIES TO FRUITS.

	FATHER	MOTHER	SON	%POSITIVE	%POSITIVE	%POSITIVE	%NEGATIVE	PER CENT OF POSITIVE FOODS			
SUBJECT	SGP	CAL-P	RRP	SGP/CAP	SGP	CAP	SGP/CAP		SGP/CAP	SGP	CAP
AGE TESTED	35	34	6	RRP	RRP	RRP	RRP	RRP	RRP	RRP	RRP
APPLE	1	1	1	1					1		
BANANA	1	1	0								
CANTALOUPE	0	1	0								
CHERRY	0	1	0								
HONEYDEW	0	1	0								
ORANGE	0	1	1			1					1
PEACH	0	1	0								
PEAR	0	1	0								
PINEAPPLE	1	1	1	1					1		
STRAWBERRY	0	1	0								
TOMATO	0	0	1					1			
VANILLA	1	1	0								
WATERMELON	0	0	0				0				
===============											
POSITIVES	4	11	4	2	0	1	1	1	2	0	1
NUMBER (N)	13	13	13	13	13	13	13	4	4	4	4
% POSITIVE	30.8	84.6	30.8	15.4	0.0	7.7	7.7	25.0	50.0	0.0	25.0

The father (SGP) tested positive for 22 of 70 (31.4%), the mother (CAP) tested positive for 31 of 70 (44.3%) and the son (RRP) was positive for 21 of 70 (30.0%) food items. Both parents' "donation" to the son appears to be stronger than the individual parent with 12 (57.1%) of RRP's 21 positives, as compared to the mother's individual contribution at 19.0% and the father's individual contribution at 4.8% (Table 5.1.2.7). By comparison with BPP's data CAL-P matches up with RRP'S positives in 16 (12 + 4) of 21 (76.2%) of the items, with BPP positives in 25 (16 + 9) of 37 (67.6%) of the 70 foods tested for indicating a strong proportionate comparison of sensitivities between mother to sons, with RRP having less total sensitivity at 21 instead of 37 Cytotoxic Test positives. The father matched up with RRP's 21 positives in 13 (12 + 1) or 61.9%, compared to BPP's 37 positives in 17 (16 + 1) or 45.9% by comparison showing only slightly less magnitude and proportionality.

TABLE 05.01.RRP.02 SGPxCAL-P (RRP) FAMILIAL DISTRIBUTION OF CYTOTOXIC TEST POSITIVE SENSITIVITIES TO GRAINS.

	FATHER	MOTHER	SON	%POSITIVE	%POSITIVE	%POSITIVE	%NEGATIVE	PER CENT OF POSITIVE FOODS			
SUBJECT	SGP	CAL-P	RRP	SGP/CAP	SGP	CAP	SGP/CAP		SGP/CAP	SGP	CAP
AGE TESTED	35	34	6	RRP	RRP	RRP	RRP	RRP	RRP	RRP	RRP
BARLEY	0	0	0				0				
CORN	1	1	1	1					1		
MALT	0	0	1					1			
OAT	1	1	1	1					1		
RICE	0	0	1					1			
RYE	0	0	0				0				
WHEAT	0	1	0								
===============											
POSITIVES	2	3	4	2	0	0	2	2	2	0	0
NUMBER (N)	7	7	7	7	7	7	7	4	4	4	4
% POSITIVE	28.6	42.9	57.1	28.6	0.0	0.0	28.6	50.0	50.0	0.0	0.0

TABLE 05.01.RRP.03 SGPxCAL-P (RRP) FAMILIAL CYTOTOXIC TEST POSITIVE SENSITIVITIES TO VEGETABLES.

	FATHER	MOTHER	SON	%POSITIVE	%POSITIVE	%POSITIVE	%NEGATIVE	PER CENT OF POSITIVE FOODS			
SUBJECT	SGP	CAL-P	RRP	SGP/CAP	SGP	CAP	SGP/CAP		SGP/CAP	SGP	CAP
AGE TESTED	35	34	6	RRP	RRP	RRP	RRP	RRP	RRP	RRP	RRP
BROCCOLI	0	1	0								
CABBAGE	0	0	0				0				
CARROT	0	1	1			1					1
CUCUMBER	1	0	0								
HOPS	0	0	0				0				
HORSERADISH	0	0	0				0				
LETTUCE	0	0	0				0				
ONION	0	0	0				0				
PEA	0	0	0				0				
PEANUT	1	1	1	1					1		
POTATO	1	1	0								
RADISH	0	0	0				0				
SOYBEAN	1	1	1	1					1		
SPINACH	0	0	0				0				
STRINGBEAN	1	0	0								
SUGAR, BEET	0	0	0				0				
SUGAR, CANE	0	0	0				0				
===============											
POSITIVES	5	5	3	2	0	1	10	0	2	0	1
NUMBER (N)	17	17	17	17	17	17	17	3	3	3	3
% POSITIVE	29.4	29.4	17.6	11.8	0.0	5.9	58.8	0.0	66.7	0.0	33.3

TABLE 05.01.RRP.04 SGPxCAL-P (RRP) FAMILIAL CYTOTOXIC TEST POSITIVE SENSITIVITIES TO FISH & SEAFOOD.

	FATHER	MOTHER	SON	%POSITIVE	%POSITIVE	%POSITIVE	%NEGATIVE	PER CENT OF POSITIVE FOODS			
SUBJECT	SGP	CAL-P	RRP	SGP/CAP	SGP	CAP	SGP/CAP		SGP/CAP	SGP	CAP
AGE TESTED	35	34	6	RRP	RRP	RRP	RRP	RRP	RRP	RRP	RRP
CARP	0	0	0				0				
CRAB	0	0	0				0				
LOBSTER	1	0	0								
SHRIMP	0	0	0				0				
SWORDFISH	0	0	0				0				
TROUT	0	0	0				0				
TUNA	1	1	0								
===============											
POSITIVES	2	1	0	0	0	0	5	0	0	0	0
NUMBER (N)	7	7	7	7	7	7	7	0	0	0	0
% POSITIVE	28.6	14.3	0.0	0.0	0.0	0.0	71.4	0.0	0.0	0.0	0.0

TABLE 05.01.RRP.05 SGPxCAL-P (RRP) FAMILIAL CYTOTOXIC TEST POSITIVE SENSITIVITIES TO MEAT, FOWL, EGG & MILK.

	FATHER	MOTHER	SON	%POSITIVE	%POSITIVE	%POSITIVE	%NEGATIVE	PER CENT OF POSITIVE FOODS			
SUBJECT	SGP	CAL-P	RRP	SGP/CAP	SGP	CAP	SGP/CAP		SGP/CAP	SGP	CAP
AGE TESTED	35	34	6	RRP	RRP	RRP	RRP	RRP	RRP	RRP	RRP
BEEF	0	0	0				0				
LAMB	0	1	0								
PORK	0	0	0				0				
VEAL	0	0	0				0				
CHICKEN	1	1	1	1					1		
TURKEY, Lt.	0	1	0								
EGG	1	0	1		1					1	
MILK	1	1	1	1					1		
CHEESE/AMER.	0	0	1					1			
CHEESE/ROQUE.	0	0	0				0				
CHEESE/SWISS	0	0	0				0				
===============											
POSITIVES	3	4	4	2	1	0	5	1	2	1	0
NUMBER (N)	11	11	11	11	11	11	11	4	4	4	4
% POSITIVE	27.3	36.4	36.4	18.2	9.1	0.0	45.5	25.0	50.0	25.0	0.0

TABLE 05.01.RRP.06 SGPxCAL-P (RRP) FAMILIAL CYTOTOXIC TEST POSITIVE SENSITIVITIES TO MISCELLANCE ITEMS.

	FATHER	MOTHER	SON	%POSITIVE	%POSITIVE	%POSITIVE	%NEGATIVE	PER CENT OF POSITIVE FOODS			
SUBJECT	SGP	CAL-P	RRP	SGP/CAP	SGP	CAP	SGP/CAP		SGP/CAP	SGP	CAP
AGE	35	34	6	RRP	RRP	RRP	RRP	RRP	RRP	RRP	RRP
CHOCOLATE	1	1	1	1					1		
COFFEE	1	1	1	1					1		
COTTONSEED	0	1	0								
GARLIC	1	0	0								
GASOLINE	0	0	0				0				
MUSHROOM	0	0	0				0				
MUSTARD	0	0	0				0				
PEPPERMINT	1	1	1	1					1		
TEA	1	1	1	1					1		
TOBACCO	0	1	1			1					1
TURPENTINE	1	0	0								
YEAST, BAKER'S	0	1	1			1					1
YEAST, BREWER'S	0	0	0				0				
ALCOHOL, 1%	0	0	0				0				
ALCOHOL, 2%	0	0	0				0				
===============											
POSITIVES	6	7	6	4	0	2	6	0	4	0	2
NUMBER (N)	15	15	15	15	15	15	15	6	6	6	6
% POSITIVE	40.0	46.7	40.0	26.7	0.0	13.3	40.0	0.0	66.7	0.0	33.3

TABLE 05.01.RRP.07 SUMMARY: SGPxCAL-P (RRP) DISTRIBUTION OF CYTOTOXIC TEST POSITIVE FOOD SENSITIVITIES.

	FATHER	MOTHER	SON	%POSITIVE	%POSITIVE	%POSITIVE	%NEGATIVE	PER CENT OF POSITIVE FOODS			
SUBJECT	SGP	CAL-P	RRP	SGP/CAP	SGP	CAP	SGP/CAP		SGP/CAP	SGP	CAP
AGE	35	34	6	RRP	RRP	RRP	RRP	RRP	RRP	RRP	RRP
POSITIVES	22	31	21	12	1	4	29	4	12	1	4
NUMBER (N)	70	70	70	70	70	70	70	21	21	21	21
% POSITIVE	31.4	44.3	30.0	17.1	1.4	5.7	41.4	19.0	57.1	4.8	19.0

RRP was Cytotoxic Test positive to four Fruits (Table 5.1.RRP.1), apple, orange, pineapple, and tomato. Every family member tested was positive (Table 05.01.MEMBERS.01) for apple including two grandmothers and a newborn brother! Pineapple was positive in all family members tested except the newborn brother. Orange was positive in the mother, maternal grandmother and all three sons, while the father and paternal grandmother were negative. Tomato was test positive in only the three sons. The mother CAL-P was positive for 11 of the 13 (84.6%) fruits tested for. The father was positive for only four of the 13 fruits tested for. Of the various grains tested for (Table 5.1.0.2) corn was positive in all but the newborn. Chicken (Table 05.01.MEMBERS.05) was positive in every family member tested and milk missed testing positive for all only in the paternal grandmother. Coffee (Table 05.01.MEMBERS.06) was positive in all family members tested and chocolate was positive in all members except the maternal grandmother and tea was positive in all members with the exception of the paternal grandmother. Tobacco tested positive in the mother, maternal grandmother, and all three sons, while the father and paternal grandmother tested negative. Not only do we see distinctive and very diagnostic test positive patterns in individuals, but also in families. The evidence points to a distinctive genetic allergy profile that is as good as a genetic fingerprint.

CASE STUDY 05.01.SSPc.01-7 SGP X CAL-P (SSP-NEWBORN): The newborn SSP was Cytotoxic Test positive for 16 of 70 (22.9%) items determined in a chord blood sample taken at the time of birth. With a newborn no second Cytotoxic Test challenge is performed, which generally adds five test positives conservatively.

Newborn SSP and his parents agreed, either positive or negative, on 38 of 70 (54.3%) of the items tested for. Every chromosome shows gene trait banding. In Calico cats, for example, the sex-linked banded trait has to appear on both X chromosomes. All Calico's are females with a very rare exception of XXY

The father and mother agreed on 16 of 70 (22.9%) positives. Of the16 positive items that both parents agreed on 8 of 16 (50.0%) were passed as positives to SSP. Of the son's 16 positives, 10 (62.5%) matched his father's positives and 11 (68.8%) matched his mom's positives, and eight (50.0%) matched both parents.

The father exclusively matched on two (garlic, egg) of 16 (12.5%) of SSP's positives. The mother exclusively matched on three (orange, cottonseed, and tobacco) of 16 (18.8%) of SSP's positives. SSP had three (18.8%) of 16 test positive items (tomato, radish, and American cheese) that did not appear in either parent.

Collectively, the father and son matched on 10 (14.3%) of 70 items tested for, while the mother and son matched on 11 (15.7%) of 70 items. In the mother to father 11 to 10 ratio (1.100) shows a "donor" advantage where the mother statistically appears to have "donated" about 10% more gene traits.

Individually, the mother to father three to two ratio (1.500) shows a "donor" advantage where the mother statistically appears to have "donated" five gene traits for every paternal gene trait "donated". If you factor in what positive food items each parent has to "offer" there is a mother to father 22 to 31 ratio (1.409) favoring the mother.

For every gene trait the father has to "offer", the mother statistically has about 40% more gene traits to "offer", advantage mother. This compares closely with his sibling brothers BPP and RRP reported in the previous cases.

SSP was positive for four fruits (Table 05.01.SSP.01), apple, orange, tomato, and vanilla. Apple was positive in SSP, and all family members tested (Table 05.01.SSP.01). Orange was positive in SSP and all family members except the father and maternal grandmother. Tomato was positive in only the three boys of the family members tested! Vanilla was positive in all family members tested except the paternal grandmother and six-year-old RRP.

The mother consumed these foods on a weekly basis during the SSP pregnancy. SSP did not test positive for any of the tested seven grains. Corn was positive in all family members tested except SSP, and oat was positive in all family members tested except the paternal grandmother and SSP. The mother also consumed these foods on a weekly basis during the SSP pregnancy.

NOTES:

TABLE 05.01.SSP.01. SGPxCAL-P (SSP) FAMILIAL DISTRIBUTION OF CYTOTOXIC TEST POSITIVE SENSITIVITIES TO FRUITS.

	FATHER	MOTHER	SON	%POSITIVE	%POSITIVE	%POSITIVE	%NEGATIVE	PER CENT OF POSITIVE FOODS			
SUBJECT	SGP	CAL-P	SSP	SGP/CAP	SGP	CAP	SGP/CAP		SGP/CAP	SGP	CAP
AGE TESTED	35	34	NEWBORN	SSP	SSP	SSP	SSP	SSP	SSP	SSP	SSP
APPLE	1	1	1	1					1		
BANANA	1	1	0								
CANTALOUPE	0	1	0								
CHERRY	0	1	0								
HONEYDEW	0	1	0								
ORANGE	0	1	1			1					1
PEACH	0	1	0								
PEAR	0	1	0								
PINEAPPLE	1	1	0								
STRAWBERRY	0	1	0								
TOMATO	0	0	1					1			
VANILLA	1	1	1	1					1		
WATERMELON	0	0	0				0				
===============											
POSITIVES	4	11	4	2	0	1	1	1	2	0	1
NUMBER (N)	13	13	13	13	13	13	13	4	4	4	4
% POSITIVE	30.8	84.6	30.8	15.4	0.0	7.7	7.7	25.0	50.0	0.0	25.0

TABLE 05.01.SSP.02 SGPxCAL-P (SSP) FAMILIAL DISTRIBUTION OF CYTOTOXIC TEST POSITIVE SENSITIVITIES TO GRAINS.

	FATHER	MOTHER	SON	%POSITIVE	%POSITIVE	%POSITIVE	%NEGATIVE	PER CENT OF POSITIVE FOODS			
SUBJECT	SGP	CAL-P	SSP	SGP/CAP	SGP	CAP	SGP/CAP		SGP/CAP	SGP	CAP
AGE TESTED	35	34	NEWBORN	SSP	SSP	SSP	SSP	SSP	SSP	SSP	SSP
BARLEY	0	0	0				0				
CORN	1	0	0								
MALT	0	0	0				0				
OAT	1	1	0								
RICE	0	0	0				0				
RYE	0	O	0				0				
WHEAT	0	1	0								
===============											
POSITIVES	2	3	0	0	0	0	4	0	0	0	0
NUMBER (N)	7	7	7	7	7	7	7	o	0	0	0
% POSITIVE	28.6	42.9	0.0	0.0	0.0	0.0	57.1				

NOTES:

TABLE 05.01.SSP.03 SGPxCAL-P (SSP) FAMILIAL CYTOTOXIC TEST POSITIVE SENSITIVITIES TO VEGETABLES.

	FATHER	MOTHER	SON	%POSITIVE	%POSITIVE	%POSITIVE	%NEGATIVE	PER CENT OF POSITIVE FOODS			
SUBJECT	SGP	CAL-P	SSP	SGP/CAP	SGP	CAP	SGP/CAP		SGP/CAP	SGP	CAP
AGE TESTED	35	34	NEWBORN	SSP	SSP	SSP	SSP	SSP	SSP	SSP	SSP
BROCCOLI	0	1	0								
CABBAGE	0	0	0				0				
CARROT	0	1	0								
CUCUMBER	1	0	0								
HOPS	0	0	0				0				
HORSERADISH	0	0	0				0				
LETTUCE	0	0	0				0				
ONION	0	0	0				0				
PEA	0	0	0				0				
PEANUT	1	1	0								
POTATO	1	1	0								
RADISH	0	0	1					1			
SOYBEAN	1	1	1	1					1		
SPINACH	0	0	0				0				
STRINGBEAN	1	0	0								
SUGAR, BEET	0	0	0				0				
SUGAR, CANE	0	0	0				0				
===============											
POSITIVES	5	5	2	1	0	0	9	1	1	0	0
NUMBER (N)	17	17	17	17	17	17	17	2	2	2	2
% POSITIVE	29.4	29.4	11.0	5.9	0.0	0.0	52.9	50.0	50.0	0.0	0.0

The unique information here is the pre-partum timing of test positive data. Conventional wisdom says the placental barrier protects the fetus from dietary antigens; therefore, food sensitivities appear in the sequence of infant diet regimes including maternal breast milk and passive transfer of maternal immune system benefits. This newborn had not yet received post-partum nourishment of any kind. Bear in mind (Chapter II.) that a positive test here reads the immune-defense system's plasma white cells and is presented here as only the first of three steps required for an immune or allergic manifestation. This allergy scenario is placed in the midst of detailed and delicate developmental processes.

Speculating for a minute that any single positive sensitizing foods react in a dose response dependent way (Chapter 3.0 Figure 03.02) potentially on susceptible end organ tissue opens the door to a model for studying a host of clinical maladies and even birth defects. Any of the above positive antigen(s) displays as a potential sigmoid dose response where very small amounts could deliver a huge blow whether it be an untimely, even localized, anaphylaxis or a very specific interference at the end organ that results in some or significant impairment.

TABLE 05.01.SSP.04 SGPxCAL-P (SSP) FAMILIAL CYTOTOXIC TEST POSITIVE SENSITIVITIES TO FISH & SEAFOOD.

	FATHER	MOTHER	SON	%POSITIVE	%POSITIVE	%POSITIVE	%NEGATIVE	PER CENT OF POSITIVE FOODS			
SUBJECT	SGP	CAL-P	SSP	SGP/CAP	SGP	CAP	SGP/CAP		SGP/CAP	SGP	CAP
AGE TESTED	35	34	NEWBORN	SSP	SSP	SSP	SSP	SSP	SSP	SSP	SSP
CARP	0	0	0				0				
CRAB	0	0	0				0				
LOBSTER	1	0	0								
SHRIMP	0	0	0				0				
SWORDFISH	0	0	0				0				
TROUT	0	0	0				0				
TUNA	1	1	0								
===============											
POSITIVES	2	1	0	0	0	0	5	0	0	0	0
NUMBER (N)	7	7	7	7	7	7	7	0	0	0	0
% POSITIVE	28.6	14.3	0.0	0.0	0.0	0.0	71.4	0.0	0.0	0.0	0.0

TABLE 05.01.SSP.05 SGPxCAL-P (SSP) FAMILIAL CYTOTOXIC TEST POSITIVE SENSITIVITIES TO MEAT, FOWL, EGG & MILK.

	FATHER	MOTHER	SON	%POSITIVE	%POSITIVE	%POSITIVE	%NEGATIVE	PER CENT OF POSITIVE FOODS			
SUBJECT	SGP	CAL-P	SSP	SGP/CAP	SGP	CAP	SGP/CAP		SGP/CAP	SGP	CAP
AGE TESTED	35	34	NEWBORN	SSP	SSP	SSP	SSP	SSP	SSP	SSP	SSP
BEEF	0	0	0				0				
LAMB	0	1	0								
PORK	0	0	0				0				
VEAL	0	0	0				0				
CHICKEN	1	1	1	1					1		
TURKEY, Lt.	0	1	0								
EGG	1	0	1		1					1	
MILK	1	1	1	1					1		
CHEESE/AMER.	0	0	1					1			
CHEESE/ROQUE.	0	0	0				0				
CHEESE/SWISS	0	0	0				0				
===============											
POSITIVES	3	4	4	2	1	0	5	1	2	1	0
NUMBER (N)	11	11	11	11	11	11	11	4	4	4	4
% POSITIVE	27.3	36.4	36.4	18.2	9.1	0.0	45.5	25.0	50.0	25.0	0.0

TABLE 05.01.SSP.06 SGPxCAL-P (SSP) FAMILIAL CYTOTOXIC TEST POSITIVE SENSITIVITIES TO MISCELLEANOUS ITEMS.

	FATHER	MOTHER	SON	%POSITIVE	%POSITIVE	%POSITIVE	%NEGATIVE	PER CENT OF POSITIVE FOODS			
SUBJECT	SGP	CAL-P	SSP	SGP/CAP	SGP	CAP	SGP/CAP		SGP/CAP	SGP	CAP
AGE	35	34	NEWBORN	SSP	SSP	SSP	SSP	SSP	SSP	SSP	SSP
CHOCOLATE	1	1	1	1					1		
COFFEE	1	1	1	1					1		
COTTONSEED	0	1	1			1					1
GARLIC	1	0	1		1					1	
GASOLINE	0	0	0				0				
MUSHROOM	0	0	0				0				
MUSTARD	0	0	0				0				
PEPPERMINT	1	1	0								
TEA	1	1	1	1					1		
TOBACCO	0	1	1			1					1
TURPENTINE	1	0	0								
YEAST, BAKER'S	0	1	0								
YEAST, BREWER'S	0	0	0				0				
ALCOHOL, 1%	0	0	0				0				
ALCOHOL, 2%	0	0	0				0				
===============											
POSITIVES	6	7	6	3	1	2	6	0	3	1	2
NUMBER (N)	15	15	15	15	15	15	15	6	6	6	6
% POSITIVE	40.0	46.7	40.0	20.0	6.7	13.3	40.0	0.0	50.0	16.7	33.3

TABLE 05.01.SSP.07 SUMMARY: SGPxCAL-P (SSP) FAMILIAL CYTOTOXIC TEST POSITIVE FOOD SENSITIVITIES.

	FATHER	MOTHER	SON	%POSITIVE	%POSITIVE	%POSITIVE	%NEGATIVE	PER CENT OF POSITIVE FOODS			
SUBJECT	SGP	CAL-P	SSP	SGP/CAP	SGP	CAP	SGP/CAP		SGP/CAP	SGP	CAP
AGE	35	34	NEWBORN	SSP	SSP	SSP	SSP	SSP	SSP	SSP	SSP
POSITIVES	22	31	16	8	2	3	30	3	8	2	3
NUMBER (N)	70	70	70	70	70	70	70	16	16	16	16
% POSITIVE	31.4	44.3	22.9	11.4	2.9	4.3	42.9	18.8	50.0	12.5	18.8

CASE STUDY 05.02.MEMBERS.01-7 SGP & Extended Family: We have looked at the SGP family in great detail.

We will now look further at SGP's two sisters HJP and DCP-G-D and the latter sister's six offspring. Of interest here are two sets of twins. Her first husband, killed in Vietnam, gave her the four oldest children and her first set of identical twins, followed by a second set of twins from a subsequent marriage. Unfortunately, no tests were run on either husband to complete the genetic mapping. Orange (Table 05.02.MEMBERS.01) tested positive with DCP-G-D and all of her offspring, yet her siblings were test negative. Oat (Table 05.02.MEMBERS.02) was positive in all members tested. Rye tested negative in the mother DCP-G-D and her siblings yet was positive in all of her children. Peanut (Table 05.02. MEMBERS.03), an often-studied antigen(s), tested positive for all members tested except the twin Laurie G. Tuna (Table 05.02.MEMBERS.04) was positive in all members except the twin Robin D. Egg (Table 05.02.MEMBERS.05) was positive in all members except the mother DCP-G-D and only the oldest son Mark G and twin Robin D were negative for chicken. Chocolate (Table 05.02..MEMBERS.06) tested positive for all members except twin Amy D and cottonseed tested negatively in the mother DCP-G-D and positive for all of her offspring. Listed here are the highlights that make the members of this group unique. We will look at the twins more closely.

TABLE 05.02.MEMBERS.01 SGP EXTENDED FAMILY CYTOTOXIC TEST POSITIVE SENSITIVITIES TO FRUITS.

	BROTHER	SISTER	SISTER	SON	DAUGHTER	DAUGHTER	DAUGHTER	DAUGHTER	DAUGHTER
SUBJECT	SGP	HJP	DCP-G-D	MARK	TERRY G	LAURIE G	KATHY G	AMY D	ROBIN D
AGE TESTED	35	30	35	17	16	14	14	9	9
APPLE	1	1	1	0	0	1	1	0	1
BANANA	1	1	0	1	1	1	0	1	1
CANTALOUPE	0	0	0	0	1	0	0	0	0
CHERRY	0	0	0	0	0	0	0	0	0
HONEYDEW	0	0	0	1	1	0	0	0	1
ORANGE	0	0	1	1	1	1	1	1	1
PEACH	0	0	1	0	1	1	0	0	0
PEAR	0	0	0	0	0	0	0	0	0
PINEAPPLE	1	1	0	0	0	0	0	0	1
STRAWBERRY	0	0	0	0	1	0	1	0	1
TOMATO	0	0	0	0	0	0	0	0	0
VANILLA	1	0	1	0	1	1	0	0	0
WATERMELON	0	0	0	0	0	0	0	0	0
===============									
POSITIVES	4	3	4	3	7	5	3	2	6
NUMBER (N)	13	13	13	13	13	13	13	13	13
% POSITIVE	30.8	23.1	30.8	23.1	53.8	38.5	23.1	15.4	46.2

TABLE 05.02.MEMBERS.02. SGP EXTENDED FAMILY CYTOTOXIC TEST POSITIVE SENSITIVITIES TO GRAINS.

	BROTHER	SISTER	SISTER	SON	DAUGHTER	DAUGHTER	DAUGHTER	DAUGHTER	DAUGHTER
SUBJECT	SGP	HJP	DCP-G-D	MARK	TERRY	LAURIE G	KATHY G	AMY D	ROBIN D
AGE TESTED	35	30	35	17	16	14	14	9	9
BARLEY	0	0	0	0	0	0	0	0	0
CORN	1	0	1	1	1	0	1	0	1
MALT	0	1	0	0	0	1	1	0	1
OAT	1	1	1	1	1	1	1	1	1
RICE	0	1	0	0	0	1	1	1	0
RYE	0	0	0	1	1	1	1	1	1
WHEAT	0	0	0	0	0	0	0	0	1
===============									
POSITIVES	2	3	2	3	3	4	5	3	5
NUMBER (N)	7	7	7	7	7	7	7	7	7
% POSITIVE	28.6	42.9	28.6	42.9	42.9	57.1	71.4	42.9	71.4

TABLE 05.02.MEMBERS.03. SGP EXTENDED FAMILY CYTOTOXIC TEST POSITIVE SENSITIVITIES TO VEGETABLES.

SUBJECT	BROTHER SGP	SISTER HJP	SISTER DCP-G-D	SON MARK	DAUGHTER TERRY	DAUGHTER LAURIE G	DAUGHTER KATHY G	DAUGHTER AMY D	DAUGHTER ROBIN D
AGE TESTED	35	30	35	17	16	14	14	9	9
BROCCOLI	0	1	0	0	0	0	0	0	0
CABBAGE	0	0	0	0	1	0	1	0	1
CARROT	0	0	0	0	0	0	0	0	0
CUCUMBER	1	0	1	0	0	0	0	0	0
HOPS	0	0	1	0	0	0	0	1	0
HORSERADISH	0	0	0	0	0	0	0	0	0
LETTUCE	0	0	0	0	0	0	0	0	0
ONION	0	0	0	0	0	0	0	0	0
PEA	0	0	0	0	0	0	0	0	0
PEANUT	1	1	1	1	1	0	1	1	1
POTATO	1	0	0	0	0	0	0	0	1
RADISH	0	0	0	0	0	0	0	0	0
SOYBEAN	1	0	1	1	0	1	1	0	1
SPINACH	0	0	0	0	0	0	0	0	0
STRINGBEAN	1	1	0	0	0	0	0	0	1
SUGAR, BEET	0	0	0	0	0	0	0	0	0
SUGAR, CANE	0	0	0	0	1	1	0	0	1
===============									
POSITIVES	5	3	4	2	3	2	3	2	6
NUMBER (N)	17	17	17	17	17	17	17	17	17
% POSITIVE	29.4	17.6	23.5	11.8	17.6	11.8	17.6	11.8	35.3

Identical twins, formed from one fertilized egg, have identical DNA. Fraternal twins, formed from two different fertilized eggs, share half of their genes similar to their common parental siblings. Identical genotypes can lead to different phenotypes, which is expression of the DNA. General appearance, size, shape, and fingerprints can be very different in identical twins. A DNA test can't determine the difference between identical twins, whereas finger or footprints on a birth certificate can.

Twins Laurie G and Kathy G (Tables 05.02.MEMBERS.1-7) agreed, either positive or negative, on 60 of 70 (85.7%) of the food items tested for. Of twin Laurie G's 17 positives, 12 (70.6%) matched twin Kathy G's positives. Of twin Kathy G's 17 positives, 12 (70.6%) matched twin Laurie G's positives. Note that five (29.4%) items did not match. The food items that did match included apple, **orange**, malt, **oat**, rice, **rye**, soybean, tuna, chicken, **egg**, chocolate, and **cottonseed**. Twin Laurie G was additionally positive for banana, peach vanilla, cane sugar, and trout, while Kathy G was negative for these five (29.4%) items. Twin Kathy G was additionally positive for strawberry, corn, cabbage, peanut, and peppermint while Laurie G was negative for these five (29.4%) items. What on the surface appeared to be identical with 17 positives each tested out to be a diagnostic discriminating difference between the two identical twins.

TABLE 05.02.MEMBERS.04. SGP EXTENDED FAMILY CYTOTOXIC TEST POSITIVE SENSITIVITIES TO FISH & SEAFOOD.

SUBJECT	BROTHER SGP	SISTER HJP	SISTER DCP-G-D	SON MARK	DAUGHTER TERRY	DAUGHTER LAURIE G	DAUGHTER KATHY G	DAUGHTER AMY D	DAUGHTER ROBIN D
AGE TESTED	35	30	35	17	16	14	14	9	9
CARP	0	0	0	0	0	0	0	0	0
CRAB	0	0	0	0	0	0	0	0	0
LOBSTER	1	0	0	0	0	0	0	0	0
SHRIMP	0	0	0	0	0	0	0	0	0
SWORDFISH	0	0	0	0	0	0	0	0	0
TROUT	0	0	0	0	0	1	0	0	0
TUNA	1	1	1	1	1	1	1	1	0
===============									
POSITIVES	2	1	1	1	1	2	1	1	0
NUMBER (N)	7	7	7	7	7	7	7	7	7
% POSITIVE	28.6	14.3	14.3	14.3	14.3	28.6	14.3	14.3	0.0

Twins Amy D and Robin D agreed, either positive or negative, on 52 of 70 (74.3%) of the items tested for. Of twin Amy D's 12 positives 8 (66.7%) matched twin Robin D's positives. Of twin Robin D's 22 positives 8 (36.4%) matched twin Amy D's positives. Note that a significant number of items did not match. The food items that did match included banana, **orange**, **oat**, **rye**, peanut, **egg**, **cottonseed,** and peppermint. Twin Amy D was additionally positive for; rice, hops, tuna, chicken, while twin Robin D was negative for these four of 12 (33.3%) items. Twin Robin D was additionally positive for; apple, honeydew, pineapple, strawberry, corn, malt, wheat, cabbage, potato, soybean, string bean, cane sugar, chocolate, and mustard, while twin Amy D was negative for these 14 of 22 (63.6%) items.

TABLE 05.02.MEMBERS.05. SGP EXTENDED FAMILY CYTOTOXIC POSITIVE SENSITIVITIES TO MEAT, FOWL, EGG & MILK.

	BROTHER	SISTER	SISTER	SON	DAUGHTER	DAUGHTER	DAUGHTER	DAUGHTER	DAUGHTER
SUBJECT	SGP	HJP	DCP-G-D	MARK	TERRY	LAURIE G	KATHY G	AMY D	ROBIN D
AGE TESTED	35	30	35	17	16	14	14	9	9
BEEF	0	0	0	0	0	0	0	0	0
LAMB	0	0	0	0	0	0	0	0	0
PORK	0	0	1	0	0	0	0	0	0
VEAL	0	0	0	0	0	0	0	0	0
CHICKEN	1	1	1	0	1	1	1	1	0
TURKEY, Lt.	0	0	0	0	0	0	0	0	0
EGG	1	1	0	1	1	1	1	1	1
MILK	1	1	0	0	0	0	0	0	0
CHEESE/AMER.	0	0	0	0	0	0	0	0	0
CHEESE/ROQUE.	0	0	0	0	0	0	0	0	0
CHEESE/SWISS	0	0	0	0	0	0	0	0	0
===============									
POSITIVES	3	3	2	1	2	2	2	2	1
NUMBER (N)	11	11	11	11	11	11	11	11	11
% POSITIVE	27.3	27.3	18.2	9.1	18.2	18.2	18.2	18.2	9.1

TABLE 05.02.MEMBERS.06 SGP EXTENDED FAMILY CYTOTOXIC TEST POSITIVE SENSITIVITIES TO MISCELLANEOUS ITEMS.

	BROTHER	SISTER	SISTER	SON	DAUGHTER	DAUGHTER	DAUGHTER	DAUGHTER	DAUGHTER
SUBJECT	SGP	HJP	DCP-G-D	MARK	TERRY	LAURIE G	KATHY G	AMY D	ROBIN D
AGE TESTED	35	30	35	17	16	14	14	9	9
CHOCOLATE	1	1	1	1	1	1	1	0	1
COFFEE	1	0	0	0	0	0	0	0	0
COTTONSEED	0	1	0	1	1	1	1	1	1
GARLIC	1	0	1	0	0	0	0	0	0
GASOLINE	0	0	0	0	0	0	0	0	0
MUSHROOM	0	0	0	0	0	0	0	0	0
MUSTARD	0	0	0	0	0	0	0	0	1
PEPPERMINT	1	1	0	0	1	0	1	1	1
TEA	1	0	1	0	0	0	0	0	0
TOBACCO	0	0	0	0	0	0	0	0	0
TURPENTINE	1	0	0	0	0	0	0	0	0
YEAST, BAKER'S	0	0	0	0	0	0	0	0	0
YEAST, BREWER'S	0	0	0	0	0	0	0	0	0
ALCOHOL, 1%	0	0	0	0	0	0	0	0	0
ALCOHOL, 2%	0	0	0	0	0	0	0	0	0
===============									
POSITIVES	6	3	3	2	3	2	3	2	4
NUMBER (N)	15	15	15	15	15	15	15	15	15
% POSITIVE	40.0	20.0	20.0	13.3	20.0	13.3	20.0	13.3	26.7

By comparison, the twins older non twin siblings Mark G and Terry G (Tables 5.1.4.1-7) agreed, either positive or negative, on 61 of 70 (87.1%) of the items tested for. Of Mark G's 12 positives 11 (91.7%) matched sibling sister Terry G positives. Of Terry G's 19 positives 11 (57.9%) matched sibling brother Mark G's positives. Note that Mark G's positives differed from Terry G by a single item while with Terry G a significant number of items did not match. The food items that did match included banana, honeydew, **orange**, corn, **oat**, **rye**, peanut, tuna, **egg**, chocolate, and **cottonseed**. Mark D was additionally positive for; soybean, while sibling Terry G was negative for these one of 11 (9.1%) items. Terry G was additionally positive for; cantaloupe, peach, strawberry, vanilla, cabbage, cane sugar, chicken, and peppermint, while sibling Mark G was negative for these eight of 19 (42.1%) items. All siblings tested positive for **orange**, **oat**, **rye,** and **cottonseed**. All of the DCP-G-D siblings (Tables 05.02.MEMBERS.01-7) agreed, either positive or negative, on 46 of 70 (65.7%) of the items tested for. Calculating the average total sensitivity for the siblings you get 16.5+/-4.1SD (24.7%CV). The high variability (coefficient of variation greater than 20%) in this family sample is probably triggered by identical twin Robin D, otherwise this data indicates great agreement. How well do related offspring match on their chi squared test positive items compared to their total test positive items. Calculating the average match on their test positive food items among the siblings in this case you get 10.9+/-1.6 SD (14.2%CV). This sample, as small as it is, has a reasonable variation that might lead to useful comparisons. Based on the pattern of positive and negative test results each sibling had a distinctive profile that could provide diagnostic discriminating differences between each sibling and group of siblings.

CASE STUDY 05.03.JLP.01-7. CRP x TAP (JLP): Fourteen-year-old JLP and her parents agreed, either positive or negative, on 44 of 70 (62.9%) of the items tested for. Of the six positive items that both parents agreed on four of six (66.7%) were passed as positives to JLP. Of the daughter's 17 positives, seven (41.2%) matched her father's positives and nine (52.9%) matched her mom's positives. The father and mother agreed on six of 70 (8.6%) positives. Both parents and JLP matched on four (cabbage, potato, egg, and coffee) of her 17 positive items. The father exclusively matched on three (malt, chocolate, and tea) of 17 (17.6%) of JLP's positives. The mother exclusively matched on five (banana, honeydew, vanilla, lettuce, and peanut) of 17 (29.4%) of JPL's positives. JLP had five (29.4%) of 17 test positive items (orange, strawberry, mustard, 1% alcohol and 2% alcohol) that did not appear in either parent.

TABLE 05.03.JLP.01. CRPXTAP (JLP) FAMILIAL DISTRIBUTION OF CYTOTOXIC TEST POSITIVE SENSITIVITIES TO FRUITS.

	FATHER	MOTHER	DAUGHTER	%POSITIVE	%POSITIVE	%POSITIVE	%NEGATIVE	PER CENT OF POSITIVE FOODS			
SUBJECT	CRP	TAP	JLP	CRP/TAP	CRP	TAP	CRP/TAP		CRP/TAP	CRP	TAP
AGE TESTED	52	47	14	JLP	JLP	JLP	JLP	JLP	JLP	JLP	JLP
APPLE	0	0	0				0				
BANANA	0	1	1			1					1
CANTALOUPE	0	0	0				0				
CHERRY	0	0	0				0				
HONEYDEW	0	1	1			1					1
ORANGE	0	0	1					1			
PEACH	0	0	0				0				
PEAR	1	0	0								
PINEAPPLE	1	0	0								
STRAWBERRY	0	0	1					1			
TOMATO	0	1	0								
VANILLA	0	1	1			1					1
WATERMELON	0	0	0				0				
===============											
POSITIVES	2	4	5	0	0	3	5	2	0	0	3
NUMBER (N)	13	13	13	13	13	13	13	5	5	5	5
% POSITIVE	15.4	30.8	38.5	0.0	0.0	23.1	38.5	40.0	0.0	0.0	60.0

TABLE 05.02.MEMBERS.07 SUMMARY: SGP EXTENDED FAMILY CYTOTOXIC TEST POSITIVE FOOD SENSITIVITIES.

	BROTHER	SISTER	SISTER	SON	DAUGHTER	DAUGHTER	DAUGHTER	DAUGHTER	DAUGHTER
SUBJECT	SGP	HJP	DCP-G-D	MARK	TERRY	LAURIE G	KATHY G	AMY D	ROBIN D
AGE TESTED	35	30	35	17	16	14	14	9	9
POSITIVES	22	16	16	12	19	17	17	12	22
NUMBER (N)	70	70	70	70	70	70	70	70	70
% POSITIVE	31.4	22.9	22.9	17.1	27.1	24.3	24.3	17.1	31.4

In general, this family had a relatively low level of Cytotoxic Test positives with CRP 17 of 70 (24.3%), TAP 14 of 70 (20.0%) and JLP 17 of 70 (24.3%). Collectively, the father and daughter matched on 7 (10.0%) of 70 items tested for, while the mother and daughter matched on 9 (12.9%) of 70 items. In the mother to father 9 to 7 ratio (1.290) shows a "donor" advantage where the mother statistically has "donated" significantly more genes traits than the father, advantage mother. If you factor in what positive food items each parent has to "offer" there is a mother to father 14 to 17 ratio (0.824) favoring the father. For every gene trait the father has to "offer", the mother statistically has about 20% less genes to "offer", advantage father. This comparison appears to be offsetting lest we consider that "donated" here list traits successfully appearing in the offspring and to "offer" only lists potentially available traits in the parents.

TABLE 05.03.JLP.02. CRP X TAP (JLP) FAMILIAL DISTRIBUTION OF CYTOTOXIC TEST POSITIVE SENSITIVITIES TO GRAINS.

	FATHER	MOTHER	DAUGHTER	%POSITIVE	%POSITIVE	%POSITIVE	%NEGATIVE	PER CENT OF POSITIVE FOODS			
SUBJECT	CRP	TAP	JLP	CRP/TAP	CRP	TAP	CRP/TAP		CRP/TAP	CRP	TAP
AGE TESTED	52	47	14	JLP	JLP	JLP	JLP	JLP	JLP	JLP	JLP
BARLEY	0	0	0				0				
CORN	0	0	0				0				
MALT	1	0	1		1					1	
OAT	0	0	0				0				
RICE	0	0	0				0				
RYE	1	0	0								
WHEAT	0	0	0				0				
===============											
POSITIVES	2	0	1	0	1	0	5	0	0	1	0
NUMBER (N)	7	7	7	7	7	7	7	1	1	1	1
% POSITIVE	28.6	0.0	14.3	0.0	14.3	0.0	71.4	0.0	0.0	100.0	0.0

NOTES:

TABLE 05.03.JLP.03. CRPxTAP (JLP) FAMILIAL CYTOTOXIC TEST POSITIVE SENSITIVITIES TO VEGETABLES.

	FATHER	MOTHER	DAUGHTER	%POSITIVE	%POSITIVE	%POSITIVE	%NEGATIVE	PER CENT OF POSITIVE FOODS			
SUBJECT	CRP	TAP	JLP	CRP/TAP	CRP	TAP	CRP/TAP		CRP/TAP	CRP	TAP
AGE TESTED	52	47	14	JLP	JLP	JLP	JLP	JLP	JLP	JLP	JLP
BROCCOLI	0	0	0				0				
CABBAGE	1	1	1	1					1		
CARROT	1	1	0								
CUCUMBER	0	0	0				0				
HOPS	1	0	0								
HORSERADISH	0	0	0				0				
LETTUCE	0	1	1			1					1
ONION	1	1	0								
PEA	0	1	0								
PEANUT	0	1	1			1					1
POTATO	1	1	1	1					1		
RADISH	0	0	0				0				
SOYBEAN	0	0	0				0				
SPINACH	0	0	0				0				
STRINGBEAN	1	0	0								
SUGAR, BEET	0	0	0				0				
SUGAR, CANE	0	0	0				0				
POSITIVES	6	7	4	2	0	2	8	0	2	0	2
NUMBER (N)	17	17	17	17	17	17	17	4	4	4	4
% POSITIVE	35.3	41.2	23.5	11.8	0.0	11.8	47.1	0.0	50.0	0.0	50.0

TABLE 05.03.JLP.04 CRP X TAP (JLP) FAMILIAL CYTOTOXIC TEST POSITIVE SENSITIVITIES TO FISH & SEAFOOD.

	FATHER	MOTHER	DAUGHTER	%POSITIVE	%POSITIVE	%POSITIVE	%NEGATIVE	PER CENT OF POSITIVE FOODS			
SUBJECT	CRP	TAP	JLP	CRP/TAP	CRP	TAP	CRP/TAP		CRP/TAP	CRP	TAP
AGE TESTED	52	47	14	JLP	JLP	JLP	JLP	JLP	JLP	JLP	JLP
CARP	0	0	0				0				
CRAB	0	0	0				0				
LOBSTER	0	0	0				0				
SHRIMP	0	1	0								
SWORDFISH	0	0	0				0				
TROUT	0	0	0				0				
TUNA	1	0	0								
===============											
POSITIVES	1	1	0	0	0	0	5	0	0	0	0
NUMBER (N)	7	7	7	7	7	7	7	0	0	0	0
% POSITIVE	14.3	14.3	0.0	0.0	0.0	0.0	71.4	0.0	0.0	0.0	0.0

TABLE 05.03.JLP.05. CRPxTAP (JLP) FAMILIAL CYTOTOXIC TEST POSITIVE SENSITIVITIES TO MEAT, FOWL, EGG & MILK.

	FATHER	MOTHER	DAUGHTER	%POSITIVE	%POSITIVE	%POSITIVE	%NEGATIVE	PER CENT OF POSITIVE FOODS			
SUBJECT	CRP	TAP	JLP	CRP/TAP	CRP	TAP	CRP/TAP		CRP/TAP	CRP	TAP
AGE TESTED	52	47	14	JLP	JLP	JLP	JLP	JLP	JLP	JLP	JLP
BEEF	0	0	0				0				
LAMB	0	0	0				0				
PORK	0	0	0				0				
VEAL	0	0	0				0				
CHICKEN	1	0	0								
TURKEY, Lt.	0	0	0				0				
EGG	1	1	1	1					1		
MILK	0	0	0				0				
CHEESE/AMER.	0	0	0				0				
CHEESE/ROQUE.	0	0	0				0				
CHEESE/SWISS	0	0	0				0				
===============											
POSITIVES	2	1	1	1	0	0	9	0	1	0	0
NUMBER (N)	11	11	11	11	11	11	11	1	1	1	1
% POSITIVE	18.2	9.1	9.1	9.1	0.0	0.0	81.8	0.0	100.0	0.0	0.0

NOTES:

TABLE 05.03.JLP.06. CRPXTAP (JLP) FAMILIAL CYTOTOXIC TEST POSITIVE SENSITIVITIES TO MISCELLANEOUS ITEMS.

	FATHER	MOTHER	DAUGHTER	%POSITIVE	%POSITIVE	%POSITIVE	%NEGATIVE	PER CENT OF POSITIVE FOODS			
SUBJECT	CRP	TAP	JLP	CRP/TAP	CRP	TAP	CRP/TAP		CRP/TAP	CRP	TAP
AGE	52	47	14	JLP	JLP	JLP	JLP	JLP	JLP	JLP	JLP
CHOCOLATE	1	0	1		1					1	
COFFEE	1	1	1	1					1		
COTTONSEED	0	0	0				0				
GARLIC	0	0	0				0				
GASOLINE	0	0	0				0				
MUSHROOM	0	0	0				0				
MUSTARD	0	0	1					1			
PEPPERMINT	0	0	0				0				
TEA	1	0	1		1						
TOBACCO	0	0	0				0				
TURPENTINE	1	0	0								
YEAST, BAKER	0	0	0				0				
YEAST, BREWER	0	0	0				0				
ALCOHOL, 1%	0	0	1					1			
ALCOHOL, 2%	0	0	1					1			
===============											
POSITIVES	4	1	6	1	2	0	8	3	1	2	0
NUMBER (N)	15	15	15	15	15	15	15	6	6	6	6
% POSITIVE	26.7	20.0	40.0	6.7	13.3	0.0	53.3	50.0	16.7	33.3	0.0

TABLE 05.03.JLP.07. SUMMARY: CRPxTAP (JLP) FAMILIAL DISTRIBUTION OF CYTOTOXIC TEST POSITIVE SENSITIVITIES.

	FATHER	MOTHER	DAUGHTER	%POSITIVE	%POSITIVE	%POSITIVE	%NEGATIVE	PER CENT OF POSITIVE FOODS			
SUBJECT	CRP	TAP	JLP	CRP/TAP	CRP	TAP	CRP/TAP		CRP/TAP	CRP	TAP
AGE	52	47	14	JLP	JLP	JLP	JLP	JLP	JLP	JLP	JLP
POSITIVES	17	14	17	4	3	5	40	5	4	3	5
NUMBER (N)	70	70	70	70	70	70	70	17	17	17	17
% POSITIVE	24.3	20.0	24.3	5.7	4.3	7.1	57.1	29.4	23.5	17.6	29.4

One of our objectives with these case studies is to determine if and how these potential genetic test positive results are distributed including what role the parents play and what clinical significance test positive results correlate with medical symptoms. This case demonstrates that both parents make significant contributions both together (66.7%) in spite of the relatively low level of total sensitivity, and there is much still to learn about this data. Let your imagination run wild here and throughout this book.

CASE STUDY05.04.MTZ.01-7 DWZ x JLZ (MTZ): Three-year-old MTZ and his parents agreed, either positive or negative, on 43 of 70 (61.4%) of the items tested for. Of the 31 positive items that both parents agreed on 25 of 31 (80.6%) were passed as positives to MTZ. Of the son's 39 positives, 29 (74.4%) matched his father's positives and 30 (76.9%) matched his mom's positives, and both parents matched on 25 of his 39 (64.1%) items. The father exclusively matched on four (rice, wheat, onion, and Swiss cheese) of 39 (10.3%) of MTZ's positives. The mother exclusively matched on five (strawberry, hops, chocolate, cottonseed, and gasoline) of 39 (12.9%) of MTZ's positives. MTZ had five (12.8%) of 39 test positive items (cantaloupe, barley, pea, milk, and Roquefort cheese) that did not appear in either parent.

In general, this family had a relatively high level of Cytotoxic Test positives with DWZ at 41 (58.6%) of 70, JLZ at 37 (52.9%) of 70 and MTZ 39 at (55.7%) of 70. Collectively, the father (DWZ) and son (MTZ) matched on 29 (41.4%) of 70 items tested for, while the mother (JLZ) and son matched on 30 (42.9%) of 70 items. In the mother to father 30 to 29 ratio (1.034) shows a "donor" advantage where the mother statistically has "donated" slightly more gene traits (3.4%). If you factor in what positive food items each parent has to "offer" there is a mother to father 37 to 41 ratio (0.902) favoring the father. For every gene the father has to "offer", the mother statistically has about 10% less genes to "offer", slight advantage father. Who has a statistical advantage, mother, or father, among all parents in these case studies?

TABLE 05.04.MTZ.01. DWZxJLZ (MTZ) FAMILIAL DISTRIBUTION OF CYTOTOXIC TEST POSITIVE SENSITIVITIES TO FRUITS.

	FATHER	MOTHER	SON	%POSITIVE	%POSITIVE	%POSITIVE	%NEGATIVE	PER CENT OF POSITIVE FOODS			
SUBJECT	DWZ	JLZ	MTZ	DMZ/JLZ	DMZ	JLZ	DMZ/JLZ		DMZ/JLZ	DMZ	JLZ
AGE TESTED	32	30	3	MTZ	MTZ	MTZ	MTZ	MTZ	MTZ	MTZ	MTZ
APPLE	1	1	1	1					1		
BANANA	1	1	1	1					1		
CANTALOUPE	0	0	1					1			
CHERRY	1	0	0								
HONEYDEW	1	0	0								
ORANGE	1	1	1	1					1		
PEACH	1	1	0								
PEAR	0	0	0				0				
PINEAPPLE	1	1	1	1					1		
STRAWBERRY	0	1	1			1					1
TOMATO	1	1	1	1					1		
VANILLA											
	1	1	1	1					1		
WATERMELON	0	0	0				0				
===============											
POSITIVES	9	8	8	6	0	1	2	1	6	0	1
NUMBER (N)	13	13	13	13	13	13	13	8	8	8	8
% POSITIVE	69.2	61.5	61.5	46.2	0.0	7.7	15.4	12.5	75.0	0.0	12.5

TABLE 05.04.MTZ.02. DWZxJLZ (MTZ) FAMILIAL DISTRIBUTION OF CYTOTOXIC TEST POSITIVE SENSITIVITIES TO GRAINS

	FATHER	MOTHER	SON	%POSITIVE	%POSITIVE	%POSITIVE	%NEGATIVE	PER CENT OF POSITIVE FOODS			
SUBJECT	DWZ	JLZ	MTZ	DMZ/JLZ	DMZ	JLZ	DMZ/JLZ		DMZ/JLZ	DMZ	JLZ
AGE TESTED	32	30	3	MTZ	MTZ	MTZ	MTZ	MTZ	MTZ	MTZ	MTZ
BARLEY	0	0	1					1			
CORN	1	1	1	1					1		
MALT	1	1	1	1					1		
OAT	1	1	1	1					1		
RICE	1	0	1		1					1	
RYE	1	1	1	1					1		
WHEAT	1	0	1		1					1	
===============											
POSITIVES	6	4	7	4	2	0	0	1	4	2	0
NUMBER (N)	7	7	7	7	7	7	7	7	7	7	7
% POSITIVE	85.7	57.1	100.0	57.1	28.6	0.0	0.0	14.3	57.1	28.6	0.0

TABLE 05.04.MTZ.03. DWZxJLZ (MTZ) FAMILIAL DISTRIBUTION OF CYTOTOXIC TEST POSITIVE SENSITIVITIES TO VEGETABLES.

	FATHER	MOTHER	SON	%POSITIVE	%POSITIVE	%POSITIVE	%NEGATIVE	PER CENT OF POSITIVE FOODS			
SUBJECT	DWZ	JLZ	MTZ	DMZ/JLZ	DMZ	JLZ	DMZ/JLZ		DMZ/JLZ	DMZ	JLZ
AGE TESTED	32	30	3	MTZ	MTZ	MTZ	MTZ	MTZ	MTZ	MTZ	MTZ
BROCCOLI	1	0	0								
CABBAGE	1	1	0								
CARROT	1	1	1	1					1		
CUCUMBER	1	0	0								
HOPS	0	1	1			1					1
HORSERADISH	0	0	0				0				
LETTUCE	1	1	0								
ONION	1	0	1		1					1	
PEA	0	0	1					1			
PEANUT	1	1	1	1					1		
POTATO	1	1	1	1					1		
RADISH	0	0	0				0				
SOYBEAN	1	1	1	1					1		
SPINACH	0	0	0				0				
STRINGBEAN	0	0	0				0				
SUGAR, BEET	0	0	0				0				
SUGAR, CANE	1	1	1	1					1		
===============											
POSITIVES	10	8	8	5	1	1	5	1	5	1	1
NUMBER (N)	17	17	17	17	17	17	17	8	8	8	8
% POSITIVE	58.8	47.1	47.1	29.4	5.9	5.9	29.4	12.5	62.5	12.5	12.5

TABLE 05.04.MTZ.04. DWZxJLZ (MTZ) FAMILIAL CYTOTOXIC TEST POSITIVE SENSITIVITIES TO FISH & SEAFOOD.

	FATHER	MOTHER	SON	%POSITIVE	%POSITIVE	%POSITIVE	%NEGATIVE	PER CENT OF POSITIVE FOODS			
SUBJECT	DWZ	JLZ	MTZ	DMZ/JLZ	DMZ	JLZ	DMZ/JLZ		DMZ/JLZ	DMZ	JLZ
AGE TESTED	32	30	3	MTZ	MTZ	MTZ	MTZ	MTZ	MTZ	MTZ	MTZ
CARP	0	0	0				0				
CRAB	0	0	0				0				
LOBSTER	0	0	0				0				
SHRIMP	1	1	0								
SWORDFISH	0	0	0				0				
TROUT	1	0	0								
TUNA	1	1	1	1					1		
===============											
POSITIVES	3	2	1	1	0	0	4	0	1	0	0
NUMBER (N)	7	7	7	7	7	7	7	1	1	1	1
% POSITIVE	42.9	28.6	14.3	14.3	0.0	0.0	57.1	0.0	100.0	0.0	0.0

TABLE 05.04.MTZ.05 DWZxJLZ (MTZ) FAMILIAL CYTOTOXIC TEST POSITIVE SENSITIVITIES TO MEAT, FOWL, EGG & MILK.

	FATHER	MOTHER	SON	%POSITIVE	%POSITIVE	%POSITIVE	%NEGATIVE	PER CENT OF POSITIVE FOODS			
SUBJECT	DWZ	JLZ	MTZ	DMZ/JLZ	DMZ	JLZ	DMZ/JLZ		DMZ/JLZ	DMZ	JLZ
AGE TESTED	32	30	3	MTZ	MTZ	MTZ	MTZ	MTZ	MTZ	MTZ	MTZ
BEEF	0	1	0								
LAMB	0	0	0				0				
PORK	1	1	0								
VEAL	0	0	0				0				
CHICKEN	1	1	1	1					1		
TURKEY, Lt.	0	0	0				0				
EGG	1	1	1	1					1		
MILK	0	0	1					1			
CHEESE/AMER.	1	1	1	1					1		
CHEESE/ROQUE.	0	0	1					1			
CHEESE/SWISS	1	0	1		1					1	
===============											
POSITIVES	5	5	6	3	1	0	3	2	3	1	0
NUMBER (N)	11	11	11	11	11	11	11	6	6	6	6
% POSITIVE	45.5	45.5	54.5	27.3	9.1	0.0	27.3	33.3	50.0	16.7	0.0

TABLE 5.3.1.6 DWZ X JLZ (MTZ) FAMILIAL DISTRIBUTION OF CYTOTOXIC TEST POSITIVE SENSITIVITIES TO MISCELLANEOUS ITEMS.

	FATHER	MOTHER	SON	%POSITIVE	%POSITIVE	%POSITIVE	%NEGATIVE	PER CENT OF POSITIVE FOODS			
SUBJECT	DWZ	JLZ	MTZ	DMZ/JLZ	DMZ	JLZ	DMZ/JLZ		DMZ/JLZ	DMZ	JLZ
AGE	32	30	3	MTZ	MTZ	MTZ	MTZ	MTZ	MTZ	MTZ	MTZ
CHOCOLATE	0	1	1			1					1
COFFEE	1	0	0								
COTTONSEED	0	1	1			1					1
GARLIC	1	1	1	1					1		
GASOLINE	0	1	1			1					1
MUSHROOM	0	0	0				0				
MUSTARD	1	1	1	1					1		
PEPPERMINT	1	1	1	1					1		
TEA	1	1	1	1					1		
TOBACCO	1	1	1	1					1		
TURPENTINE	1	1	0								
YEAST, BAKER	1	1	1	1					1		
YEAST, BREWER	0	0	0				0				
ALCOHOL, 1%	0	0	0				0				
ALCOHOL, 2%	0	0	0				0				
===============											
POSITIVES	8	10	9	6	0	3	4	0	6	0	3
NUMBER (N)	15	15	15	15	15	15	15	9	9	9	9
% POSITIVE	53.3	66.7	60.0	40.0	0.0	20.0	26.7	0.0	66.7	0.0	33.3

TABLE 05.04.MTZ.07 DWZxJLZ (MTZ) SUMMARY: FAMILIAL CYTOTOXIC TEST POSITIVE SENSITIVITIES.

	FATHER	MOTHER	SON	%POSITIVE	%POSITIVE	%POSITIVE	%NEGATIVE	PER CENT OF POSITIVE FOODS			
SUBJECT	DWZ	JLZ	MTZ	DMZ/JLZ	DMZ	JLZ	DMZ/JLZ		DMZ/JLZ	DMZ	JLZ
AGE	32	30	3	MTZ	MTZ	MTZ	MTZ	MTZ	MTZ	MTZ	MTZ
POSITIVES	41	37	39	25	4	5	18	5	25	4	5
NUMBER (N)	70	70	70	70	70	70	70	39	39	39	39
% POSITIVE	58.6	52.9	55.7	35.7	5.7	7.1	25.7	12.8	64.1	10.3	12.8

One of our objectives with these case studies is to determine how these potential genetic test positive results are distributed. This case demonstrates that both parents make significant contributions to their son's test positive profile both together and individually. Of the 31 food items that both parents agreed on 25 (80.6%) were passed to the offspring. Of the offspring's 39 positive food items about 75% match with the father or mother. All members of this family listed multiple clinical complaints thought to be of a nonspecific allergic origin. Given the information in Tables 05.03.MTZ.01-7 menus are formulated, and clinical outcomes tracked. The family profile was separate and unique compared to the other family profiles reported in this Program.

CASE STUDY 05.05.MLK-W.01-7 EWK x HCK (MLK-W): Twenty-nine-year-old MLK-W and her parents agreed, either positive or negative, on 19 of 70 (27.1%) of the items tested for. The father and mother agreed on only six of 70 (8.6%) positives. Of the six positive items that both parents agreed on five of six (83.3%) were passed as positives to MLK-W. Of her 44 positives, 19 (43.2%) matched her father's positives and 16 (36.4%) matched her mom's positives. Both parents and MLK-W matched on five of her 44 (11.4%) items. The father exclusively matched on 14 (apple, cantaloupe, corn, rice, chicken, milk, American cheese, Roquefort cheese, Swiss cheese, peppermint, tobacco, lettuce, cane sugar and tuna) of 44 (31.8%) of MLK-W's positives. The mother exclusively matched on 11 (pineapple, vanilla, barley, oat, rye, wheat, hops, trout, egg, chocolate, and baker's yeast) of 44 (25.0%) of MLK-W's positives. MLK-W had 14 (31.8%) of 44 test positive items (banana, honeydew, orange, strawberry, watermelon, pork, coffee, mustard, tea, brewer's yeast, cabbage, pea, beet sugar and lobster) that did not appear in either parent.

Collectively, the father (EWK) and daughter (MLK-W) matched on 19 (27.1%) of 70 items tested for, while the mother (HCK) and daughter matched on 16 (22.9%) of 70 items. In the mother to father 16 to 19 ratio (0.842) shows a "donor" advantage where the mother statistically has "donated" about 15% fewer gene traits. Both parents had 23 positive food items to "offer". Collectively for every gene trait the father has "donated" the mother has "donated" about 15% less gene traits but had an equal number of gene traits to "offer". Individually, the father and daughter agreed on 14 (20.0%) of 70 items tested for, whereas the mother and daughter agreed on 11 (15.7%) of 70 items. In the 11 to 14 ratio (0.786) there is a slight "donor" advantage for the father. Individually for every gene the father has "donated" the mother has "donated" about 20% less genes but had an equal number of genes to "offer", slight advantage father.

TABLE 05.05.MLK-W.01. EWKxHCK (MLK-W) FAMILIAL CYTOTOXIC TEST POSITIVE SENSITIVITIES TO FRUITS.

	FATHER	MOTHER	DAUGHTER	%POSITIVE	%POSITIVE	%POSITIVE	%NEGATIVE	PER CENT OF POSITIVE FOODS			
SUBJECT	EWK	HCK	MLK-W	EWK/HCK	EWK	HCK	EWK/HCK		EWK/HCK	EWK	HCK
AGE TESTED	61	58	29	MLW	MLW	MLW	MLW	MLW	MLW	MLW	MLW
APPLE	1	0	1		1					1	
BANANA	0	0	1					1			
CANTALOUPE	1	0	1		1					1	
CHERRY	0	1	0								
HONEYDEW	0	0	1					1			
ORANGE	0	0	1					1			
PEACH	1	1	1	1					1		
PEAR	0	1	0								
PINEAPPLE	0	1	1			1					1
STRAWBERRY	0	0	1					1			
TOMATO	1	1	0								
VANILLA	0	1	1			1					1
WATERMELON	0	0	1					1			
===============											
POSITIVES	4	6	10	1	2	2	0	5	1	2	2
NUMBER (N)	13	13	13	13	13	13	13	10	10	10	10
% POSITIVE	30.8	46.2	76.9	7.7	15.4	15.4	0.0	50.0	10.0	20.0	20.0

TABLE 05.05.MLK-W.02. EWKxHCK (MLK-W) FAMILIAL DISTRIBUTION OF CYTOTOXIC TEST POSITIVE SENSITIVITIES TO GRAINS.

	FATHER	MOTHER	DAUGHTER	%POSITIVE	%POSITIVE	%POSITIVE	%NEGATIVE	PER CENT OF POSITIVE FOODS			
SUBJECT	EWK	HCK	MLK-W	EWK/HCK	EWK	HCK	EWK/HCK		EWK/HCK	EWK	HCK
AGE TESTED	61	58	29	MLW	MLW	MLW	MLW	MLW	MLW	MLW	MLW
BARLEY	0	1	1			1					1
CORN	1	0	1		1					1	
MALT	1	1	1	1					1		
OAT	0	1	1			1					1
RICE	1	0	1		1					1	
RYE	0	1	1			1					1
WHEAT	0	1	1			1					1
===============											
POSITIVES	3	5	7	1	2	4	0	0	1	2	4
NUMBER (N)	7	7	7	7	7	7	7	7	7	7	7
% POSITIVE	42.9	71.4	100.0	14.3	28.6	57.1	0.0	0.0	14.3	28.6	57.1

TABLE 05.05.MLK-W.03. EWKxHCK (MLK-W) FAMILIAL CYTOTOXIC TEST POSITIVE SENSITIVITIES TO VEGETABLES.

	FATHER	MOTHER	DAUGHTER	%POSITIVE	%POSITIVE	%POSITIVE	%NEGATIVE	PER CENT OF POSITIVE FOODS			
SUBJECT	EWK	HCK	MLK-W	EWK/HCK	EWK	HCK	EWK/HCK		EWK/HCK	EWK	HCK
AGE TESTED	61	58	29	MLW	MLW	MLW	MLW	MLW	MLW	MLW	MLW
BROCCOLI	0	1	0								
CABBAGE	0	0	1					1			
CARROT	1	1	1	1					1		
CUCUMBER	0	0	0				0				
HOPS	0	1	1			1					1
HORSERADISH	0	0	0				0				
LETTUCE	1	0	1		1					1	
ONION	1	0	0								
PEA	0	0	1					1			
PEANUT	0	0	0				0				
POTATO	1	1	1	1					1		
RADISH	0	0	0				0				
SOYBEAN	1	1	1	1					1		
SPINACH	0	0	0				0				
STRINGBEAN	0	0	0				0				
SUGAR, BEET	0	0	1					1			
SUGAR, CANE	1	0	1		1					1	
===============											
POSITIVES	6	5	9	3	2	1	6	3	3	2	1
NUMBER (N)	17	17	17	17	17	17	17	9	9	9	9
% POSITIVE	35.3	29.4	52.9	17.6	11.8	5.9	35.3	33.3	33.3	22.2	11.1

TABLE 05.05.MLK-W.04. EWKxHCK (MLK-W) FAMILIAL CYTOTOXIC TEST POSITIVE SENSITIVITIES TO FISH & SEAFOOD.

	FATHER	MOTHER	DAUGHTER	%POSITIVE	%POSITIVE	%POSITIVE	%NEGATIVE	PER CENT OF POSITIVE FOODS			
SUBJECT	EWK	HCK	MLK-W	EWK/HCK	EWK	HCK	EWK/HCK		EWK/HCK	EWK	HCK
AGE TESTED	61	58	29	MLW	MLW	MLW	MLW	MLW	MLW	MLW	MLW
CARP	0	0	0				0				
CRAB	0	0	0				0				
LOBSTER	0	0	1					1			
SHRIMP	0	0	0				0				
SWORDFISH	0	1	0								
TROUT	0	1	1			1					1
TUNA	1	0	1		1					1	
===============											
POSITIVES	1	2	3	0	1	1	3	1	0	1	1
NUMBER (N)	7	7	7	7	7	7	7	3	3	3	3
% POSITIVE	14.3	28.6	42.9	0.0	14.3	14.3	42.9	33.3	0.0	33.3	33.3

TABLE 05.05.MLK-W.05. EWKXHCK (MLK-W) FAMILIAL CYTOTOXIC TEST POSITIVE SENSITIVITIES TO MEAT, FOWL, EGG & MILK.

	FATHER	MOTHER	DAUGHTER	%POSITIVE	%POSITIVE	%POSITIVE	%NEGATIVE	PER CENT OF POSITIVE FOODS			
SUBJECT	EWK	HCK	MLK-W	EWK/HCK	EWK	HCK	EWK/HCK		EWK/HCK	EWK	HCK
AGE TESTED	61	58	29	MLW	MLW	MLW	MLW	MLW	MLW	MLW	MLW
BEEF	0	0	0				0				
LAMB	0	0	0				0				
PORK	0	0	1					1			
VEAL	0	1	0								
CHICKEN	1	0	1		1					1	
TURKEY, Lt.	1	0	0								
EGG	0	1	1			1					1
MILK	1	0	1		1					1	
CHEESE/AMER	1	0	1		1					1	
CHEESE/ROQUE	1	0	1		1					1	
CHEESE/SWISS	1	0	1		1					1	
===============											
POSITIVES	6	2	7	0	5	1	2	1	0	5	1
NUMBER (N)	11	11	11	11	11	11	11	7	7	7	7
% POSITIVE	54.5	18.2	63.6	0.0	45.5	9.1	18.2	14.3	0.0	71.4	14.3

TABLE 05.05.MLK-W.06. EWKxHCK (MLK-W) FAMILIAL CYTOTOXIC TEST POSITIVE SENSITIVITIES TO MISCELLANEOUS ITEMS.

	FATHER	MOTHER	DAUGHTER	%POSITIVE	%POSITIVE	%POSITIVE	%NEGATIVE	PER CENT OF POSITIVE FOODS			
SUBJECT	EWK	HCK	MLK-W	EWK/HCK	EWK	HCK	EWK/HCK		EWK/HCK	EWK	HCK
AGE	61	58	29	MLW	MLW	MLW	MLW	MLW	MLW	MLW	MLW
CHOCOLATE	0	1	1			1					1
COFFEE	0	0	1					1			
COTTONSEED	0	1	0								
GARLIC	0	0	0				0				
GASOLINE	1	0	0								
MUSHROOM	0	0	0				0				
MUSTARD	0	0	1					1			
PEPPERMINT	1	0	1		1					1	
TEA	0	0	1					1			
TOBACCO	1	0	1		1					1	
TURPENTINE	0	0	0				0				
YEAST, BAKER	0	1	1			1					1
YEAST, BREWER	0	0	1					1			
ALCOHOL, 1%	0	0	0				0				
ALCOHOL, 2%	0	0	0				0				
===============											
POSITIVES	3	3	8	0	2	2	5	4	0	2	2
NUMBER (N)	15	15	15	15	15	15	15	8	8	8	8
% POSITIVE	20.0	20.0	53.3	0.0	13.3	13.3	33.3	50.0	0.0	25.0	25.0

TABLE 05.05.MLK-W.07. SUMMARY: EWKxHCK (MLK-W) FAMILIAL CYTOTOXIC TEST POSITIVE FOOD SENSITIVITIES.

	FATHER	MOTHER	DAUGHTER	%POSITIVE	%POSITIVE	%POSITIVE	%NEGATIVE	PER CENT OF POSITIVE FOODS			
SUBJECT	EWK	HCK	MLK-W	EWK/HCK	EAK	HCK	EWK/HCK		EWK/HCK	EAK	HCK
AGE	61	58	29	MLW	MLW	MLW	MLW	MLW	MLW	MLW	MLW
POSITIVES	23	23	44	5	14	11	16	14	5	14	11
NUMBER (N)	70	70	70	70	70	70	70	44	44	44	44
% POSITIVE	32.9	32.9	62.9	7.1	20.0	15.7	22.9	31.8	11.4	31.8	25.0

The parents exhibited relatively low levels of Cytotoxic Test positives at 32.9% of the foods tested for, while the daughter exhibited relatively high 44 (62.9%) of 70 level of sensitivity. One of our objectives with a case study is to determine if and how these test positive results are specifically distributed genetically including what role the parents play and how clinically significant test positive results correlate with medical histories. This family submitted to this case study program to help understand the daughter's involvement with persistent clinical complaints that appear to be of allergic origin. We provide the information above (Tables 05.05.MLK-W.01-7) listing positive foods, negative foods and foods not tested. The patient formulates menus from the list of negative foods avoiding foods that have not been tested. Benefits are generally seen immediately, while carefully adding foods back identifying specific foods with undesirable clinical complaints.

MLK-W was a 29-year-old five-foot five-inch 112-pound female with a history of headaches and migraines. Additional intermittent complaints included diarrhea, constipation and stomach upset. Under the category of allergy, she listed sneezing, nasal blocking, and hay fever which she attributed to spring pollens. She listed no skin allergies, itching, or mental symptoms such as sleepiness, fatigue, or depression. She listed no use of alcohol, tobacco, drugs or over the counter remedies. MLK-W listed

University of Missouri – Columbia

5400 Arsenal Street
St. Louis, Mo. 63139

SCHOOL OF MEDICINE
Department of Psychiatry St. Louis Division
Missouri Institute of Psychiatry

Telephone
314 644-2400

September 11, 1973

Charles Norland, M. D.
114 North Taylor
St. Louis, Mo. 63110

Dear Chuck:

We completed the preliminary analysis of the results on ______ today. These conclusions were based on a summary of the three cytotoxic tests that we have done -- two in the fasting state and one we consider a provocative reaction done after the ingestion of foods to which she proved to be sensitive on our first test. I am enclosing three lists: 1) foods to which she is sensitive; 2) foods that proved sensitizing when the test was done at the height of the white blood cell response, and 3) foods which she proved not sensitive to under all testing conditions.

The graph is of white cell counts done every one-half hour over a four hour period, indicating percent rise from her resting baseline of white blood cells to a peak reaction as shown on the enclosed graph. This occurred after the ingestion of three cups of whole milk which produced some immediate eructation, hiccups and temporal headache. There is little doubt in my mind but what her symptoms were due to milk.

I would feel that she should certainly be instructed to avoid milk and also those foods to which she showed a positive sensitivity.

I trust that the results of our testing will prove helpful. Please let us know if we can be of further assistance.

Sincerely yours,

George Ulett M.D.

George A. Ulett, M. D.
Chairman and Professor

GAU:ecs

Enclosures

bc: Dr. Perry ✓

house dust, damp air, trees (April and May) and grasses as factors increasing her allergic symptoms.

CASE STUDY 05.06.MEMBERS.01-7 JES x ERS (FAMILY): This family case study (Tables 05.06. MEMBERS.01-7) reviews 49-year-old parents and four children ranging in age from 12 to 22 years of age. The parents and all four children agreed either positive or negative on 19 (27.1%) of 70 items tested for. Both parents and all four children were Cytotoxic Test positive for eight (11.4%) of 70 items (apple, vanilla, corn, oat, rye, potato, coffee, and tobacco) tested for. These comparisons would be significant between parents and a single child but are exceptional considering such agreement involving parents and all four children. The comparisons are statistical and do not always include the same positive mix of food items. This family displayed mixed levels of positive food sensitivity. The father (JES) was Cytotoxic Test positive for 34 (48.6%) of 70 items and the mother was positive for 29 (41.4%) of 70 items. The 22-year son (GWS) was positive for 26 (37.1%) of 70 items, while among the three daughters; LMS was positive for 41 (58.6%), KES was positive for 26 (37.1%) and JES was positive for 39 (55.7%) of the 70 items tested for. Together, the parents matched on 21 (30.0%) of 70 items tested for. Of the 21 items that both parents agreed on eight (38.1%) items were also positive in all the children.

In table 05.06.MEMBERS.01 (Fruits) apple and vanilla were positive for all family members. Orange was positive for all family members except KES. Tomato was positive for all children and negative in both parents. The father JES (61.5%) and daughters LMS (53.8%) and JES (69.2%) showed high levels of sensitivities to fruits. In table 05.06.MEMBERS.02 (Grains) the whole family tested greater than 70.0% positive with the mother ERS positive for all grains tested. All family members tested positive for corn, oat, and rye, and high to this family, was the sensitivity to milk and milk products. In table 05.06. MEMBERS.03 (Vegetables) the whole family was positive for potato with daughters LMS (58.8%) and JES (52.9%) showing relatively high levels of sensitivity to vegetables. Fish and seafood (Table 05.06. MEMBERS.04) sensitivities were relatively minimal for all members of the family.

TABLE 05.06.MEMBERS.01 JESxERS FAMILIAL CYTOTOXIC TEST POSITIVE SENSITIVITIES TO FRUITS.

	FATHER	MOTHER	SON	DAUGHTER	DAUGHTER	DAUGHTER
SUBJECT	JES	ERS	GWS	LMS	KES	JES
AGE TESTED	49	49	22	19	17	12
APPLE	1	1	1	1	1	1
BANANA	0	0	0	1	0	1
CANTALOUPE	1	0	0	0	0	1
CHERRY	0	0	0	1	0	1
HONEYDEW	1	0	0	0	0	0
ORANGE	1	1	1	1	0	1
PEACH	1	0	0	0	1	0
PEAR	0	0	0	0	0	0
PINEAPPLE	1	0	0	0	1	1
STRAWBERRY	1	0	0	1	1	1
TOMATO	0	0	1	1	1	1
VANILLA	1	1	1	1	1	1
WATERMELON	0	0	0	0	0	0
===============						
POSITIVES	8	3	4	7	6	9
NUMBER (N)	13	13	13	13	13	13
% POSITIVE	61.5	23.1	30.8	53.8	46.2	69.2

TABLE 05.06.MEMBERS.02. JESxERS FAMILIAL CYTOTOXIC TEST POSITIVE SENSITIVITIES TO GRAINS.

	FATHER	MOTHER	SON	DAUGHTER	DAUGHTER	DAUGHTER
SUBJECT	JES	ERS	GWS	LMS	KES	JES
AGE TESTED	49	49	22	19	17	12
BARLEY	1	1	1	1	0	1
CORN	1	1	1	1	1	1
MALT	0	1	1	0	1	0
OAT	1	1	1	1	1	1
RICE	0	1	0	1	1	0
RYE	1	1	1	1	1	1
WHEAT	1	1	0	1	1	1
===============						
POSITIVES	5	7	5	6	6	5
NUMBER (N)	7	7	7	7	7	7
% POSITIVE	71.4	100.0	71.4	85.7	85.7	71.4

TABLE 05.06.MEMBERS.03. JESxERS FAMILIAL CYTOTOXIC TEST POSITIVE SENSITIVITIES TO VEGETABLES.

	FATHER	MOTHER	SON	DAUGHTER	DAUGHTER	DAUGHTER
SUBJECT	JES	ERS	GWS	LMS	KES	JES
AGE TESTED	49	49	22	19	17	12
BROCCOLI	0	1	0	0	0	1
CABBAGE	1	0	0	0	1	1
CARROT	1	1	1	1	1	0
CUCUMBER	0	1	0	0	0	0
HOPS	0	0	1	1	1	0
HORSERADISH	0	0	0	0	0	0
LETTUCE	1	1	1	1	0	1
ONION	0	1	0	1	0	1
PEA	0	0	0	0	0	0
PEANUT	1	0	1	1	0	1
POTATO	1	1	1	1	1	1
RADISH	0	1	0	1	0	0
SOYBEAN	1	1	0	1	0	1
SPINACH	0	0	0	0	0	0
STRINGBEAN	0	0	1	0	0	1
SUGAR, BEET	0	0	0	1	0	0
SUGAR, CANE	1	0	0	1	1	1
===============						
POSITIVES	7	8	6	10	5	9
NUMBER (N)	17	17	17	17	17	17
% POSITIVE	41.2	47.1	35.3	58.8	29.4	52.9

TABLE 05.06.MEMBERS.04. JESxERS FAMILIAL CYTOTOXIC POSITIVE SENSITIVITIES TO FISH & SEAFOOD.

	FATHER	MOTHER	SON	DAUGHTER	DAUGHTER	DAUGHTER
SUBJECT	JES	ERS	GWS	LMS	KES	JES
AGE TESTED	49	49	22	19	17	12
CARP	0	0	0	0	0	0
CRAB	0	0	0	0	0	0
LOBSTER	0	0	0	1	0	0
SHRIMP	0	0	1	0	1	0
SWORDFISH	0	0	0	0	0	0
TROUT	0	0	0	0	0	0
TUNA	1	1	0	1	0	0
===============						
POSITIVES	1	1	1	2	1	0
NUMBER (N)	7	7	7	7	7	7
% POSITIVE	14.3	14.3	14.3	28.6	14.3	0.0

TABLE 05.06.MEMBERS.05. JESxERS CYTOTOXIC POSITIVE ALLERGENS TO MEAT, FOWL, EGG, & MILK.

	FATHER	MOTHER	SON	DAUGHTER	DAUGHTER	DAUGHTER
SUBJECT	JES	ERS	GWS	LMS	KES	JES
AGE	49	49	22	19	17	12
BEEF	1	0	0	1	1	1
LAMB	1	1	1	1	1	1
PORK	0	1	1	1	1	0
VEAL	0	0	1	1	1	1
	0	0	0	1	1	1
CHICKEN	0	0	0	1	0	0
TURKEY Lt.	1	0	1	1	0	1
	1	1	0	0	1	1
TEA	0	1	0	1	0	1
TOBACCO	1	1	1	1	1	1
TURPENTINE	0	0	1	1	0	0
YEAST, BAKER	1	1	0	0	0	1
YEAST, BREWER	1	0	0	0	0	0
ALCOHOL, 1%	0	0	0	1	0	0
ALCOHOL, 2%	0	0	0	1	0	0
===============						
POSITIVES	7	6	6	12	7	9
NUMBER (N)	15	15	15	15	15	15
% POSITIVE	46.7	40.0	40.0	80.0	46.7	60.0

In table 05.06.MEMBERS.05 (Meat, Fowl, Egg, and Milk) the number of test positive items was low with no items test positive for all members of the family, however the father JES (54.5%) and youngest daughter JES (63.6%) tested relatively high to this group of items, particularity noteworthy was the sensitivity to milk and milk products. In table 05.06.MEMBERS.06 (Miscellaneous Items) the entire family tested positive for coffee and tobacco with otherwise minimal sensitivities toward the other items in this table. Daughters LMS (80.0%) and JES (60.0%) posted individually high levels of sensitivities to these miscellaneous items.

TABLE 05.06.MEMBERS.06. JESxERS FAMILIAL CYTOTOXIC POSITIVE SENSITIVITIES TO MISCELLANEOUS ITEMS.

	FATHER	MOTHER	SON	DAUGHTER	DAUGHTER	DAUGHTER
SUBJECT	JES	ERS	GWS	LMS	KES	JES
AGE	49	49	22	19	17	12
CHOCOLATE	1	0	0	1	1	1
COFFEE	1	1	1	1	1	1
COTTONSEED	0	1	1	1	1	0
GARLIC	0	0	1	1	1	1
GASOLINE	0	0	0	1	1	1
MUSHROOM	0	0	0	1	0	0
MUSTARD	1	0	1	1	0	1
PEPPERMINT	1	1	0	0	1	1
TEA	0	1	0	1	0	1
TOBACCO	1	1	1	1	1	1
TURPENTINE	0	0	1	1	0	0
YEAST, BAKER	1	1	0	0	0	1
YEAST, BREWER	1	0	0	0	0	0
ALCOHOL, 1%	0	0	0	1	0	0
ALCOHOL, 2%	0	0	0	1	0	0
===============						
POSITIVES	7	6	6	12	7	9
NUMBER (N)	15	15	15	15	15	15
% POSITIVE	46.7	40.0	40.0	80.0	46.7	60.0

TABLE 05.06.MEMBERS.07 SUMMARY: JESxERS FAMILIAL CYTOTOXIC TEST POSITIVE FOOD SENSITIVITIES.

	FATHER	MOTHER	SON	DAUGHTER	DAUGHTER	DAUGHTER
SUBJECT	JES	ERS	GWS	LMS	KES	JES
AGE	49	49	22	19	17	12
POSITIVES	34	29	26	41	26	39
NUMBER (N)	70	70	70	70	70	70
% POSITIVE	48.6	41.4	37.1	58.6	37.1	55.7

Clearly the father JES (48.6%) and two daughters LMS (58.6%) and JES (55.7%) displayed the highest levels of Cytotoxic Test positives items (Table 05.06.MEMBERS.01.7). The mother ERS (41.1%) and two children GWS (37.1%) and KES (37.1%) tested out with relatively low levels of sensitivity and no major clinical complaints. The father JES medical history included Allergy Skin Tests and treatment that provided recommendations for a special diet from his Allergist. The Cytotoxic Test procedure uses the same allergen reagent preparations that are used by Allergists for Skin Tests. The father JES listed headaches, catches colds easily, rhinorrhea, upset stomach, chest pain, hay fever, sneezing, nasal blocking, and eczema with itching hands as symptoms associated with what was perceived as allergy was prescribed ACTH for relief. The 19-year-old daughter LMS listed variable clinical complaints that included catches colds easily, diarrhea, constipation, dizziness, sneezing, nasal blocking, cough, itching eyes, sleepiness, fatigue, and eczema. House dust, hot or warm air and grass were listed as factors that increased the symptoms. LMS also listed tobacco use and was Cytotoxic Test positive for tobacco. Antihistamines and cortisones were prescribed for relief.

The 11-year-old daughter RES listed clinical complaints that included itching eyes, eczema, and food allergy. She listed house dust as a factor that increased her symptoms. Cortisone was prescribed to manage her symptoms. Given a high level of sensitivity any one allergen could trigger serious clinical problems. Multiple allergens could lead to many symptoms difficult to diagnose and nearly impossible to treat specifically.

CASE STUDY 05.06.GWSa.01. JES x ERS (GWS): Twenty-two-year-old GWS and his parents agreed, either positive or negative, on 36 of 70 (51.4%) of the items tested for. The father and mother agreed on 21 of 70 (30.0%) positives. Of the 21 positive items that both parents agreed on 15 of 21 (21.4%) were passed as positive to GWS. Of GWS's 26 positives, 17 (65.4%) matched his father's positives, 17 (65.4%) matched his mom's positives and 15 (57.7%) matched both parents. The father exclusively matched on only two (peanut and mustard) of 26 (7.7%) of GWS's positives. The mother exclusively matched on only two (malt and cottonseed) of 26 (7.7%) of GWS's positives. GWS's had seven (26.9%) of 26 test positive items (tomato, hops, string bean, pork, garlic, and turpentine) that did not appear in either parent. Collectively, the father (JES) and son (GWS) matched on 17 (24.3%) of 70 items tested for, while the mother (ERS) and son matched on 17 (24.3%) of 70 items. In the mother to father 17 to 17 ratio (1.0) shows no "donor" difference between the father and mother. For every gene trait the father has "donated" the mother has "donated" an equal number of genes. Individually, the father and daughter agreed on two (2.9%) of 70 items tested for, whereas the mother and daughter agreed on two (2.9%) of 70 items. In the two-to-two ratio (1.0) there is no "donor" advantage for the father, for every gene the father has "donated" the mother has "donated" an equal number of genes. If you factor in what positive food items each parent has to "offer" there is a mother to father 29 to 34 ratios (0.853) favoring the father. For every gene the father has to "offer", the mother statistically has about 15% less genes to "offer". One of our objectives with these case studies is to determine if and how these potential genetic test positive results are distributed including what role the parents play and how clinically significance test positive results correlate with medical symptoms. This case demonstrates that collectively both parents make significant contributions to their son's test positive items together (57.7%) but not individually (7.7%).

TABLE 05.06.GWSa.01 JESxERS (GWS) FAMILIAL DISTRIBUTION OF CYTOTOXIC TEST POSITIVE SENSITIVITIES TO FRUITS.

	FATHER	MOTHER	SON	%POSITIVE	%POSITIVE	%POSITIVE	%NEGATIVE	PER CENT OF POSITIVE FOODS			
SUBJECT	JES	ERS	GWS	JES/ERS	JES	ERS	JES/ERS		JES/ERS	JES	ERS
AGE TESTED	49	49	22	GWS	GWS	GWS	GWS	GWS	GWS	GWS	GWS
APPLE	1	1	1	1					1		
BANANA	0	0	0				0				
CANTALOUPE	1	0	0								
CHERRY	0	0	0				0				
HONEYDEW	1	0	0								
ORANGE	1	1	1	1					1		
PEACH	1	0	0								
PEAR	0	0	0				0				
PINEAPPLE	1	0	0								
STRAWBERRY	1	0	0								
TOMATO	0	0	1					1			
VANILLA	1	1	1	1					1		
WATERMELON	0	0	0				0				
===============											
POSITIVES	8	3	4	3	0	0	4	1	3	0	0
NUMBER (N)	13	13	13	13	13	13	13	4	4	4	4
% POSITIVE	61.5	23.1	30.8	23.1	0.0	0.0	30.8	25.0	75.0	0.0	0.0

TABLE 05.06.GWSa.02 JESxERS (GWS) FAMILIAL DISTRIBUTION OF CYTOTOXIC TEST POSITIVE SENSITIVITIES TO GRAINS.

	FATHER	MOTHER	SON	%POSITIVE	%POSITIVE	%POSITIVE	%NEGATIVE	PER CENT OF POSITIVE FOODS			
SUBJECT	JES	ERS	GWS	JES/ERS	JES	ERS	JES/ERS		JES/ERS	JES	ERS
AGE TESTED	49	49	22	GWS	GWS	GWS	GWS	GWS	GWS	GWS	GWS
BARLEY	1	1	1	1					1		
CORN	1	1	1	1					1		
MALT	0	1	1			1					1
OAT	1	1	1	1					1		
RICE	0	1	0								
RYE	1	1	1	1					1		
WHEAT	1	1	0								
===============											
POSITIVES	5	7	5	4	0	1	0	0	4	0	1
NUMBER (N)	7	7	7	7	7	7	7	5	5	5	5
% POSITIVE	71.4	100.0	71.4	57.1	0.0	14.3	0.0	0.0	80.0	0.0	20.0

TABLE 05.06.GWSa.03 JESxERS (GWS) FAMILIAL DISTRIBUTION OF CYTOTOXIC POSITIVE SENSITIVITIES TO VEGETABLES.

	FATHER	MOTHER	SON	%POSITIVE	%POSITIVE	%POSITIVE	%NEGATIVE	PER CENT OF POSITIVE FOODS			
SUBJECT	JES	ERS	GWS	JES/ERS	JES	ERS	JES/ERS		JES/ERS	JES	ERS
AGE TESTED	49	49	22	GWS	GWS	GWS	GWS	GWS	GWS	GWS	GWS
BROCCOLI	0	1	0								
CABBAGE	1	0	0								
CARROT	1	1	1	1					1		
CUCUMBER	0	1	0								
HOPS	0	0	1					1			
HORSERADISH	0	0	0				0				
LETTUCE	1	1	1	1					1		
ONION	0	1	0								
PEA	0	0	0				0				
PEANUT	1	0	1		1					1	
POTATO	1	1	1	1					1		
RADISH	0	1	0								
SOYBEAN	1	1	0								
SPINACH	0	0	0				0				
STRINGBEAN	0	0	1					1			
SUGAR, BEET	0	0	0				0				
SUGAR, CANE	1	0	0								
===============											
POSITIVES	7	8	6	3	1	0	4	2	3	1	0
NUMBER (N)	17	17	17	17	17	17	17	6	6	6	6
% POSITIVE	41.2	47.1	35.3	17.6	5.9	0.0	23.5	33.3	50.0	16.7	0.0

TABLE 05.06.GWSa.04 JESxERS (GWS) FAMILIAL CYTOTOXIC TEST POSITIVE SENSITIVITIES TO FISH & SEAFOOD.

	FATHER	MOTHER	SON	%POSITIVE	%POSITIVE	%POSITIVE	%NEGATIVE	PER CENT OF POSITIVE FOODS			
SUBJECT	JES	ERS	GWS	JES/ERS	JES	ERS	JES/ERS		JES/ERS	JES	ERS
AGE TESTED	49	49	22	GWS	GWS	GWS	GWS	GWS	GWS	GWS	GWS
CARP	0	0	0				0				
CRAB	0	0	0				0				
LOBSTER	0	0	0				0				
SHRIMP	0	0	1					1			
SWORDFISH	0	0	0				0				
TROUT	0	0	0				0				
TUNA	1	1	0								
===============											
POSITIVES	1	1	1	0	0	0	5	1	0	0	0
NUMBER (N)	7	7	7	7	7	7	7	1	1	1	1
% POSITIVE	14.3	14.3	14.3	0.0	0.0	0.0	71.4	100.0	0.0	0.0	0.0

TABLE 05.06.GWSa.05 JESxERS (GWS) FAMILIAL CYTOTOXIC TEST POSITIVE SENSITIVITIES TO MEAT, FOWL, EGG & MILK.

	FATHER	MOTHER	SON	%POSITIVE	%POSITIVE	%POSITIVE	%NEGATIVE	PER CENT OF POSITIVE FOODS			
SUBJECT	JES	ERS	GWS	JES/ERS	JES	ERS	JES/ERS		JES/ERS	JES	ERS
AGE TESTED	49	49	22	GWS	GWS	GWS	GWS	GWS	GWS	GWS	GWS
BEEF	0	0	0				0				
LAMB	0	0	0				0				
PORK	0	0	1					1			
VEAL	0	0	0				0				
CHICKEN	1	1	1	1					1		
TURKEY, Lt.	1	1	0								
EGG	1	1	1	1					1		
MILK	1	0	0								
CHEESE/AMER.	1	1	1	1					1		
CHEESE/ROQUE.	0	0	0				0				
CHEESE/SWISS	1	0	0								
===============											
POSITIVES	6	4	4	3	0	0	4	1	3	0	0
NUMBER (N)	11	11	11	11	11	11	11	4	4	4	4
% POSITIVE	54.5	36.4	36.4	27	0	0	36	25	75	0	0

TABLE 05.06.GWSa.06 JESxERS (GWS) FAMILIAL CYTOTOXIC TEST POSITIVE SENSITIVITIES TO MISCELLANEOUS ITEMS.

	FATHER	MOTHER	SON	%POSITIVE	%POSITIVE	%POSITIVE	%NEGATIVE	PER CENT OF POSITIVE FOODS			
SUBJECT	JES	ERS	GWS	JES/ERS	JES	ERS	JES/ERS		JES/ERS	JES	ERS
AGE	49	49	22	GWS	GWS	GWS	GWS	GWS	GWS	GWS	GWS
CHOCOLATE	1	0	0								
COFFEE	1	1	1	1					1		
COTTONSEED	0	1	1			1					1
GARLIC	0	0	1					1			
GASOLINE	0	0	0				0				
MUSHROOM	0	0	0				0				
MUSTARD	1	0	1		1					1	
PEPPERMINT	1	1	0								
TEA	0	1	0								
TOBACCO	1	1	1	1					1		
TURPENTINE	0	0	1					1			
YEAST, BAKER	1	1	0								
YEAST, BREWER	1	0	0								
ALCOHOL, 1%	0	0	0				0				
ALCOHOL, 2%	0	0	0				0				
==============											
POSITIVES	7	6	6	2	1	1	4	2	2	1	1
NUMBER (N)	15	15	15	15	15	15	15	6	6	6	6
% POSITIVE	46.7	40.0	40.0	13.3	6.7	6.7	26.7	33.3	33.3	16.7	16.7

TABLE 05.06.GWSa.07 SUMMARY: JESxERS (GWS) FAMILIAL CYTOTOXIC TEST POSITIVE FOOD SENSITIVITIES.											
	FATHER	MOTHER	SON	%POSITIVE	%POSITIVE	%POSITIVE	%NEGATIVE	PER CENT OF POSITIVE FOODS			
SUBJECT	JES	ERS	GWS	JES/ERS	JES	ERS	JES/ERS		JES/ERS	JES	ERS
AGE	49	49	22	GWS	GWS	GWS	GWS	GWS	GWS	GWS	GWS
POSITIVES	34	29	26	15	2	2	21	7	15	2	2
NUMBER (N)	70	70	70	70	70	70	70	26	26	26	26
% POSITIVE	48.6	41.4	37.1	21.4	2.9	2.9	30.0	26.9	57.7	7.7	7.7

CASE STUDY 05.06.LMSb.01-7. JES x ERS (LMS): Nineteen-year-old LMS and her parents agreed, either positive or negative, on 30 of 70 (42.9%) of the items tested for. The father (JES) and mother (ERS) agreed on 21 of 70 (30.0%) positives. Of the 21 positive items that both parents agreed on 16 of 21 (76.2%) were passed as positive to LMS. Of LMS's 41 positives, 22 (53.7%) matched her father's positives, 21 (51.2%) matched her mom's positives and 16 (39.0%) matched both parents The father exclusively matched on six (strawberry, peanut cane sugar, Swiss cheese chocolate and mustard) of 41 (14.6%) of LMS's positives. The mother exclusively matched on five (rice, onion, radish, cottonseed ,and tea) of 41 (12.2%) of LMS's positives. LMS had 14 (34.1%) of 41 test positive items (banana, cherry, tomato, hops, beet sugar, lobster, beef, pork, garlic gasoline, mushroom, turpentine, and 1&2% alcohol) that did not appear in either parent. Collectively, the father (JES) and daughter (LMS) matched on 22 (31.4%) of 70 items tested for, while the mother (ERS) and daughter matched on 21 (30.0%) of 70 items. In the mother to father 21 to 22 ratio (0.955) shows no "donor" difference between the father and mother. If you factor in what positive food items each parent has to "offer" there is a mother to father 29 to 34 ratio (0.853) favoring the father. For every gene the father has "donated" the mother has "donated" an equal number of genes but had 15% less genes to "offer". Individually, the father and daughter agreed on six (8.6%) of 70 stuffs tested for, whereas the mother and daughter agreed on five (7.1%) of 70 items. In the five to six ratio (0.833) there appears to be a slight "donor" advantage for the father, for every gene the father has "donated" the mother has "donated" about 15% less as many. For every gene the father has to "offer", the mother statistically has about 15% less genes "donated", but 15% less genes to "offer" individually.

TABLE 05.06.LMSb.01. JESxERS (LMS) FAMILIAL DISTRIBUTION OF CYTOTOXIC TEST POSITIVE SENSITIVITIES TO FRUITS.											
	FATHER	MOTHER	DAUGHTER	%POSITIVE	%POSITIVE	%POSITIVE	%NEGATIVE	PER CENT OF POSITIVE FOODS			
SUBJECT	JES	ERS	LMS	JES/ERS	JES	ERS	JES/ERS		JES/ERS	JES	ERS
AGE	49	49	19	LMS	LMS	LMS	LMS	LMS	LMS	LMS	LMS
APPLE	1	1	1	1					1		
BANANA	0	0	1					1			
CANTALOUPE	1	0	0								
CHERRY	0	0	1					1			
HONEYDEW	1	0	0								
ORANGE	1	1	1	1					1		
PEACH	1	0	0								
PEAR	0	0	0				0				
PINEAPPLE	1	0	0								
STRAWBERRY	1	0	0		1					1	
TOMATO	0	0	0					1			
VANILLA	1	1	1	1					1		
WATERMELON	0	0	0				0				
===============											
POSITIVES	8	3	5	3	1	0	2	3	3	1	0
NUMBER (N)	13	13	13	13	13	13	13	7	7	7	7
% POSITIVE	61.5	23.1	38.5	23.1	7.7	0.0	15.4	42.9	42.9	14.3	0.0

TABLE 05.06.LMSb.02 JESxERS (LMS) FAMILIAL DISTRIBUTION OF CYTOTOXIC TEST POSITIVE SENSITIVITIES TO GRAINS.

	FATHER	MOTHER	DAUGHTER	%POSITIVE	%POSITIVE	%POSITIVE	%NEGATIVE	PER CENT OF POSITIVE FOODS			
SUBJECT	JES	ERS	LMS	JES/ERS	JES	ERS	JES/ERS		JES/ERS	JES	ERS
AGE TESTED	49	49	19	LMS	LMS	LMS	LMS	LMS	LMS	LMS	LMS
BARLEY	1	1	1	1					1		
CORN	1	1	1	1					1		
MALT	0	1	0								
OAT	1	1	1	1					1		
RICE	0	1	1			1					1
RYE	1	1	1	1					1		
WHEAT	1	1	1	1					1		
===============											
POSITIVES	5	7	6	5	0	1	0	0	5	0	1
NUMBER (N)	7	7	7	7	7	7	7	6	6	6	6
% POSITIVE	71.4	100.0	85.7	71.4	0.0	14.3	0.0	0.0	83.3	0.0	16.7

TABLE 05.06.LMSb.03 JESxERS (LMS) FAMILIAL DISTRIBUTION OF CYTOTOXIC TEST POSITIVE SENSITIVITIES TO VEGETABLES.

	FATHER	MOTHER	DAUGHTER	%POSITIVE	%POSITIVE	%POSITIVE	%NEGATIVE	PER CENT OF POSITIVE FOODS			
SUBJECT	JES	ERS	LMS	JES/ERS	JES	ERS	JES/ERS		JES/ERS	JES	ERS
AGE TESTED	49	49	19	LMS	LMS	LMS	LMS	LMS	LMS	LMS	LMS
BROCCOLI	0	1	0								
CABBAGE	1	0	0								
CARROT	1	1	1	1					1		
CUCUMBER	0	1	0								
HOPS	0	0	1					1			
HORSERADISH	0	0	0				0				
LETTUCE	1	1	1	1					1		
ONION	0	1	1			1					1
PEA	0	0	0				0				
PEANUT	1	0	1		1					1	
POTATO	1	1	1	1					1		
RADISH	0	1	1			1					1
SOYBEAN	1	1	1	1					1		
SPINACH	0	0	0				0				
STRINGBEAN	0	0	0				0				
SUGAR, BEET	0	0	1					1			
SUGAR, CANE	1	0	1		1					1	
===============											
POSITIVES	7	8	10	4	2	2	4	2	4	2	2
NUMBER (N)	17	17	17	17	17	17	17	10	10	10	10
% POSITIVE	41.2	47.1	58.8	23.5	11.8	11.8	23.5	20.0	40.0	20.0	20.0

TABLE 05.06.LMSb.04 JESxERS (LMS) FAMILIAL CYTOTOXIC TEST POSITIVE SENSITIVITIES TO FISH & SEAFOOD.											
	FATHER	MOTHER	DAUGHTER	%POSITIVE	%POSITIVE	%POSITIVE	%NEGATIVE	PER CENT OF POSITIVE FOODS			
SUBJECT	JES	ERS	LMS	JES/ERS	JES	ERS	JES/ERS		JES/ERS	JES	ERS
AGE TESTED	49	49	19	LMS	LMS	LMS	LMS	LMS	LMS	LMS	LMS
CARP	0	0	0				0				
CRAB	0	0	0				0				
LOBSTER	0	0	1					1			
SHRIMP	0	0	0				0				
SWORDFISH	0	0	0				0				
TROUT	0	0	0				0				
TUNA	1	1	1	1					1		
===============											
POSITIVES	1	1	2	1	0	0	5	1	1	0	0
NUMBER (N)	7	7	7	7	7	7	7	2	2	2	2
% POSITIVE	14.3	14.3	28.6	14.3	0.0	0.0	71.4	50.0	50.0	0.0	0.0

TABLE 05.06.LMSb.05 JESxERS (LMS) FAMILIAL CYTOTOXIC TEST POSITIVE SENSITIVITIES TO MEAT, FOWL, EGG & MILK.											
	FATHER	MOTHER	DAUGHTER	%POSITIVE	%POSITIVE	%POSITIVE	%NEGATIVE	PER CENT OF POSITIVE FOODS			
SUBJECT	JES	ERS	LMS	JES/ERS	JES	ERS	JES/ERS		JES/ERS	JES	ERS
AGE TESTED	49	49	19	LMS	LMS	LMS	LMS	LMS	LMS	LMS	LMS
BEEF	0	0	1					1			
LAMB	0	0	0				0				
PORK	0	0	1					1			
VEAL	0	0	0				0				
CHICKEN	1	1	1	1					1		
TURKEY, Lt.	1	1	0								
EGG	1	1	0								
MILK	1	0	0								
CHEESE/AMER.	1	1	0								
CHEESE/ROQUE.	0	0	0				0				
CHEESE/SWISS	1	0	1		1					1	
===============											
POSITIVES	6	4	4	1	1	0	3	2	1	1	0
NUMBER (N)	11	11	11	11	11	11	11	4	4	4	4
% POSITIVE	54.5	36.4	36.4	9	9	0	27	50	25	25	0

TABLE 05.06.LMSb.06 JESxERS (LMS) FAMILIAL CYTOTOXIC TEST POSITIVE SENSITIVITIES TO MISCELLANEOUS ITEMS.

	FATHER	MOTHER	DAUGHTER	%POSITIVE	%POSITIVE	%POSITIVE	%NEGATIVE	PER CENT OF POSITIVE FOODS			
SUBJECT	JES	ERS	LMS	JES/ERS	JES	ERS	JES/ERS		JES/ERS	JES	ERS
AGE	49	49	19	LMS	LMS	LMS	LMS	LMS	LMS	LMS	LMS
CHOCOLATE	1	0	1		1					1	
COFFEE	1	1	1	1					1		
COTTONSEED	0	1	1			1					1
GARLIC	0	0	1					1			
GASOLINE	0	0	1					1			
MUSHROOM	0	0	1					1			
MUSTARD	1	0	1		1					1	
PEPPERMINT	1	1	0								
TEA	0	1	1			1					1
TOBACCO	1	1	1	1					1		
TURPENTINE	0	0	1					1			
YEAST, BAKER	1	1	0								
YEAST, BREWER	1	0	0								
ALCOHOL, 1%	0	0	1					1			
ALCOHOL, 2%	0	0	1					1			
===============											
POSITIVES	7	6	12	2	2	2	0	6	2	2	2
NUMBER (N)	15	15	15	15	15	15	15	12	12	12	12
% POSITIVE	46.7	40.0	80.0	13.3	13.3	13.3	0.0	50.0	16.7	16.7	16.7

TABLE 05.06.LMSb.07 SUMMARY: JESxERS (LMS) FAMILIAL DISTRIBUTION OF CYTOTOXIC TEST POSITIVE FOOD SENSITIVITIES.

	FATHER	MOTHER	DAUGHTER	%POSITIVE	%POSITIVE	%POSITIVE	%NEGATIVE	PER CENT OF POSITIVE FOODS			
SUBJECT	JES	ERS	LMS	JES/ERS	JES	ERS	JES/ERS		JES/ERS	JES	ERS
AGE	49	49	19	LMS	LMS	LMS	LMS	LMS	LMS	LMS	LMS
POSITIVES	34	29	41	16	6	5	14	14	16	6	5
NUMBER (N)	70	70	70	70	70	70	70	41	41	41	41
% POSITIVE	48.6	41.4	58.6	22.9	8.6	7.1	20.0	34.1	39.0	14.6	12.2

CASE STUDY 05.06.KESc.01-7 JES x ERS (KES): Seventeen-year-old KES and her parents agreed, either positive or negative, on 35 of 70 (50.0%) of the items tested for. The father and mother agreed on 21 of 70 (30.0%) positives. Of the 21 positive items that both parents agreed on 12 of 21 (57.1%) were passed as positive to KES. Of the daughter's 26 positives, 18 (69.2%) matched her father's positives, 15 (57.7%) matched her mom's positives and 12 (46.2%) matched both parents

TABLE 05.06.KESc.01 JESxERS (KES) FAMILIAL DISTRIBUTION OF CYTOTOXIC TEST POSITIVE SENSITIVITIES TO FRUITS.

	FATHER	MOTHER	DAUGHTER	%POSITIVE	%POSITIVE	%POSITIVE	%NEGATIVE	PER CENT OF POSITIVE FOODS			
SUBJECT	JES	ERS	KES	JES/ERS	JES	ERS	JES/ERS		JES/ERS	JES	ERS
AGE TESTED	49	49	17	KES	KES	KES	KES	KES	KES	KES	KES
APPLE	1	1	1	1					1		
BANANA	0	0	0				0				
CANTALOUPE	1	0	0								
CHERRY	0	0	0				0				
HONEYDEW	1	0	0								
ORANGE	1	1	0								
PEACH	1	0	1		1					1	
PEAR	0	0	0				0				
PINEAPPLE	1	0	1		1					1	
STRAWBERRY	1	0	1		1					1	
TOMATO	0	0	1					1			
VANILLA	1	1	1	1					1		
WATERMELON	0	0	0				0				
===============											
POSITIVES	8	3	6	2	3	0	4	1	2	3	0
NUMBER (N)	13	13	13	13	13	13	13	6	6	6	6
% POSITIVE	61.5	23.1	46.2	15.4	23.1	0.0	30.8	16.7	33.3	50.0	0.0

TABLE 05.06.KESc.02 JESxERS (KES) FAMILIAL DISTRIBUTION OF CYTOTOXIC TEST POSITIVE SENSITIVITIES TO GRAINS.

	FATHER	MOTHER	DAUGHTER	%POSITIVE	%POSITIVE	%POSITIVE	%NEGATIVE	PER CENT OF POSITIVE FOODS			
SUBJECT	JES	ERS	KES	JES/ERS	JES	ERS	JES/ERS		JES/ERS	JES	ERS
AGE TESTED	49	49	17	KES	KES	KES	KES	KES	KES	KES	KES
BARLEY	1	1	0								
CORN	1	1	1	1					1		
MALT	0	1	1			1					1
OAT	1	1	1	1					1		
RICE	0	1	1			1					1
RYE	1	1	1	1					1		
WHEAT	1	1	1	1					1		
===============											
POSITIVES	5	7	6	4	0	2	0	0	4	0	2
NUMBER (N)	7	7	7	7	7	7	7	6	6	6	6
% POSITIVE	71.4	100.0	85.7	57.1	0.0	28.6	0.0	0.0	66.7	0.0	33.3

TABLE 05.06.KESc.03 JESxERS (KES) FAMILIAL DISTRIBUTION OF CYTOTOXIC TEST POSITIVE SENSITIVITIES TO VEGETABLES.

	FATHER	MOTHER	DAUGHTER	%POSITIVE	%POSITIVE	%POSITIVE	%NEGATIVE	PER CENT OF POSITIVE FOODS			
SUBJECT	JES	ERS	KES	JES/ERS	JES	ERS	JES/ERS		JES/ERS	JES	ERS
AGE TESTED	49	49	17	KES	KES	KES	KES	KES	KES	KES	KES
BROCCOLI	0	1	0								
CABBAGE	1	0	1			1					1
CARROT	1	1	1	1					1		
CUCUMBER	0	1	0								
HOPS	0	0	1					1			
HORSERADISH	0	0	0				0				
LETTUCE	1	1	0								
ONION	0	1	0								
PEA	0	0	0				0				
PEANUT	1	0	0								
POTATO	1	1	1	1					1		
RADISH	0	1	0								
SOYBEAN	1	1	0								
SPINACH	0	0	0				0				
STRINGBEAN	0	0	0				0				
SUGAR, BEET	0	0	0				0				
SUGAR, CANE	1	0	1			1					1
===============											
POSITIVES	7	8	5	2	0	2	5	1	2	0	2
NUMBER (N)	17	17	17	17	17	17	17	5	5	5	5
% POSITIVE	41.2	47.1	29.4	11.8	0.0	11.8	29.4	20.0	40.0	0.0	40.0

TABLE 05.06.KESc.04 JESxERS (KES) FAMILIAL CYTOTOXIC TEST POSITIVE SENSITIVITIES TO FISH & SEAFOOD.

	FATHER	MOTHER	DAUGHTER	%POSITIVE	%POSITIVE	%POSITIVE	%NEGATIVE	PER CENT OF POSITIVE FOODS			
SUBJECT	JES	ERS	KES	JES/ERS	JES	ERS	JES/ERS		JES/ERS	JES	ERS
AGE TESTED	49	49	17	KES	KES	KES	KES	KES	KES	KES	KES
CARP	0	0	0				0				
CRAB	0	0	0				0				
LOBSTER	0	0	0				0				
SHRIMP	0	0	1					1			
SWORDFISH	0	0	0				0				
TROUT	0	0	0				0				
TUNA	1	1	0								
===============											
POSITIVES	1	1	1	0	0	0	5	1	0	0	0
NUMBER (N)	7	7	7	7	7	7	7	1	1	1	1
% POSITIVE	14.3	14.3	14.3	0.0	0.0	0.0	71.4	100.0	0.0	0.0	0.0

TABLE 05.06.KESc.05 JESxERS (KES) FAMILIAL CYTOTOXIC TEST POSITIVE SENSITIVITIES TO MEAT, FOWL, EGG & MILK.

	FATHER	MOTHER	DAUGHTER	%POSITIVE	%POSITIVE	%POSITIVE	%NEGATIVE	PER CENT OF POSITIVE FOODS			
SUBJECT	JES	ERS	KES	JES/ERS	JES	ERS	JES/ERS		JES/ERS	JES	ERS
AGE TESTED	49	49	17	KES	KES	KES	KES	KES	KES	KES	KES
BEEF	0	0	0				0				
LAMB	0	0	0				0				
PORK	0	0	0				0				
VEAL	0	0	0				0				
CHICKEN	1	1	0								
TURKEY, Lt.	1	1	0								
EGG	1	1	0								
MILK	1	0	0								
CHEESE/AMER.	1	1	1	1					1		
CHEESE/ROQUE.	0	0	0				0				
CHEESE/SWISS	1	0	0								
===============											
POSITIVES	6	4	1	1	0	0	5	0	1	0	0
NUMBER (N)	11	11	11	11	11	11	11	1	1	1	1
% POSITIVE	54.5	36.4	9.1	9.1	0.0	0.0	45.5	0.0	100.0	0.0	0.0

TABLE 05.06.KESc.06 JESxERS (KES) FAMILIAL CYTOTOXIC TEST POSITIVE SENSITIVITIES TO MISCELLANEOUS ITEMS.

	FATHER	MOTHER	DAUGHTER	%POSITIVE	%POSITIVE	%POSITIVE	%NEGATIVE	PER CENT OF POSITIVE FOODS			
SUBJECT	JES	ERS	KES	JES/ERS	JES	ERS	JES/ERS		JES/ERS	JES	ERS
AGE	49	49	17	KES	KES	KES	KES	KES	KES	KES	KES
CHOCOLATE	1	0	1		1					1	
COFFEE	1	1	1	1					1		
COTTONSEED	0	1	1			1					1
GARLIC	0	0	1					1			
GASOLINE	0	0	1					1			
MUSHROOM	0	0	0				0				
MUSTARD	1	0	0								
PEPPERMINT	1	1	1	1					1		
TEA	0	1	0								
TOBACCO	1	1	1	1					1		
TURPENTINE	0	0	0				0				
YEAST, BAKER	1	1	0								
YEAST, BREWER	1	0	0								
ALCOHOL, 1%	0	0	0				0				
ALCOHOL, 2%	0	0	0				0				
===============											
POSITIVES	7	6	7	3	1	1	4	2	3	1	1
NUMBER (N)	15	15	15	15	15	15	15	7	7	7	7
% POSITIVE	46.7	40.0	46.7	20.0	6.7	6.7	26.7	28.6	42.9	14.3	14.3

The father exclusively matched KES on six (peach, pineapple, strawberry, cabbage, cane sugar and chocolate) of 26 (23.1%) positives. The mother exclusively matched KES on three (malt, rice, and cottonseed) of 26 (11.5%) positives. KES had five (19.2%) of 26 test positive items (tomato, hops, shrimp, garlic, and gasoline) that did not appear in either parent. Collectively, the father (JES) and daughter (KES) matched on 18 (25.7%) of 70 items tested for, whereas the mother and daughter on matched on 15 (21.4%) of 70 items. In the mother to father 15 to 18 ratio (0.833) there appears to be a "donor" difference between the father and mother.

For every gene the father has "donated" the mother has "donated" about eight tenths as many. Individually, the father and daughter agreed on 6 (8.6%) of 70 items tested for, whereas the mother and daughter agreed on three (4.3%) of 70 items. In the three to six ratio (0.50) there appears to be a "donor" advantage for the father, for every gene the father has "donated" the mother has "donated" than half as many. If you factor in what positive food items each parent has to "offer" there is a mother to father 29 to 34 ratio (0.853) favoring the father. For every gene the father has to "offer", the mother statistically has about eight tenths of a gene to "offer". Individually the genes that showed up or were "donated" showed a discrepancy with the mother "donating" half as many genes for KES, whereas collectively the ratio was genes "donated" 0.833 to genes "offered" 0.853 indicating that the mother had donated fewer gene-traits, but proportionally had fewer gene-traits to offer.

TABLE 05.06.KESc.07 SUMMARY: JESxERS (KES) FAMILIAL CYTOTOXIC TEST POSITIVE FOOD SENSITIVITIES.

	FATHER	MOTHER	DAUGHTER	%POSITIVE	%POSITIVE	%POSITIVE	%NEGATIVE	PER CENT OF POSITIVE FOODS			
SUBJECT	JES	ERS	KES	JES/ERS	JES	ERS	JES/ERS		JES/ERS	JES	ERS
AGE	49	49	17	KES	KES	KES	KES	KES	KES	KES	KES
POSITIVES	34	29	26	12	4	5	23	5	12	4	5
NUMBER (N)	70	70	70	70	70	70	70	26	26	26	26
% POSITIVE	48.6	41.4	37.1	17.1	5.7	7.1	32.9	19.2	46.2	15.4	19.2

CASE STUDY 05.06.JESd.01-7 JES x ERS (JES): Eleven-year-old JES and her parents agreed, either positive or negative, on 38 of 70 (54.3%) of the items tested for.

The father and mother agreed on 21 of 70 (30.0%) positives. Of the 21 positive items that both parents agreed on 18 of 21 (85.7%) were passed as positives to JES.

Of JES's 39 positives, 28 (71.8%) matched her father's positives, 21 (53.8%) matched her mom's positives and 18 (46.2%) matched both parents. The father exclusively matched on 10 (cantaloupe, pineapple, strawberry, cabbage, peanut, cane sugar, milk, Swiss cheese, chocolate, and mustard) of 39 (25.6%) of JES's positives. The mother exclusively matched on three (broccoli, onion ,and tea) of 39 (7.7%) of JES's positives.

JES had eight (20.5%) of 39 test positive items (banana, cherry, tomato, string bean, pork, Roquefort cheese, garlic, and gasoline) that did not appear in either parent. The father (RES) and 12-year-old daughter (RES) listed the clinical responsible for their visit to this program. Collectively, the father (JES) and daughter (JES) agreed on 28 (40.0%) of 70 items tested for, whereas the mother and daughter on agreed on 21 (30.0%) of 70 items.

In the mother to father 21 to 28 ratio (0.750) there appears to be a "donor" difference between the father and mother. For every gene-trait the father has "donated" the mother has "donated" about three fourths as many. Individually, the father and daughter agreed on 10 (14.3%) of 70 items tested for, whereas the mother and daughter agreed on only three (4.3%) of 70 items. In the three to 10 ratio (0.30) there appears to be a "donor" advantage for the father, for every gene the father has "donated" the mother has "donated" less than one third as many. That this individual ratio differs from the collective ratio is driven in large part by the number of allergens that both parents agree on collectively (18) and the three to ten individual ratio in favor of the father.

If you factor in what positive items each parent has to "offer" there is a mother to father 29 to 34 ratio (0.853) favoring the father. For every gene-trait the father has to "offer", the mother statistically has a little less than one gene to "offer", whereas the genes that showed up or were "donated" showed a discrepancy with the mother donating one third as many genes in this offspring. Either way, the capacity (specific gene or trigger) to be sensitized may be transmitted not the sensitivity itself. The trigger initiates the three-step model proposed by Ulett and Perry (1975).

TABLE 05.06.JESd.01. JESxERS (JES) FAMILIAL DISTRIBUTION OF CYTOTOXIC TEST POSITIVE SENSITIVITIES TO FRUITS.

	FATHER	MOTHER	DAUGHTER	%POSITIVE	%POSITIVE	%POSITIVE	%NEGATIVE	PER CENT OF POSITIVE FOODS			
SUBJECT	JES	ERS	JES	JES/ERS	JES	ERS	JES/ERS		JES/ERS	JES	ERS
AGE TESTED	49	49	12	JES	JES	JES	JES	JES	JES	JES	JES
APPLE	1	1	1	1					1		
BANANA	0	0	1					1			
CANTALOUPE	1	0	1		1					1	
CHERRY	0	0	1					1			
HONEYDEW	1	0	0								
ORANGE	1	1	1	1					1		
PEACH	1	0	0								
PEAR	0	0	0				0				
PINEAPPLE	1	0	1		1					1	
STRAWBERRY	1	0	1		1					1	
TOMATO	0	0	1					1			
VANILLA	1	1	1	1					1		
WATERMELON	0	0	0				0				
===============											
POSITIVES	8	3	9	3	3	0	2	3	3	3	0
NUMBER (N)	13	13	13	13	13	13	13	9	9	9	9
% POSITIVE	61.5	23.1	69.2	23.1	23.1	0.0	15.4	33.3	33.3	33.3	0.0

TABLE 05.06.JESd.02. JESxERS (JES) FAMILIAL DISTRIBUTION OF CYTOTOXIC TEST POSITIVE SENSITIVITIES TO GRAINS.

	FATHER	MOTHER	DAUGHTER	%POSITIVE	%POSITIVE	%POSITIVE	%NEGATIVE	PER CENT OF POSITIVE FOODS			
SUBJECT	JES	ERS	JES	JES/ERS	JES	ERS	JES/ERS		JES/ERS	JES	ERS
AGE TESTED	49	49	12	JES	JES	JES	JES	JES	JES	JES	JES
BARLEY	1	1	1	1					1		
CORN	1	1	1	1					1		
MALT	0	1	0								
OAT	1	1	1	1					1		
RICE	0	1	0								
RYE	1	1	1	1					1		
WHEAT	1	1	1	1					1		
===============											
POSITIVES	5	7	5	5	0	0	0	0	5	0	0
NUMBER (N)	7	13	13	13	13	13	13	5	5	5	5
% POSITIVE	71.4	53.8	38.5	38.5	0.0	0.0	0.0	0.0	100.0	0.0	0.0

TABLE 05.06.JESd.03. JESxERS (JES) FAMILIAL DISTRIBUTION OF CYTOTOXIC TEST POSITIVE SENSITIVITIES TO VEGETABLES.

	FATHER	MOTHER	DAUGHTER	%POSITIVE	%POSITIVE	%POSITIVE	%NEGATIVE	PER CENT OF POSITIVE FOODS			
SUBJECT	JES	ERS	JES	JES/ERS	JES	ERS	JES/ERS		JES/ERS	JES	ERS
AGE TESTED	49	49	12	JES	JES	JES	JES	JES	JES	JES	JES
BROCCOLI	0	1	1			1					1
CABBAGE	1	0	1		1					1	
CARROT	1	1	0								
CUCUMBER	0	1	0								
HOPS	0	0	0				0				
HORSERADISH	0	0	0				0				
LETTUCE	1	1	1	1					1		
ONION	0	1	1			1					1
PEA	0	0	0				0				
PEANUT	1	0	1		1					1	
POTATO	1	1	1	1					1		
RADISH	0	1	0								
SOYBEAN	1	1	1	1					1		
SPINACH	0	0	0				0				
STRINGBEAN	0	0	1					1			
SUGAR, BEET	0	0	0				0				
SUGAR, CANE	1	0	1		1					1	
===============											
POSITIVES	7	8	9	3	3	2	5	1	3	3	2
NUMBER (N)	17	17	17	17	17	17	17	9	9	9	9
% POSITIVE	41.2	47.1	52.9	17.6	17.6	11.8	29.4	11.1	33.3	33.3	22.2

	FATHER	MOTHER	DAUGHTER	%POSITIVE	%POSITIVE	%POSITIVE	%NEGATIVE	PER CENT OF POSITIVE FOODS			
SUBJECT	JES	ERS	JES	JES/ERS	JES	ERS	JES/ERS		JES/ERS	JES	ERS
AGE TESTED	49	49	12	JES	JES	JES	JES	JES	JES	JES	JES
CARP	0	0	0				0				
CRAB	0	0	0				0				
LOBSTER	0	0	0				0				
SHRIMP	0	0	0				0				
SWORDFISH	0	0	0				0				
TROUT	0	0	0				0				
TUNA	1	1	0								
===============											
POSITIVES	1	1	0	0	0	0	6	0	0	0	0
NUMBER (N)	7	7	7	7	7	7	7	0	0	0	0
% POSITIVE	14.3	14.3	0.0	0.0	0.0	0.0	85.7	0.0	0.0	0.0	0.0

TABLE 05.06.JESd.05. JESxERS (JES) FAMILIAL CYTOTOXIC TEST POSITIVE SENSITIVITIES TO MEAT, FOWL, EGG & MILK.

	FATHER	MOTHER	DAUGHTER	%POSITIVE	%POSITIVE	%POSITIVE	%NEGATIVE	PER CENT OF POSITIVE FOODS			
SUBJECT	JES	ERS	JES	JES/ERS	JES	ERS	JES/ERS		JES/ERS	JES	ERS
AGE TESTED	49	49	12	JES	JES	JES	JES	JES	JES	JES	JES
BEEF	0	0	0				0				
LAMB	0	0	0				0				
PORK	0	0	1					1			
VEAL	0	0	0				0				
CHICKEN	1	1	1	1					1		
TURKEY, Lt.	1	1	0								
EGG	1	1	1	1					1		
MILK	1	0	1		1					1	
CHEESE/AMER.	1	1	1	1					1		
CHEESE/ROQUE.	0	0	1					1			
CHEESE/SWISS	1	0	1		1					1	
POSITIVES	6	4	7	3	2	0	3	2	3	2	0
NUMBER (N)	11	11	11	11	11	11	11	7	7	7	7
% POSITIVE	54.5	36.4	63.6	27.3	18.2	0.0	27.3	28.6	42.9	28.6	0.0

NOTES:

TABLE 05.06.JESd.06. JESxERS (JES) FAMILIAL CYTOTOXIC TEST POSITIVE SENSITIVITIES TO MISCELLANEOUS ITEMS.

	FATHER	MOTHER	DAUGHTER	%POSITIVE	%POSITIVE	%POSITIVE	%NEGATIVE	PER CENT OF POSITIVE FOODS			
SUBJECT	JES	ERS	JES	JES/ERS	JES	ERS	JES/ERS		JES/ERS	JES	ERS
AGE	49	49	12	JES	JES	JES	JES	JES	JES	JES	JES
CHOCOLATE	1	0	1		1					1	
COFFEE	1	1	1	1					1		
COTTONSEED	0	1	0								
GARLIC	0	0	1					1			
GASOLINE	0	0	1					1			
MUSHROOM	0	0	0				0				
MUSTARD	1	0	1		1					1	
PEPPERMINT	1	1	1	1					1		
TEA	0	1	1			1					1
TOBACCO	1	1	1	1					1		
TURPENTINE	0	0	0				0				
YEAST, BAKER	1	1	1	1					1		
YEAST, BREWER	1	0	0								
ALCOHOL, 1%	0	0	0				0				
ALCOHOL, 2%	0	0	0				0				
POSITIVES	7	6	9	4	2	1	4	2	4	2	1
NUMBER (N)	15	15	15	15	15	15	15	9	9	9	9
% POSITIVE	46.7	40.0	60.0	26.7	13.3	6.7	26.7	22.2	44.4	22.2	11.1

TABLE 05.06.JESd.07 SUMMARY: JESXERS (JES) FAMILIAL CYTOTOXIC TEST POSITIVE FOOD SENSITIVITIES.

	FATHER	MOTHER	DAUGHTER	%POSITIVE	%POSITIVE	%POSITIVE	%NEGATIVE	PER CENT OF POSITIVE FOODS			
SUBJECT	JES	ERS	JES	JES/ERS	JES	ERS	JES/ERS		JES/ERS	JES	ERS
AGE	49	49	12	JES	JES	JES	JES	JES	JES	JES	JES
POSITIVES	34	29	39	18	10	3	20	8	18	10	3
NUMBER (N)	70	76	76	76	76	76	76	39	39	39	39
% POSITIVE	48.6	38.2	51.3	23.7	13.2	3.9	26.3	20.5	46.2	25.6	7.7

CASE STUDY 05.07.GPS.01-7 JGS x SDP-S (GPS): In this case we see a son with a very low level (17.1%) of Cytotoxic Test positive results, a mother with moderate levels (28.6%) and a father with a relative high level (55.7%) of sensitivity.

Three-year-old GPS and his parents agreed, either positive or negative, on 21 of 70 (30.0%) of the items tested for. The father and mother agreed on 12 of 70 (17.1%) positives. Of the 12 positive items that both parents agreed on only two of 12 (16.7%) were passed as positives to GPS. Of GPS's 12 positives, seven (58.3%) matched his father's positives, four (33.3%) matched his mom's positives and only two (16.7%) matched both parents.

The father exclusively matched on five (peach, lettuce, onion, potato, and egg) of 12 (41.7%) of GPS's positives. The mother exclusively matched on only two (orange and strawberry) of 12 (16.7%) of GPS's positives. GPS had three (25.0%) of 12 test positive items (corn, rye, and cottonseed) that did not appear in either parent.

Collectively, the father and son agreed on seven (10.0%) of 70 items tested for, whereas the mother and son agreed on only four (5.7%) of 70 items. In the mother to father four to seven ratio (0.571) there appears to be a "donor" difference between the father and mother. For every gene the father has "donated" the mother has "donated" about half as many.

Individually, the father and son agreed on five (7.1%) of 70 items tested for, whereas the mother and son agreed on only two (2.9%) of 70 items. In the two to five ratio (0.40) there appears to be a "donor" advantage for the father, for every gene the father has "donated" the mother has "donated" less than half as many. If you factor in what positive items each parent has to "offer" there is a mother to father 20 to 39 ratio (0.513) favoring the father. For every gene the father has to "offer", the mother statistically has about half as many genes to offer. Simply stated the father has more to "offer", advantage father.

TABLE 05.07.GPS.01 JGSxSDP-S (GPS) FAMILIAL DISTRIBUTION OF CYTOTOXIC TEST POSITIVE SENSITIVITIES TO FRUITS.

	FATHER	MOTHER	SON	%POSITIVE	%POSITIVE	%POSITIVE	%NEGATIVE	PER CENT OF POSITIVE FOODS			
SUBJECT	JGS	SDP-S	GPS	JGS/SDS	JGS	SDS	JGS/SDS		JGS/SDS	JGS	SDS
AGE TESTED	43	28	3	GPS	GPS	GPS	GPS	GPS	GPS	GPS	GPS
APPLE	1	1	0								
BANANA	0	0	0				0				
CANTALOUPE	0	0	0				0				
CHERRY	0	1	0								
HONEYDEW	0	0	0				0				
ORANGE	0	1	1			1					1
PEACH	1	0	1		1					1	
PEAR	0	1	0								
PINEAPPLE	1	0	0								
STRAWBERRY	0	1	1			1					1
TOMATO	1	0	0								
VANILLA	0	0	0				0				
WATERMELON	1	0	0								
===============											
POSITIVES	5	5	3	0	1	2	4	0	0	1	2
NUMBER (N)	13	13	13	13	13	13	13	3	3	3	3
% POSITIVE	38.5	38.5	23.1	0.0	7.7	15.4	30.8	0.0	0.0	33.3	66.7

NOTES:

TABLE 05.07.GPS.02 JGSxSDP-S (GPS) FAMILIAL DISTRIBUTION OF CYTOTOXIC POSITIVE SENSITIVITIES TO GRAINS.

	FATHER	MOTHER	SON	%POSITIVE	%POSITIVE	%POSITIVE	%NEGATIVE	PER CENT OF POSITIVE FOODS			
SUBJECT	JGS	SDP-S	GPS	JGS/SDS	JGS	SDS	JGS/SDS		JGS/SDS	JGS	SDS
AGE TESTED	43	28	3	GPS	GPS	GPS	GPS	GPS	GPS	GPS	GPS
BARLEY	1	0	0								
CORN	0	0	1					1			
MALT	0	0	0				0				
OAT	1	1	1	1					1		
RICE	1	1	0								
RYE	0	0	1					1			
WHEAT	1	0	0								
===============											
POSITIVES	4	2	3	1	0	0	1	2	1	0	0
NUMBER (N)	7	7	7	7	7	7	7	3	3	3	3
% POSITIVE	57.1	28.6	42.9	14.3	0.0	0.0	14.3	66.7	33.3	0.0	0.0

	FATHER	MOTHER	SON	%POSITIVE	%POSITIVE	%POSITIVE	%NEGATIVE	PER CENT OF POSITIVE FOODS			
SUBJECT	JGS	SDP-S	GPS	JGS/SDS	JGS	SDS	JGS/SDS		JGS/SDS	JGS	SDS
AGE TESTED	43	28	3	GPS	GPS	GPS	GPS	GPS	GPS	GPS	GPS
BROCCOLI	0	1	0								
CABBAGE	0	0	0				0				
CARROT	1	1	1	1					1		
CUCUMBER	0	0	0				0				
HOPS	1	0	0								
HORSERADISH	0	0	0				0				
LETTUCE	1	0	1		1					1	
ONION	1	0	1		1					1	
PEA	1	0	0								
PEANUT	0	1	0								
POTATO	1	0	1		1					1	
RADISH	1	0	0								
SOYBEAN	1	0	0								
SPINACH	0	0	0				0				
STRINGBEAN	1	0	0								
SUGAR, BEET	0	0	0				0				
SUGAR, CANE	0	1	0								
===============											
POSITIVES	9	4	4	1	3	0	5	0	1	3	0
NUMBER (N)	17	17	17	17	17	17	17	4	4	4	4
% POSITIVE	52.9	23.5	23.5	5.9	17.6	0.0	29.4	0.0	25.0	75.0	0.0

TABLE 05.07.GPS.05 JGSxSDP-S (GPS) FAMILIAL CYTOTOXIC TEST POSITIVE SENSITIVITIES TO MEAT, POULTRY & DAIRY.

	FATHER	MOTHER	SON	%POSITIVE	%POSITIVE	%POSITIVE	%NEGATIVE	PER CENT OF POSITIVE FOODS			
SUBJECT	JGS	SDP-S	GPS	JGS/SDS	JGS	SDS	JGS/SDS		JGS/SDS	JGS	SDS
AGE TESTED	43	28	3	GPS	GPS	GPS	GPS	GPS	GPS	GPS	GPS
BEEF	0	0	0				0				
LAMB	1	0	0								
PORK	1	1	0								
VEAL	1	0	0								
CHICKEN	1	1	0								
TURKEY, Lt.	1	1	0								
EGG	1	0	1		1					1	
MILK	1	0	0								
CHEESE/AMER.	1	0	0								
CHEESE/ROQUE.	1	0	0								
CHEESE/SWISS	0	0	0				0				
===============											
POSITIVES	9	3	1	0	1	0	2	0	0	1	0
NUMBER (N)	11	11	11	11	11	11	11	1	1	1	1
% POSITIVE	81.8	27.3	9.1	0.0	9.1	0.0	18.2	0.0	0.0	100.0	0.0

TABLE 05.07.GPS.06 JGSxSDP-S (GPS) FAMILIAL CYTOTOXIC TEST POSITIVE SENSITIVITIES TO MISCELLANEOUS ITEMS.

	FATHER	MOTHER	SON	%POSITIVE	%POSITIVE	%POSITIVE	%NEGATIVE	PER CENT OF POSITIVE FOODS			
SUBJECT	JGS	SDP-S	GPS	JGS/SDS	JGS	SDS	JGS/SDS		JGS/SDS	JGS	SDS
AGE	43	28	3	GPS	GPS	GPS	GPS	GPS	GPS	GPS	GPS
ALCOHOL, 1%	1	0	0								
ALCOHOL, 2%	1	0	0								
CHOCOLATE	1	1	0								
COFFEE	1	1	0								
COTTONSEED	0	0	1					1			
GARLIC	0	1	0								
GASOLINE	0	0	0				0				
MUSHROOM	1	0	0								
MUSTARD	0	0	0				0				
PEPPERMINT	0	0	0				0				
TEA	1	0	0								
TOBACCO	1	0	0								
TURPENTINE	0	0	0				0				
YEAST, BAKER	1	1	0								
YEAST, BREWER	0	0	0				0				
===============											
POSITIVES	8	4	1	0	0	0	5	1	0	0	0
NUMBER (N)	15	15	15	15	15	15	15	1	1	1	1
% POSITIVE	53.3	26.7	6.7	0.0	0.0	0.0	33.3	100.0	0.0	0.0	0.0

NOTES:

TABLE 05.07.GJS.07 SUMMARY: JGSxSDP-S (GPS) FAMILIAL CYTOTOXIC TEST POSITIVE FOOD SENSITIVITIES.

	FATHER	MOTHER	SON	%POSITIVE	%POSITIVE	%POSITIVE	%NEGATIVE	PER CENT OF POSITIVE FOODS			
SUBJECT	JGS	SDP-S	GPS	JGS/SDS	JGS	SDS	JGS/SDS		JGS/SDS	JGS	SDS
AGE TESTED	43	28	3	GPS	GPS	GPS	GPS	GPS	GPS	GPS	GPS
TOTAL POSITIVES	39	20	12	2	5	2	20	3	2	5	2
NUMBER(N)	70	70	70	70	70	70	70	12	12	12	12
% POSITIVE	55.7	28.6	17.1	2.9	7.1	2.9	28.6	25.0	16.7	41.7	16.7

CASE STUDY 05.08.MEMBERS.01-7 JMR x RJR (FAMILY): This case study reviews the familial food sensitivities for a 36-year-old father (JMR), 34-year-old mother (RJR), a 12-year-old daughter (KKR) and a six-year-old son (JSR). These parents volunteered for the Family Study Program because they suspected that food allergy may be a contributing factor to their son's general medical discomforts and inability to concentrate in school. JSR was described by his parents as an "allergic child" and the mother identified herself as an "allergic person", while the father and daughter had no complaints of allergic origin to their knowledge. The son had a history of headaches, otic sensitivity to loud noise, mental fatigue, crabby disposition as well as frequent stomach upset and pain. Allergic reactions, most prevalent in the spring and fall of the year, included sneezing, nasal blocking, itching in the eyes and nose, and swelling in the lymph glands. The mother listed a history of headache, stomach upset, frequent colds, dizziness, chest pains, and mastoid sinuses pains. Allergic signs included a persistent cough, itchy eyes nose and throat, sinus congestions and lack of physical strength.

Both parents and the two children agreed either positive or negative on 25 (35.7%) of 70 items tested for. All family members tested positive for corn (Table 05.07.KKR.02), pork and chicken (Table 05.07. KKR.05), in addition to coffee and peppermint (Table 05.07.MEMBERS.06). This family displayed mixed levels of sensitivity. The father (JMR) was Cytotoxic Test positive for 20 (28.6%) of 70 items and the mother was positive for 34 (48.6%) of 70 items. The 12-year daughter (KKR) was positive for 24 (34.3%) of 70 items, while the six-year-old son was positive for 36 (51.4%) of the 70 items tested for. Together, the parents agreed positive on ten (14.3%) of 70 items tested for. Of the ten items that both parents agreed on five (50.0%) items were also positive in both children. When we analyze these parents and each individual child we will carefully look at the contribution of both parents as opposed to that of each individual parent regards the test positive profile of each offspring

TABLE 05.08.MEMBERS.01. JMRxRJR FAMILIAL CYTOTOXIC POSITIVE SENSITIVITIES TO FRUITS.

	FATHER	MOTHER	DAUGHTER	SON
SUBJECT	JMR	RJR	KKR	JSR
AGE TESTED	36	34	12	6
APPLE	1	0	1	1
BANANA	0	1	1	1
CANTALOUPE	1	0	0	1
CHERRY	0	1	1	0
HONEYDEW	0	1	0	1
ORANGE	1	0	1	1
PEACH	1	1	0	0
PEAR	0	0	0	0
PINEAPPLE	0	1	1	1
STRAWBERRY	0	1	0	0
TOMATO	0	1	1	1
VANILLA	1	1	0	1
WATERMELON	0	0	0	0
===============				
POSITIVES	5	8	6	8
NUMBER (N)	13	13	13	13
% POSITIVE	38.5	61.5	46.2	61.5

TABLE 05.08.MEMBERS.02 JMRxRJR FAMILIAL CYTOTOXIC POSITIVE SENSITIVITIES TO GRAINS.

	FATHER	MOTHER	DAUGHTER	SON
SUBJECT	JMR	RJR	KKR	JSR
AGE TESTED	36	34	12	6
BARLEY	0	0	0	0
CORN	1	1	1	1
MALT	0	0	0	0
OAT	0	1	1	1
RICE	1	0	1	1
RYE	0	1	1	1
WHEAT	0	1	1	1
===============				
POSITIVES	2	4	5	5
NUMBER (N)	7	7	7	7
% POSITIVE	28.6	57.1	71.4	71.4

TABLE 05.08.MEMBERS.03 JMRxRJR FAMILIAL CYTOTOXIC POSITIVE SENSITIVITIES TO VEGETABLES.

	FATHER	MOTHER	DAUGHTER	SON
SUBJECT	JMR	RJR	KKR	JSR
AGE TESTED	36	34	12	6
BROCCOLI	0	1	0	0
CABBAGE	0	0	0	0
CARROT	0	0	1	0
CUCUMBER	0	1	0	0
HOPS	0	0	1	0
HORSERADISH	0	1	0	0
LETTUCE	0	0	1	1
ONION	1	1	0	1
PEA	0	0	0	0
PEANUT	0	0	0	1
POTATO	0	1	0	0
RADISH	0	0	0	0
SOYBEAN	1	0	1	1
SPINACH	0	0	0	0
STRINGBEAN	1	0	0	0
SUGAR, BEET	0	1	1	1
SUGAR, CANE	0	1	1	1
===============				
POSITIVES	3	7	6	6
NUMBER (N)	17	17	17	17
% POSITIVE	17.6	41.2	35.3	35.3

TABLE 05.08.MEMBERS.04 JMRXRJR CYTOTOXIC POSITIVE SENSITIVITIES TO FISH & SEAFOOD.

	FATHER	MOTHER	DAUGHTER	SON
SUBJECT	JMR	RJR	KKR	JSR
AGE TESTED	36	34	12	6
CARP	0	1	0	0
CRAB	0	0	0	0
LOBSTER	1	1	0	1
SHRIMP	1	0	0	0
SWORDFISH	0	0	0	0
TROUT	0	1	0	0
TUNA	1	1	0	1
===============				
POSITIVES	3	4	0	2
NUMBER (N)	7	7	7	7
% POSITIVE	42.9	57.1	0.0	28.6

TABLE 05.08.MEMBERS.05 JMRxRJR CYTOTOXIC POSITIVE SENSITIVITIES TO MEAT/FOWL/EGG & MILK.				
	FATHER	MOTHER	DAUGHTER	SON
SUBJECT	JMR	RJR	KKR	JSR
AGE TESTED	36	34	12	6
BEEF	1	0	0	1
LAMB	0	0	0	0
PORK	1	1	1	1
VEAL	0	0	0	0
CHICKEN	1	1	1	1
TURKEY, Lt.	0	0	0	0
EGG	1	0	1	1
MILK	0	1	0	1
CHEESE/AMER.	0	1	0	1
CHEESE/ROQUE.	0	0	0	1
CHEESE/SWISS	0	0	0	1
===============				
POSITIVES	4	4	3	8
NUMBER (N)	11	11	11	11
% POSITIVE	36.4	36.4	27.3	72.7

TABLE 05.08.MEMBERS.06 JMRxRJR FAMILIAL CYTOTOXIC TEST POSITIVE SENSITIVITIES TO MISC. ITEMS.				
	FATHER	MOTHER	DAUGHTER	SON
SUBJECT	JMR	RJR	KKR	JSR
AGE	36	34	12	6
CHOCOLATE	0	1	1	1
COFFEE	1	1	1	1
COTTONSEED	0	1	0	0
GARLIC	0	0	0	0
GASOLINE	0	0	0	0
MUSHROOM	0	0	0	0
MUSTARD	0	0	0	0
PEPPERMINT	1	1	1	1
TEA	0	1	0	1
TOBACCO	1	0	0	1
TURPENTINE	0	0	0	0
YEAST, BAKER	0	1	1	1
YEAST, BREWER	0	1	0	1
ALCOHOL, 1%	0	0	0	0
ALCOHOL, 2%	0	0	0	0
==============				
POSITIVES	3	7	4	7
NUMBER (N)	15	15	15	15
% POSITIVE	20.0	46.7	26.7	46.7

TABLE 05.08.MEMBERS.07 SUMMARY: JMRxRJR FAMILIAL CYTOTOXIC POSITIVE FOOD SENSITIVITIES.				
	FATHER	MOTHER	DAUGHTER	SON
SUBJECT	JMR	RJR	KKR	JSR
AGE	36	34	12	6
POSITIVES	20	34	24	36
NUMBER (N)	70	70	70	70
% POSITIVE	28.6	48.6	34.3	51.4

Of the 70 items surveyed 20 (28.6%) were not positive in any family member (Table 05..08.MEMBERS.08).

It is from this list of "grocery items" that two-, four- or eight-week menus are constructed. On these menus, primary clinical complaints disappear quickly with the sometimes-added benefit of weight loss, disappearance of acne, and normalized sleep patterns. As foods are added back to their diets you carefully watch for offending clinical signs.

Of the ten items that both parents agreed on five (50.0%) appeared in the daughter, and nine (90.0%) appeared in the son.

The father and both children agreed on nine (12.9%) of 70 items tested for, whereas the mother agreed on 15 (21.4%) of 70 items. In the 15 to 9 ratio (1.7) there appears to be a "donor" difference between the father and mother, however if you factor in what positive items each parent has to "offer" there is a 34 to 20 ratio (1.7) to balance out what appears to be that difference.

Table 05.08.MEMBERS.08 summarizes the family. Look for the notorious food allergens (i.e., peanut, the grains, sea foods, milk, chocolate) to see who has them and to see where they came from. One might also look into the role food allergens might have with cases of food addiction (coffee, tobacco etc.).

NOTES:

TABLE 05.08.MEMBERS.08 JMRxRJR FAMILIAL DISTRIBUTION OF CYTOTOXIC TEST POSITIVES TO 70 FOODS.

A. Test Results Positive In Both Parents

Father	Mother	Daughter	Son
JMR	RJR	KKR	JSR
36	34	12	6
PEACH	PEACH		
VANILLA	VANILLA		VANILLA
CORN	CORN	CORN	CORN
ONION	ONION		ONION
LOBSTER	LOBSTER		LOBSTER
TUNA	TUNA		TUNA
PORK	PORK	PORK	PORK
CHICKEN	CHICKEN	CHICKEN	CHICKEN
COFFEE	COFFEE	COFFEE	COFFEE
PEPPERMINT	PEPPERMINT	PEPPERMINT	PEPPERMINT

B. Test Results Positive In Various Family Members

Father	Mother	Daughter	Son
JMR	RJR	KKR	JSR
36	34	12	6
APPLE		APPLE	APPLE
	BANANA	BANANA	BANANA
CANTALOUPE			CANTALOUPE
	CHERRY	CHERRY	
	HONEYDEW		HONEYDEW
ORANGE		ORANGE	ORANGE
	PINEAPPLE	PINEAPPLE	PINEAPPLE
	TOMATO	TOMATO	TOMATO
	OAT	OAT	OAT
RICE		RICE	RICE
	RYE	RYE	RYE
	WHEAT	WHEAT	WHEAT
		LETTUCE	LETTUCE
SOYBEAN		SOYBEAN	SOYBEAN
	BEET SUGAR	BEET SUGAR	BEET SUGAR
	CANE SUGAR	CANE SUGAR	CANE SUGAR
BEEF			BEEF
EGG		EGG	EGG
	MILK		MILK
	AMER CHEESE		AMER CHEESE
	CHOCOLATE	CHOCOLATE	CHOCOLATE
	TEA		TEA
TOBACCO			TOBACCO
	BAKE YEAST	BAKE YEAST	BAKE YEAST
	BREW YEAST		BREW YEAST

C. Test Results Negative In All Family Members

PEAR	LAMB
WATERMELON	VEAL
BARLEY	TURKEY, Lt
MALT	GARLIC
CABBAGE	GASOLINE
PEA	MUSHROOM
RADISH	MUSTARD
SPINACH	TURPENTINE
CRAB	1%ALCOHOL
SWORDFISH	2%ALCOHOL

D. Exclusive to Individual Family Members

Father	Mother	Daughter	Son
SHRIMP	BROCCOLI	CARROT	ROQUE CHEESE
STRINGBEAN	CARP	HOPS	SWISS CHEESE
	COTTONSEED		PEANUT
	CUCUMBER		
	HORSERADISH		
	POTATO		
	STRAWBERRY		
	TROUT		

CASE STUDY 05.08.KKRa.01-7 JMR x RJR (KKR): For the KKR case, the father (JMR) tested positive for 20 of 70 (28.6%), the mother (RJR) tested positive for 34 of 70 (48.6%) and the daughter (KKR) was positive for 24 of 70 (34.3%). The twelve-year-old KKR and her parents agreed, either positive or negative, on 28 of 70 (40.0%) of the items tested for. The father and mother agreed on 10 of 70 (14.3%) positives. Of the 10 positive items that both parents agreed on five of 10 (50.0%) were passed as positives to KKR. Of

KKR's 24 positives, 10 (41.7%) matched her father's positives, 16 (66.7%) matched her mom's positives and five (20.8%) matched both parents. The father exclusively matched on five (apple, orange, rice, soybean, and egg) of 24 (20.8%) of KKR's positives. The mother exclusively matched on 11 (banana, cherry, pineapple, tomato, oat, rye, wheat, beet sugar, cane sugar, chocolate, and baker's yeast) of 24 (45.8%) of KKR's positives. KKR had one test positive item (lettuce) that did not appear in either parent.

KKR was positive (Table 05.08.KKR.01) for six of 13 fruits (apple, banana, cherry, orange, pineapple ,and tomato). Apple, and orange was positive in KKR and her father, and the mother was positive for banana, cherry pineapple, and tomato. KKR tested positive for five (corn, oat, rice, rye, and wheat) of the seven grains tested (Table 05.07.KKR.02). Corn was positive in all family members tested; the mother was positive for five grains excluding only barley, malt, and rice with KKR. Of the vegetables (Table 05.08. KKR.03) tested for KKR, carrot, hops, lettuce, beet sugar and cane sugar were positive. KKR did not test positive for any of the fish and sea foods items (Table 05.08.KKR.04). Various sources of protein (Table 05.08.KKR.05) that tested positive for KKR included pork, chicken, and egg. Among the miscellaneous items tested for (Table 05.08.KKR.06) KKR was positive for chocolate, coffee, peppermint, baker's yeast. Both parents also agreed on coffee and peppermint where the mother tested positive for all four items.

TABLE 05.08.KKRa.01 JMRxRJR (KKR) FAMILIAL DISTRIBUTION OF CYTOTOXIC TEST POSITIVE SENSITIVITIES TO FRUITS.

	FATHER	MOTHER	DAUGHTER	%POSITIVE	%POSITIVE	%POSITIVE	%NEGATIVE	PER CENT OF POSITIVE FOODS			
SUBJECT	JMR	RJR	KKR	JMR/RJR	JMR	RJR	JMR/RJR		JMR/RJR	JMR	RJR
AGE TESTED	36	34	12	KKR	KKR	KKR	KKR	KKR	KKR	KKR	KKR
APPLE	1	0	1		1					1	
BANANA	0	1	1			1					1
CANTALOUPE	1	0	0								
CHERRY	0	1	1			1					1
HONEYDEW	0	1	0								
ORANGE	1	0	1		1					1	
PEACH	1	1	0								
PEAR	0	0	0				0				
PINEAPPLE	0	1	1			1					1
STRAWBERRY	0	1	0								
TOMATO	0	1	1			1					1
VANILLA	1	1	0								
WATERMELON	0	0	0				0				
===============											
POSITIVES	5	8	6	0	2	4	2	0	0	2	4
NUMBER (N)	13	13	13	13	13	13	13	6	6	6	6
% POSITIVE	38.5	61.5	46.2	0.0	15.4	30.8	15.4	0.0	0.0	33.3	66.7

TABLE 05.08.KKRa.02 JMRxRJR (KKR) FAMILIAL DISTRIBUTION OF CYTOTOXIC TEST POSITIVE SENSITIVITIES TO GRAINS.

	FATHER	MOTHER	DAUGHTER	%POSITIVE	%POSITIVE	%POSITIVE	%NEGATIVE	PER CENT OF POSITIVE FOODS			
SUBJECT	JMR	RJR	KKR	JMR/RJR	JMR	RJR	JMR/RJR		JMR/RJR	JMR	RJR
AGE TESTED	36	34	12	KKR	KKR	KKR	KKR	KKR	KKR	KKR	KKR
BARLEY	0	0	0				0				
CORN	1	1	1	1					1		
MALT	0	0	0				o				
OAT	0	1	1			1					1
RICE	1	0	1		1					1	
RYE	0	1	1			1					1
WHEAT	0	1	1			1					1
===============											
POSITIVES	2	4	5	1	1	3	2	0	1	1	3
NUMBER (N)	7	7	7	7	7	7	7	5	5	5	5
% POSITIVE	28.6	57.1	71.4	14.3	14.3	42.9	28.6	0.0	20.0	20.0	60.0

	FATHER	MOTHER	DAUGHTER	%POSITIVE	%POSITIVE	%POSITIVE	%NEGATIVE	PER CENT OF POSITIVE FOODS			
SUBJECT	JMR	RJR	KKR	JMR/RJR	JMR	RJR	JMR/RJR		JMR/RJR	JMR	RJR
AGE TESTED	36	34	12	KKR	KKR	KKR	KKR	KKR	KKR	KKR	KKR
BROCCOLI	0	1	0								
CABBAGE	0	0	0				0				
CARROT	0	0	1								
CUCUMBER	0	1	0								
HOPS	0	0	1								
HORSERADISH	0	1	0								
LETTUCE	0	0	1								
ONION	1	1	0								
PEA	0	0	0				0				
PEANUT	0	0	0				0				
POTATO	0	1	0								
RADISH	0	0	0				0				
SOYBEAN	1	0	1		1					1	
SPINACH	0	0	0				0				
STRINGBEAN	1	0	0								
SUGAR, BEET	0	1	1			1					1
SUGAR, CANE	0	1	1			1					1
===============											
POSITIVES	3	7	6	0	1	2	5	0	0	1	2
NUMBER (N)	17	17	17	17	17	17	17	6	6	6	6
% POSITIVE	17.6	41.2	35.3	0.0	5.9	11.8	29.4	0.0	0.0	16.7	33.3

TABLE 05.08.KKRa.04 JMRxRJR (KKR) FAMILIAL CYTOTOXIC TEST POSITIVE SENSITIVITIES TO FISH AND SEAFOOD.

	FATHER	MOTHER	DAUGHTER	%POSITIVE	%POSITIVE	%POSITIVE	%NEGATIVE	PER CENT OF POSITIVE FOODS			
SUBJECT	JMR	RJR	KKR	JMR/RJR	JMR	RJR	JMR/RJR		JMR/RJR	JMR	RJR
AGE TESTED	36	34	12	KKR	KKR	KKR	KKR	KKR	KKR	KKR	KKR
CARP	0	1	0								
CRAB	0	0	0				0				
LOBSTER	1	1	0								
SHRIMP	1	0	0								
SWORDFISH	0	0	0				0				
TROUT	0	1	0								
TUNA	1	1	0								
===============											
POSITIVES	3	4	0	0	0	0	2	0	0	0	0
NUMBER (N)	7	7	7	7	7	7	7	0	0	0	0
% POSITIVE	42.9	57.1	0.0	0.0	0.0	0.0	28.6	0.0	0.0	0.0	0.0

TABLE 05.08.KKRa.05 JMRxRJR (KKR) FAMILIAL CYTOTOXIC TEST POSITIVE SENSITIVITIES TO MEAT, FOWL, EGG & MILK.

	FATHER	MOTHER	DAUGHTER	%POSITIVE	%POSITIVE	%POSITIVE	%NEGATIVE	PER CENT OF POSITIVE FOODS			
SUBJECT	JMR	RJR	KKR	JMR/RJR	JMR	RJR	JMR/RJR		JMR/RJR	JMR	RJR
AGE TESTED	36	34	12	KKR	KKR	KKR	KKR	KKR	KKR	KKR	KKR
BEEF	1	0	0								
LAMB	0	0	0				0				
PORK	1	1	1	1					1		
VEAL	0	0	0				0				
CHICKEN	1	1	1	1					1		
TURKEY, Lt.	0	0	0				0				
EGG	1	0	1		1					1	
MILK	0	1	0								
CHEESE/AMER.	0	1	0								
CHEESE/ROQUE.	0	0	0				0				
CHEESE/SWISS	0	0	0				0				
===============											
POSITIVES	4	4	3	2	1	0	5	0	2	1	0
NUMBER (N)	11	11	11	11	11	11	11	3	3	3	3
% POSITIVE	36.4	36.4	27.3	18.2	9.1	0.0	45.5	0.0	66.7	33.3	0.0

TABLE 05.08.KKRa.06 JMRxRJR (KKR) FAMILIAL CYTOTOXIC TEST POSITIVE SENSITIVITIES TO MISCELLANEOUS ITEMS.

	FATHER	MOTHER	DAUGHTER	%POSITIVE	%POSITIVE	%POSITIVE	%NEGATIVE	PER CENT OF POSITIVE FOODS			
SUBJECT	JMR	RJR	KKR	JMR/RJR	JMR	RJR	JMR/RJR		JMR/RJR	JMR	RJR
AGE	36	34	12	KKR	KKR	KKR	KKR	KKR	KKR	KKR	KKR
CHOCOLATE	0	1	1			1					1
COFFEE	1	1	1	1					1		
COTTONSEED	0	1	0								
GARLIC	0	0	0				0				
GASOLINE	0	0	0				0				
MUSHROOM	0	0	0				0				
MUSTARD	0	0	0				0				
PEPPERMINT	1	1	1	1					1		
TEA	0	1	0								
TOBACCO	1	0	0								
TURPENTINE	0	0	0				0				
YEAST, BAKER	0	1	1			1					1
YEAST, BREWER	0	1	0								
ALCOHOL, 1%	0	0	0				0				
ALCOHOL, 2%	0	0	0				0				
===============											
POSITIVES	3	7	4	2	0	2	7	0	2	0	2
NUMBER (N)	15	15	15	15	15	15	15	4	4	4	4
% POSITIVE	20.0	46.7	26.7	13.3	0.0	13.3	46.7	0.0	50.0	0.0	50.0

TABLE 05.08.KKRa.07 SUMMARY: JMRXRJR (KKR) FAMILIAL CYTOTOXIC TEST POSITIVE FOOD SENSITIVITIES.

	FATHER	MOTHER	DAUGHTER	%POSITIVE	%POSITIVE	%POSITIVE	%NEGATIVE	PER CENT OF POSITIVE FOODS			
SUBJECT	JMR	RJR	KKR	JMR/RJR	JMR	RJR	JMR/RJR		JMR/RJR	JMR	RJR
AGE	36	34	12	KKR	KKR	KKR	KKR	KKR	KKR	KKR	KKR
POSITIVES	20	34	24	5	5	11	23	0	5	5	11
NUMBER (N)	70	70	70	70	70	70	70	24	24	24	24
% POSITIVE	28.6	48.6	34.3	7.1	7.1	15.7	32.9	0.0	20.8	20.8	45.8

Of the ten items that both parents agreed on five (50.0%) appeared in the daughter. The father and daughter agreed on ten (14.3%) of 70 items tested for, whereas the mother agreed with the daughter on 16 (22.9%) of 70 items. In the 16 to 10 ratio (1.6) there appears to be a "donor" difference between the father and mother, however if you factor in what positive items each parent has to "offer" there is a 34 to 20 ratio (1.7) to balance out what appears to be that close difference. It appears that genetically they are proportionally passing what they got!

CASE STUDY 05.08.JSRb.01-7. JMR x RJR (JSR): For the JSR case, the father (JMR) tested positive for 20 of 70 (28.6%), the mother (RJR) tested positive for 34 of 70 (48.6%), the sister (KKR) was positive for 24 of 70 (34.3%), and the son (JSR) tested positive for 36 of 70 potential allergens (51.4%). Six-year-old JSR and his parents (Tables 05.08.JSR.1-6) agreed, either positive or negative, on 31 of 70 (44.3%) of the items tested for. The father and mother agreed on 10 of 70 (14.3%) positives. Of the 10 positive items that both parents agreed on nine of 10 (90.0%) were passed as positives to JSR! Of JSR's 36 positives, 17 (47.2%) matched his father's positives, 24 (66.7%) matched his mom's positives and nine (25.0%) matched both parents. The father exclusively matched on eight (apple, cantaloupe, orange, rice, soybean, beef, egg, and tobacco) of 36 (22.2%) of JSR's positives. The mother exclusively matched on 15 (banana, honeydew, pineapple, tomato, oat, rye, wheat, beet sugar, cane sugar, milk, American cheese, chocolate, tea, baker's yeast, and brewer's yeast) of 36 (41.7%) of GPS's positives. JSR had four (11.1%) of 36 test positive items (lettuce, peanut, Roquefort cheese and Swiss cheese) that did not appear in either parent. JSR was positive (Table 05.08.JSR.01) for eight of 13 fruits (apple, banana, cantaloupe, honeydew, orange, pineapple, tomato, and vanilla. Vanilla was positive in all family members tested. Apple and orange were positive in JSR and his father, and the mother was positive for banana, honeydew, pineapple, and tomato. JSR tested positive for five (corn, oat, rice, rye, and wheat) of the seven grains tested (Table 05.08.JSR.02). Corn was positive in all family members. The mother matched JSR on four of five grains excluding only rice. Of the vegetables (Table 05.08.JSR.03) tested for JSR six of 17 were positive and only three matched with the mother. JSR tested positive for lobster and tuna matching both parents (Table 05.08.JSR.04). Various sources of protein (Table 05.08.JSR.05) that tested positive for JSR included beef, pork, chicken, egg, milk, American cheese, Roquefort cheese and Swiss cheese. Among the miscellaneous items tested for (Table 05.08.JSR.06) JSR was positive for chocolate, coffee, peppermint tea, tobacco, baker's yeast, and brewer's yeast. The mother matched all but tea, which the father did match. Coffee and peppermint were positive in all family members. The diet management program was highly successful for JSR.

TABLE 05.08.JSRb.01. JMRXRJR (JSR) FAMILIAL DISTRIBUTION OF CYTOTOXIC TEST POSITIVE SENSITIVITIES TO FRUITS.

	FATHER	MOTHER	SON	%POSITIVE	%POSITIVE	%POSITIVE	%NEGATIVE	PER CENT OF POSITIVE FOODS			
SUBJECT	JMR	RJR	JSR	JMR/RJR	JMR	RJR	JMR/RJR		JMR/RJR	JMR	RJR
AGE TESTED	36	34	6	JSR	JSR	JSR	JSR	JSR	JSR	JSR	JSR
APPLE	1	0	1		1					1	
BANANA	0	1	1			1					1
CANTALOUPE	1	0	1		1					1	
CHERRY	0	1	0								
HONEYDEW	0	1	1			1					1
ORANGE	1	0	1		1					1	
PEACH	1	1	0								
PEAR	0	0	0				0				
PINEAPPLE	0	1	1			1					1
STRAWBERRY	0	1	0								
TOMATO	0	1	1			1					1
VANILLA	1	1	1	1					1		
WATERMELON	0	0	0				0				
==============											
POSITIVES	5	8	8	1	3	4	2	0	1	3	4
NUMBER (N)	13	13	13	13	13	13	13	8	8	8	8
% POSITIVE	38.5	61.5	61.5	7.7	23.1	30.8	15.4	0.0	12.5	37.5	50.0

TABLE 05.08.JSRb.02. JMRXRJR (JSR) FAMILIAL DISTRIBUTION OF CYTOTOXIC TEST POSITIVE SENSITIVITIES TO GRAINS.

	FATHER	MOTHER	SON	%POSITIVE	%POSITIVE	%POSITIVE	%NEGATIVE	PER CENT OF POSITIVE FOODS			
SUBJECT	JMR	RJR	JSR	JMR/RJR	JMR	RJR	JMR/RJR		JMR/RJR	JMR	RJR
AGE TESTED	36	34	6	JSR	JSR	JSR	JSR	JSR	JSR	JSR	JSR
BARLEY	0	0	0				0				
CORN	1	1	1	1					1		
MALT	0	0	0				0				
OAT	0	1	1			1					1
RICE	1	0	1		1					1	
RYE	0	1	1			1					1
WHEAT	0	1	1			1					1
===============											
POSITIVES	2	4	5	1	1	3	2	0	1	1	3
NUMBER (N)	7	7	7	7	7	7	7	5	5	5	5
% POSITIVE	28.6	57.1	71.4	14.3	14.3	42.9	28.6	0.0	20.0	20.0	60.0

TABLE 05.08.JSRb.03. JMRXRJR (JSR) FAMILIAL CYTOTOXIC TEST POSITIVE SENSITIVITIES TO VEGETABLES.

	FATHER	MOTHER	SON	%POSITIVE	%POSITIVE	%POSITIVE	%NEGATIVE	PER CENT OF POSITIVE FOODS			
SUBJECT	JMR	RJR	JSR	JMR/RJR	JMR	RJR	JMR/RJR		JMR/RJR	JMR	RJR
AGE TESTED	36	34	6	JSR	JSR	JSR	JSR	JSR	JSR	JSR	JSR
BROCCOLI	0	1	0								
CABBAGE	0	0	0				0				
CARROT	0	0	0				0				
CUCUMBER	0	1	0								
HOPS	0	0	0				0				
HORSERADISH	0	1	0								
LETTUCE	0	0	1					1			
ONION	1	1	1	1					1		
PEA	0	0	0				0				
PEANUT	0	0	1					1			
POTATO	0	1	0								
RADISH	0	0	0				0				
SOYBEAN	1	0	1		1					1	
SPINACH	0	0	0				0				
STRINGBEAN	1	0	0								
SUGAR, BEET	0	1	1			1					1
SUGAR, CANE	0	1	1			1					1
===============											
POSITIVES	3	7	6	1	1	2	6	2	1	1	2
NUMBER (N)	17	17	17	17	17	17	17	6	6	6	6
% POSITIVE	17.6	41.2	35.3	5.9	5.9	11.8	35.3	33.3	16.7	16.7	33.3

TABLE 05.08.JSRb.04. JMRXRJR (JSR) FAMILIAL CYTOTOXIC TEST POSITIVE SENSITIVITIES TO FISH & SEAFOOD.

	FATHER	MOTHER	SON	%POSITIVE	%POSITIVE	%POSITIVE	%NEGATIVE	PER CENT OF POSITIVE FOODS			
SUBJECT	JMR	RJR	JSR	JMR/RJR	JMR	RJR	JMR/RJR		JMR/RJR	JMR	RJR
AGE TESTED	36	34	6	JSR	JSR	JSR	JSR	JSR	JSR	JSR	JSR
CARP	0	1	0								
CRAB	0	0	0				0				
LOBSTER	1	1	1	1					1		
SHRIMP	1	0	0								
SWORDFISH	0	0	0				0				
TROUT	0	1	0								
TUNA	1	1	1	1					1		
===============											
POSITIVES	3	4	2	2	0	0	2	0	2	0	0
NUMBER (N)	7	7	7	7	7	7	7	2	2	2	2
% POSITIVE	42.9	57.1	28.6	28.6	0.0	0.0	28.6	0.0	100.0	0.0	0.0

TABLE 05.08.JSRb.05. JMRxRJR (JSR) FAMILIAL CYTOTOXIC TEST POSITIVE SENSITIVITIES TO MEAT, FOWL, EGG & MILK.

	FATHER	MOTHER	SON	%POSITIVE	%POSITIVE	%POSITIVE	%NEGATIVE	PER CENT OF POSITIVE FOODS			
SUBJECT	JMR	RJR	JSR	JMR/RJR	JMR	RJR	JMR/RJR		JMR/RJR	JMR	RJR
AGE TESTED	36	34	6	JSR	JSR	JSR	JSR	JSR	JSR	JSR	JSR
BEEF	1	0	1		1					1	
LAMB	0	0	0				0				
PORK	1	1	1	1					1		
VEAL	0	0	0				0				
CHICKEN	1	1	1	1					1		
TURKEY, Lt.	0	0	0				0				
EGG	1	0	1		1					1	
MILK	0	1	1			1					1
CHEESE/AMER.	0	1	1			1					1
CHEESE/ROQUE.	0	0	1					1			
CHEESE/SWISS	0	0	1					1			
===============											
POSITIVES	4	4	8	2	2	2	3	2	2	2	2
NUMBER (N)	11	11	11	11	11	11	11	8	8	8	8
% POSITIVE	36.4	36.4	72.7	18.2	18.2	18.2	27.3	25.0	25.0	25.0	25.0

TABLE 05.08.JSRb.06. JMRxRJR (JSR) FAMILIAL CYTOTOXIC TEST POSITIVE SENSITIVITIES TO MISCELLANEOUS ITEMS.

	FATHER	MOTHER	SON	%POSITIVE	%POSITIVE	%POSITIVE	%NEGATIVE	PER CENT OF POSITIVE FOODS			
SUBJECT	JMR	RJR	JSR	JMR/RJR	JMR	RJR	JMR/RJR		JMR/RJR	JMR	RJR
AGE	36	34	6	JSR	JSR	JSR	JSR	JSR	JSR	JSR	JSR
CHOCOLATE	0	1	1			1					1
COFFEE	1	1	1	1					1		
COTTONSEED	0	1	0								
GARLIC	0	0	0				0				
GASOLINE	0	0	0				0				
MUSHROOM	0	0	0				0				
MUSTARD	0	0	0				0				
PEPPERMINT	1	1	1	1					1		
TEA	0	1	1			1					1
TOBACCO	1	0	1		1					1	
TURPENTINE	0	0	0				0				
YEAST, BAKER	0	1	1			1					1
YEAST, BREWER	0	1	1			1					1
ALCOHOL, 1%	0	0	0				0				
ALCOHOL, 2%	0	0	0				0				
===============											
POSITIVES	3	7	7	2	1	4	7	0	2	1	4
NUMBER (N)	15	15	15	15	15	15	15	7	7	7	7
% POSITIVE	20.0	46.7	46.7	13.3	6.7	26.7	46.7	0.0	28.6	14.3	57.1

TABLE 05.08.JSRb.07. SUMMARY: JMRxRJR (JSR) FAMILIAL CYTOTOXIC TEST POSITIVE FOOD SENSITIVITIES.

	FATHER	MOTHER	SON	%POSITIVE	%POSITIVE	%POSITIVE	%NEGATIVE	PER CENT OF POSITIVE FOODS			
SUBJECT	JMR	RJR	JSR	JMR/RJR	JMR	RJR	JMR/RJR		JMR/RJR	JMR	RJR
AGE	36	34	6	JSR	JSR	JSR	JSR	JSR	JSR	JSR	JSR
POSITIVES	20	34	36	9	8	15	22	4	9	8	15
NUMBER (N)	70	70	70	70	70	70	70	36	36	36	36
% POSITIVE	28.6	48.6	51.4	12.9	11.4	21.4	31.4	11.1	25.0	22.2	41.7

Of the ten items that both parents agreed on nine (90.0%) appeared in the son. Collectively, the father and son agreed on 17 (24.3%) of 70 items tested for, whereas the mother and son agreed on 25 (35.7%) of 70 items. In the 25 to 17 ratio (1.47) there appears to be a "donor" difference between the father and mother that is less than we saw with their daughter (1.60). Individually, the father and son agreed on 8 (11.4%) of 70 items tested for, whereas the mother and son agreed on 15 (21.4%) of 70 items. In the 15 to 8 ratio (1.88) there appears to be a "donor" advantage for the mother that was less than we saw in their daughter (1.22). If you factor in what positive items each parent has to "offer" there is a 34 to 20 ratio (1.70) that somewhat off sets the mother matching up with the son more frequently than the father. Proportionally each parent appears to be passing what they got. Simply stated, individually the mother has more a few more allergens to "offer", advantage mother. Remember (Ulett and Perry: 1974, 1975) there are three steps to manifesting an allergy (having the trait, initiating immunochemical steps, and then manifesting a response in an end organ such as skin, cardiovascular system, intestine, or nervous system).

CASE STUDY 05.09.EJG.01-7 LRG x SNG (EJG) Newborn: For the EJG case Table 05.09.EJG.07 summarizes (Tables 05.09.EJG.01-6) the totals for all the items tested for EJG. The father (LRG) tested positive for 23 of 70 (32.9%), the mother (SNG) tested positive for 49 of 70 (70.0%) and the newborn daughter (EJG) was already positive for 19 of 70 (27.1%). Newborn EJG and her parents agreed, either positive or negative, on 23 of 70 (32.9%) of the items tested for. The father and mother agreed on 19 of 70 (27.1%) positives. Of the 19 positive items that both parents agreed on seven of 19 (36.8%) were passed as positives to EJG. Of EJG's 19 positives, eight (42.1%) matched her father's positives, 17 (89.5%) matched her mom's positives and seven (36.8%) matched both parents. The father exclusively matched on only one (spinach) of 19 (5.3%) of EJG's positives. The mother exclusively matched on 10 (apple, cantaloupe, pineapple, malt, rice, broccoli, hops, soybean, beet sugar, and Swiss cheese) of 19 (52.6%) of EJG's positives. EJG had only one (string bean) of 19 (5.3%) test positive items that were negative or did not appear in either parent. The newborn EJG was Cytotoxic Test positive for 19 of 70 (27.1%) determined from a chord blood sample taken at the time of birth. EJG was positive for four fruits (Table 5.8.1.1), apple, banana, cantaloupe, and pineapple. Banana was positive in EJG and both parents, and the mother was positive for all four fruits. EJG tested positive for three (corn, malt, and rice) of the seven grains tested (Table 5.8.1.2). Corn was positive in all family members tested; the mother was positive for all three grains. Of the vegetables tested for EJG broccoli, hops, soybean, and beet sugar were positive. The mother tested positive for all four items. EJG did not test positive for any of the tested fish and sea foods. Various sources of protein (Table 5.8.1.5) that tested positive for EJG included pork, egg, milk, American cheese, Roquefort cheese and Swiss cheese. The mother tested positive for all of these items as did the father with the exception of Swiss cheese. Among the miscellaneous items tested for (Table 5.8.1.6) EJG was negative for all items where the mother tested positive for 13 (86.7%) of 15 and the father was positive for only two (13.3%) of 15 items.

TABLE 05.09.EJG.01 LRGXSNG (EJG) FAMILIAL DISTRIBUTION OF CYTOTOXIC TEST POSITIVE SENSITIVITIES TO FRUITS.

	FATHER	MOTHER	DAUGHTER	%POSITIVE	%POSITIVE	%POSITIVE	%NEGATIVE	PER CENT OF POSITIVE FOODS			
SUBJECT	LRG	SNG	EJG	LRG/SNG	LRG	SNG	LRG/SNG		LRG/SNG	LRG	SNG
AGE TESTED	35	26	NEWBORN	EJG	EJG	EJG	EJG	EJG	EJG	EJG	EJG
APPLE	0	1	1			1					1
BANANA	1	1	1	1					1		
CANTALOUPE	0	1	1			1					1
CHERRY	0	0	0				0				
HONEYDEW	1	0	0								
ORANGE	1	1	0								
PEACH	0	1	0								
PEAR	0	0	0				0				
PINEAPPLE	0	1	1			1					1
STRAWBERRY	1	1	0								
TOMATO	1	1	0								
VANILLA	1	1	0								
WATERMELON	0	0	0				0				
===============											
POSITIVES	6	9	4	1	0	3	3	0	1	0	3
NUMBER (N)	13	13	13	13	13	13	13	4	4	4	4
% POSITIVE	46.2	69.2	30.8	7.7	0.0	23.1	23.1	0.0	25.0	0.0	75.0

TABLE 05.09.EJG.02 LRGXSNG (EJG) FAMILIAL DISTRIBUTION OF CYTOTOXIC TEST POSITIVE SENSITIVITIES TO GRAINS.

	FATHER	MOTHER	DAUGHTER	%POSITIVE	%POSITIVE	%POSITIVE	%NEGATIVE	PER CENT OF POSITIVE FOODS			
SUBJECT	LRG	SNG	EJG	LRG/SNG	LRG	SNG	LRG/SNG		LRG/SNG	LRG	SNG
AGE TESTED	35	26	NEWBORN	EJG	EJG	EJG	EJG	EJG	EJG	EJG	EJG
BARLEY	1	1	0								
CORN	1	1	1	1					1		
MALT	0	1	1			1					1
OAT	1	1	0								
RICE	0	1	1			1					1
RYE	0	1	0								
WHEAT	0	1	0								
===============											
POSITIVES	3	7	3	1	0	2	0	0	1	0	2
NUMBER (N)	7	7	7	7	7	7	7	3	3	3	3
% POSITIVE	42.9	100.0	42.9	14.3	0.0	28.6	0.0	0.0	33.3	0.0	66.7

TABLE 05.09.EJG.03 LRGxSNG (EJG) FAMILIAL DISTRIBUTION OF CYTOTOXIC TEST POSITIVE SENSITIVITIES TO VEGETABLES.

	FATHER	MOTHER	DAUGHTER	%POSITIVE	%POSITIVE	%POSITIVE	%NEGATIVE	PER CENT OF POSITIVE FOODS			
SUBJECT	LRG	SNG	EJG	LRG/SNG	LRG	SNG	LRG/SNG		LRG/SNG	LRG	SNG
AGE TESTED	35	26	NEWBORN	EJG	EJG	EJG	EJG	EJG	EJG	EJG	EJG
BROCCOLI	0	1	1			1					1
CABBAGE	1	0	0								
CARROT	1	1	0								
CUCUMBER	0	0	0				0				
HOPS	0	1	1			1					1
HORSERADISH	0	0	0				0				
LETTUCE	0	1	0								
ONION	0	1	0								
PEA	0	1	0								
PEANUT	1	1	0								
POTATO	1	1	0								
RADISH	0	0	0				0				
SOYBEAN	0	1	1			1					1
SPINACH	1	0	1		1					1	
STRINGBEAN	0	0	1					1			
SUGAR, BEET	0	1	1			1					1
SUGAR, CANE	1	1	0								
===============											
POSITIVES	6	11	6	0	1	4	3	1	0	1	4
NUMBER (N)	17	17	17	17	17	17	17	6	6	6	6
% POSITIVE	35.3	64.7	35.3	0.0	5.9	23.5	17.6	16.7	0.0	16.7	66.7

TABLE 05.09.EJG.04 LRGxSNG (EJG) FAMILIAL CYTOTOXIC TEST POSITIVE SENSITIVITIES TO VARIOUS FISH & SEAFOOD.

	FATHER	MOTHER	DAUGHTER	%POSITIVE	%POSITIVE	%POSITIVE	%NEGATIVE	PER CENT OF POSITIVE FOODS			
SUBJECT	LRG	SNG	EJG	LRG/SNG	LRG	SNG	LRG/SNG		LRG/SNG	LRG	SNG
AGE TESTED	35	26	NEWBORN	EJG	EJG	EJG	EJG	EJG	EJG	EJG	EJG
CARP	0	0	0				0				
CRAB	0	0	0				0				
LOBSTER	0	1	0								
SHRIMP	0	1	0								
SWORDFISH	0	0	0				0				
TROUT	1	0	0								
TUNA	0	0	0				0				
===============											
POSITIVES	1	2	0	0	0	0	4	0	0	0	0
NUMBER (N)	7	7	7	7	7	7	7	0	0	0	0
% POSITIVE	14.3	28.6	0.0	0.0	0.0	0.0	57.1	0.0	0.0	0.0	0.0

TABLE 05.09.EJG.05 LRGxSNG (EJG) FAMILIAL CYTOTOXIC TEST POSITIVE SENSITIVITIES TO MEAT, POULTRY & DAIRY.

	FATHER	MOTHER	DAUGHTER	%POSITIVE	%POSITIVE	%POSITIVE	%NEGATIVE	PER CENT OF POSITIVE FOODS			
SUBJECT	LRG	SNG	EJG	LRG/SNG	LRG	SNG	LRG/SNG		LRG/SNG	LRG	SNG
AGE TESTED	35	26	NEWBORN	EJG	EJG	EJG	EJG	EJG	EJG	EJG	EJG
BEEF	0	0	0				0				
LAMB	0	0	0				0				
PORK	1	1	1	1					1		
VEAL	0	0	0				0				
CHICKEN	0	1	0								
TURKEY, Lt.	0	0	0				0				
EGG	1	1	1	1					1		
MILK	1	1	1	1					1		
CHEESE/AMER.	1	1	1	1					1		
CHEESE/ROQUE.	1	1	1	1					1		
CHEESE/SWISS	0	1	1			1					1
===============											
POSITIVES	5	7	6	5	0	1	4	0	5	0	1
NUMBER (N)	11	11	11	11	11	11	11	6	6	6	6
% POSITIVE	45.5	63.6	54.5	45.5	0.0	9.1	36.4	0.0	83.3	0.0	16.7

TABLE 05.09.EJG.06 LRGxSNG (EJG) FAMILIAL CYTOTOXIC TEST POSITIVE SENSITIVITIES TO MISCELLANEOUS ITEMS.

	FATHER	MOTHER	DAUGHTER	%POSITIVE	%POSITIVE	%POSITIVE	%NEGATIVE	PER CENT OF POSITIVE FOODS			
SUBJECT	LRG	SNG	EJG	LRG/SNG	LRG	SNG	LRG/SNG		LRG/SNG	LRG	SNG
AGE	35	26	NEWBORN	EJG	EJG	EJG	EJG	EJG	EJG	EJG	EJG
ALCOHOL, 1%	0	0	0				0				
ALCOHOL, 2%	0	0	0				0				
CHOCOLATE	1	1	0								
COFFEE	1	1	0								
COTTONSEED	0	1	0								
GARLIC	0	1	0								
GASOLINE	0	1	0								
MUSHROOM	0	1	0								
MUSTARD	0	1	0								
PEPPERMINT	0	1	0								
TEA	0	1	0								
TOBACCO	0	1	0								
TURPENTINE	0	1	0								
YEAST, BAKER	0	1	0								
YEAST, BREWER	0	1	0								
===============											
POSITIVES	2	13	0	0	0	0	2	0	0	0	0
NUMBER (N)	15	15	15	15	15	15	15	0	0	0	0
% POSITIVE	13.3	86.7	0.0	0.0	0.0	0.0	13.3	0.0	0.0	0.0	0.0

TABLE 05.09.EJG.07 SUMMARY: LRGxSNG (EJG) FAMILIAL DISTRIBUTION OF CYTOTOXIC TEST POSITIVE FOOD SENSITIVITIES.

	FATHER	MOTHER	DAUGHTER	%POSITIVE	%POSITIVE	%POSITIVE	%NEGATIVE	PER CENT OF POSITIVE FOODS			
SUBJECT	LRG	SNG	EJG	LRG/SNG	LRG	SNG	LRG/SNG		LRG/SNG	LRG	SNG
AGE TESTED	35	26	NEWBORN	EJG	EJG	EJG	EJG	EJG	EJG	EJG	EJG
TOTAL POSITIVES	23	49	19	7	1	10	16	1	7	1	10
NUMBER (N)	70	70	70	70	70	70	70	19	19	19	19
% POSITIVE	32.9	70.0	27.1	10.0	1.4	14.3	22.9	5.3	36.8	5.3	52.6

As with Case Study 05.01.SGP.01-7 SGP x CAL-P (SSP), newborn male, the unique information here is the pre-partum timing of test positive data. Conventional wisdom says the placental barrier protects the fetus from dietary antigens; therefore, food sensitivities appear in the sequence of infant diet regimes including maternal breast milk and passive transfer of maternal immune system benefits. This newborn had not yet received post-partum nourishment of any kind. Bear in mind (Chapter II.) that a positive test here reads the immune-defense system's white plasma cells and is presented here as only the first of three steps required for an immune or allergic manifestation. This allergy scenario is placed in the midst of detailed and delicate developmental processes. Speculating for a minute that any single positive sensitizing foods react in a dose response dependent way (Chapter 3.0 Figure 3.2) potentially on susceptible end organ tissue opens the door to a model for studying a host of clinical maladies and even birth defects. Any of the above positive antigen(s) displays as a potential sigmoid dose response where very small amounts could deliver a huge blow whether it be an untimely, even localized, anaphylaxis or a very specific interference at the end organ that results in some or significant impairment.

CASE STUDY 05.10.SBS.01-7 TBS x KMS (SBS): Three-year-old SBS and his parents agreed, either positive or negative, on 37 of 70 (52.9%) of the items tested for. The father and mother agreed on only three of 70 (4.3%) positives. Of the three positive items that both parents agreed on, only two (cabbage and carrot) of three (66.7%) were passed as positives to SBS. Of SBS's 27 positives, nine (33.3%) matched his father's positives, 12 (44.4%) matched his mom's positives and only two (7.4%) matched both parents. The father exclusively matched on seven (vanilla, oat, hops, peanut, soybean, chicken, and Swiss cheese) of 27 (25.9%) of SBS's positives. The mother exclusively matched on 10 (banana, cherry, pineapple, tomato, oat, rye, wheat, beet sugar, cane sugar, and baker's yeast) of 27 (37.0%) of SBS's positives. SBS had eight (29.6%) of 27 test positive items (cherry, strawberry, malt, milk, Roquefort cheese, baker's yeast, and 1-2% alcohol) that did not appear in either parent.

Both parents exhibited a relatively low level of sensitivity. The father (TBS) tested positive for 12 of 70 (17.1%), the mother (KMS) tested positive for 18 of 70 (25.7%), and SBS was positive for 27 of 70 (38.6%) items that they were tested for. SBS exhibited a moderate level of sensitivity (ranked 9th of 15 off-springs in this study) in spite of both parents agreeing on only two test positive items passed to SBS. Father listed no complaints or treatment for allergy, while the mother did list complaints and has received seasonal allergy shots every week during the fall and every three weeks during the winter. She was also skin tested for mold spores, grasses, trees, and house dust, but not tested for animals or foods. She continued to suffer from "Hay fever" symptoms. SBS was being seen by a Doctor for a host of symptoms that refused to go away. These test results were sent to his office to be used as part of a diet and allergen management program. The recommendation is to specifically avoid all test positive food items as well as foods that were not tested for, and feed only foods that were test negative for a period of two weeks. Generally, this results in the disappearance of many undesirable symptoms as well as a bonus loss of undesirable weight and acne. Given success then foods are added back gradually looking for the offending food(s).

TABLE 05.10.SBS.01 TBSxKMS (SBS) FAMILIAL DISTRIBUTION OF CYTOTOXIC TEST POSITIVE SENSITIVITIES TO FRUITS.

	FATHER	MOTHER	SON	%POSITIVE	%POSITIVE	%POSITIVE	%NEGATIVE	PER CENT OF POSITIVE FOODS			
SUBJECT	TBS	KMS	SBS	TBS/KMS	TSB	KMS	TBS/KMS		TBS/KMS	TSB	KMS
AGE TESTED	28	25	3	SBS	SBS	SBS	SBS	SBS	SBS	SBS	SBS
APPLE	0	1	1			1					1
BANANA	0	0	0				0				
CANTALOUPE	0	0	0				0				
CHERRY	0	0	1					1			
HONEYDEW	0	0	0				0				
ORANGE	0	1	1			1					1
PEACH	0	0	0				0				
PEAR	0	0	0				0				
PINEAPPLE	1	1	0								
STRAWBERRY	0	0	1					1			
TOMATO	0	0	0				0				
VANILLA	1	0	1		1					1	
WATERMELON	0	0	0				0				
===============											
POSITIVES	2	3	5	0	1	2	7	2	0	1	2
NUMBER (N)	13	13	13	13	13	13	13	5	5	5	5
% POSITIVE	15.4	23.1	38.5	0.0	7.7	15.4	53.8	40.0	0.0	20.0	40.0

TABLE 05.10.SBS.02 TBSxKMS (SBS) FAMILIAL DISTRIBUTION OF CYTOTOXIC TEST POSITIVE SENSITIVITIES TO GRAINS.

								PER CENT OF POSITIV FOODS			
SUBJECT	TBS	KMS	SBS	TBS/KMS	TSB	KMS	TBS/KMS		TBS/KMS	TSB	KMS
AGE TESTED	28	25	3	SBS	SBS	SBS	SBS	SBS	SBS	SBS	SBS
BARLEY	0	0	0				0				
CORN	0	1	1			1					1
MALT	0	0	1					1			
OAT	1	0	1		1					1	
RICE	0	1	0								
RYE	0	1	1			1					1
WHEAT	0	0	0				0				
===============											
POSITIVES	1	3	4	0	1	2	2	1	0	1	2
NUMBER (N)	7	7	7	7	7	7	7	4	4	4	4
% POSITIVE	14.3	42.9	57.1	0.0	14.3	28.6	28.6	25.0	0.0	25.0	50.0

TABLE 05.10.SBS.03 TBSxKMS (SBS) FAMILIAL DISTRIBUTION OF CYTOTOXIC TEST POSITIVE SENSITIVITIES TO VEGETABLES.

	FATHER	MOTHER	SON	%POSITIVE	%POSITIVE	%POSITIVE	%NEGATIVE	PER CENT OF POSITIVE FOODS			
SUBJECT	TBS	KMS	SBS	TBS/KMS	TSB	KMS	TBS/KMS		TBS/KMS	TSB	KMS
AGE TESTED	28	25	3	SBS	SBS	SBS	SBS	SBS	SBS	SBS	SBS
BROCCOLI	0	0	0				0				
CABBAGE	1	1	1	1					1		
CARROT	1	1	1	1					1		
CUCUMBER	0	0	0				0				
HOPS	1	0	1		1					1	
HORSERADISH	0	0	0				0				
LETTUCE	0	0	0				0				
ONION	0	0	0				0				
PEA	0	0	0				0				
PEANUT	1	0	1		1					1	
POTATO	0	1	0								
RADISH	0	0	0				0				
SOYBEAN	1	0	1		1					1	
SPINACH	0	0	0				0				
STRINGBEAN	0	0	0				0				
SUGAR, BEET	0	0	0				0				
SUGAR, CANE	0	1	1			1					1
===============											
POSITIVES	5	4	6	2	3	1	10	0	2	3	1
NUMBER (N)	17	17	17	17	17	17	17	6	6	6	6
% POSITIVE	29.4	23.5	35.3	11.8	17.6	5.9	58.8	0.0	33.3	50.0	16.7

TABLE 05.10.SBS.04 TBSxKMS (SBS) FAMILIAL CYTOTOXIC TEST POSITIVE SENSITIVITIES TO VARIOUS FISH & SEAFOOD.

	FATHER	MOTHER	SON	%POSITIVE	%POSITIVE	%POSITIVE	%NEGATIVE	PER CENT OF POSITIVE FOODS			
SUBJECT	TBS	KMS	SBS	TBS/KMS	TBS	KMS	TBS/KMS		TBS/KMS	TBS	KMS
AGE TESTED	28	25	3	SBS	SBS	SBS	SBS	SBS	SBS	SBS	SBS
CARP	0	0	0				0				
CRAB	0	0	0				0				
LOBSTER	1	0	0								
SHRIMP	0	0	0								
SWORDFISH	0	0	0				0				
TROUT	0	0	0				0				
TUNA	1	0	0				0				
===============											
POSITIVES	2	0	0	0	0	0	5		0		0
NUMBER (N)	7	7	7	7	7	7	7	0	0	0	0
% POSITIVE	28.6	0.0	0.0	0.0	0.0	0.0	71.4	0.0	0.0	0.0	0.0

TABLE 05.10.SBS.05 TBSxKMS (SBS) FAMILIAL CYTOTOXIC TEST POSITIVE SENSITIVITIES TO MEAT, POULTRY & DAIRY.

	FATHER	MOTHER	SON	%POSITIVE	%POSITIVE	%POSITIVE	%NEGATIVE	PER CENT OF POSITIVE FOODS			
SUBJECT	TBS	KMS	SBS	LRG/SNG	TBS	KMS	TBS/KMS		TBS/KMS	TBS	KMS
AGE TESTED	28	25	3	SBS	SBS	SBS	SBS	SBS	SBS	SBS	SBS
BEEF	0	0	0				0				
LAMB	0	0	0				0				
PORK	0	0	0				0				
VEAL	0	0	0				0				
CHICKEN	1	0	1		1					1	
TURKEY, Lt.	0	0	0				0				
EGG	0	1	1			1					1
MILK	0	0	1					1			
CHEESE/AMER.	0	1	1			1					1
CHEESE/ROQUE.	0	0	1					1			
CHEESE/SWISS	1	0	1		1					1	
===============											
POSITIVES	2	2	6	0	2	2	5	2	0	2	2
NUMBER (N)	11	11	11	11	11	11	11	6	6	6	6
% POSITIVE	18.2	18.2	54.5	0.0	18.2	18.2	45.5	33.3	0.0	33.3	33.3

TABLE 05.10.SBS.06 TBSxKMS (SBS) FAMILIAL CYTOTOXIC TEST POSITIVE SENSITIVITIES TO MISCELLANEOUS ITEMS.

	FATHER	MOTHER	SON	%POSITIVE	%POSITIVE	%POSITIVE	%NEGATIVE	PER CENT OF POSITIVE FOODS			
SUBJECT	TBS	KMS	SBS	TBS/KMS	TSB	KMS	TBS/KMS		TBS/KMS	TSB	KMS
AGE	28	25	3	SBS	SBS	SBS	SBS	SBS	SBS	SBS	SBS
ALCOHOL, 1%	0	0	1					1			
ALCOHOL, 2%	0	0	1					1			
CHOCOLATE	0	1	1			1					1
COFFEE	0	1	0								
COTTONSEED	0	0	0				0				
GARLIC	0	0	0				0				
GASOLINE	0	1	0								
MUSHROOM	0	0	0				0				
MUSTARD	0	0	0				0				
PEPPERMINT	0	1	0								
TEA	0	1	1			1					1
TOBACCO	0	1	1			1					1
TURPENTINE	0	0	0				0				
YEAST, BAKER	0	0	1					1			
YEAST, BREWER	0	0	0				0				
===============											
POSITIVES	0	6	6	0	0	3	6	3	0	0	3
NUMBER (N)	15	15	15	15	15	15	15	6	6	6	6
% POSITIVE	0.0	40.0	40.0	0.0	0.0	20.0	40.0	50.0	0.0	0.0	50.0

TABLE 05.10.SBS.07 SUMMARY: TBSxKMS(SBS) FAMILIAL DISTRIBUTION OF CYTOTOXIC TEST POSITIVE FOOD SENSITIVITIES.

	FATHER	MOTHER	SON	%POSITIVE	%POSITIVE	%POSITIVE	NEGATIVE	PER CENT OF POSITIVE FOODS			
SUBJECT	TBS	KMS	SBS	TBS/KMS	TSB	KMS	TBS/KMS		TBS/KMS	TSB	KMS
AGE TESTED	28	25	3	SBS	SBS	SBS	SBS	SBS	SBS	SBS	SBS
TOTAL POSITIVES	12	18	27	2	7	10	35	8	2	7	10
TOTAL NUMBER	70	70	70	70	70	70	70	27	27	27	27
% POSITIVE	17.1	25.7	38.6	2.9	10.0	14.3	50.0	29.6	7.4	25.9	37.0

CASE STUDY SUMMARY 5.10: Each case study recorded above revealed details about the offspring, the parents, and in some cases siblings and grandparents. Table 5.11 provides a data summary for 15 children detailing their results compared to their parents and to each other. The data in Table 5.11 was sorted and ranked by the children's ascending percent Cytotoxic Test positive items. The data was further analyzed for those that had less than 40% vs. greater than 40% test positive percentages. The children's overall average percent Cytotoxic Test positive items was 40.0+/-11.8SD (29.3%CV), compared to 37.8+/-10.7SD (28.3%CV) for the fathers and 41.7+/-12.9SD (30.9%CV) for mothers for this very small database of 15 offspring. This database represents the first case study undertaken. Ideally a database with a coefficient of variation greater than 20% demands more data and/or more selective study design. In Chapter 6.0, Population Analysis, a database of 130 participants' total sensitivity averaged 37.3+/-5.9SD (15.8%CV) with additional evidence that age and sex showed no change over a range of two to 82 years of age. The premise has been advanced here earlier that by two years of age we have been exposed to the items tested for in this study and genetically have responded. We continue, to look for more evidence.

Where both parents were Cytotoxic Test positive, 67.1+/-21.5SD (32.1%CV) per cent of these positives were passed to the offspring! Analyzing this data further, those with less than 40% sensitivity averaged 54.5+/-19.4SD (35.7%CV) reflecting a high degree of variability in this data, however those with greater than 40% sensitivity averaged 86.0+/-9.7SD (11.3%CV). This difference proved to be significantly (P<0.025) different even in this small database. Generally speaking, the greater than 40% population is an at-risk population.

Of the 70 items the children were tested for an average of 6.9% matched the father's positives, 9.5% matched the mother's positives and 14.9% matched both parents. The individual difference for both the father (6.9%) and the mother (9.5%) are statistically significantly different from the data for both parents (14.9%) at the 95% or better confidence level. The difference between the individual father and mother was not significant. It appears that when all 70 items are considered that both parents combined exert more input than the individual parent when it comes to passing the traits.

When the children are sorted to analyze those with less than 40% sensitivity, 4.8% matched the father's positives, 8.3% matched the mother's positives and 10.6% matched both parents. The individual difference for the fathers (4.8%) were statistically significantly different from the data for both parents (10.6%) at the 95% or better confidence level, however the difference for the mothers (8.3%) was not significant. The difference between the individual fathers and mothers was significant at the 95% confidence level. Restated, it appears that there is no difference between the influence exerted by the mothers and that of both parents as opposed to the fathers having a somewhat separate influence in determining the items passed to the children with less than 40% sensitivity.

When the children are sorted to analyze those with greater than 40% sensitivity, 10.2% matched the father's positives, 11.4% matched the mothers' positives and 21.2% matched both parents. The individual difference for both the fathers (10.2%) and the mothers (11.4%) are statistically significantly different from the data for both parents (21.2%) at the 95% or better confidence level. The individual difference for the fathers at less than 40% sensitivity (4.8%) compared to that at greater than 40% sensitivity (10.2%) was significant at the 95% or better confidence level, whereas that same comparison for the mothers 8.3% vs 11.4% did not prove to be significant. The difference for both the fathers and mothers at less than 40% sensitivity (10.6%) compared to that at greater than 40% sensitivity (21.2%) was significant at the 95% or better confidence level. Recall that the data for the children were sorted for those with less than or greater than 40% Cytotoxic Test positives (sensitivities) and the parents went along for the ride. Through this the mothers hang with the more significant data for both parents while the fathers appear to get hung out there somewhat when all 70 items are considered.

TABLE 5.11 ANALYSIS OF FAMILIAL CASE STUDIES OFFSPRING'S GENETIC DISTRIBUTIONS OF CYTOTOXIC TEST POSITIVE FOOD SENSITIVITIES: LESS THAN 40% vs GREATER THAN 40%.

				%POSITIVE	%POSITIVE	%POSITIVE	PERCENT OF CHILD POSITIVES			
	% POSITIVE (70 FOODS)			PARENTS	FATHER	MOTHER	FROM	PARENTS	Father's	Mother's
CASE	FATHERS	MOTHERS	CHILDREN	CHILD	CHILD	CHILD	PARENTS	CHILD	CHILD	CHILD
GPS(5.6.1)	55.7	28.6	17.1	2.9	7.1	2.9	16.7	16.7	58.3	33.3
SSP(5.1.3)	31.4	44.3	22.9	11.4	2.9	4.3	50.0	50.0	62.5	68.8
JLP(5.2.1)	24.3	20.0	24.3	5.7	4.3	7.1	66.7	23.5	41.2	52.9
EJG(5.8.1)	32.9	70.0	27.1	10.0	1.4	14.3	36.8	36.8	42.1	89.5
RRP(5.1.2)	31.4	44.3	30.0	17.1	1.4	5.7	75.0	57.1	61.9	76.2
KKR(5.7.2)	28.6	48.6	34.3	7.1	7.1	15.7	50.0	20.8	41.7	66.7
KES(5.5.3)	48.6	41.4	37.1	17.1	5.7	7.1	57.1	46.2	69.2	57.7
GWS(5.5.1)	48.6	41.4	37.1	21.4	2.9	2.9	71.4	57.7	65.4	65.4
SBS(5.9.1)	17.1	25.7	38.6	2.9	10.0	14.3	66.7	7.4	33.3	44.4
JSR(5.7.1)	28.6	48.6	51.4	12.9	11.4	21.4	90.0	25.0	47.2	66.7
BPP(5.1.1)	31.4	44.3	52.9	22.9	1.4	12.9	100.0	43.2	43.2	67.6
MTZ(5.3.1)	58.6	52.9	55.7	35.7	5.7	7.1	80.6	64.1	74.4	76.9
JES(5.5.4)	48.6	41.4	55.7	25.7	14.3	4.3	85.7	46.2	71.8	53.8
LMS(5.5.2)	48.6	41.4	58.6	22.9	8.6	7.1	76.2	39.0	53.7	51.2
MLK-W(5.4.1)	32.9	32.9	62.9	7.1	20.0	15.7	83.3	11.4	31.8	36.4
AVERAGE	37.8	41.7	40.4	14.9	6.9	9.5	67.1	36.3	53.2	60.5
SUM	567.3	625.8	605.7	222.8	104.2	142.8	1006.2	545.1	797.7	907.5
SUMSQ	23625.8	28126.5	27496.7	4573.1	1115.4	1814.3	74200.5	24282.4	45199.8	58257.2
NUMBER (N)	15.0	15.0	15.0	15.0	15.0	15.0	15.0	15.0	15.0	15.0
SQRT OF N	3.9	3.9	3.9	3.9	3.9	3.9	3.9	3.9	3.9	3.9
MAXIMUM	58.6	70.0	62.9	35.7	20.0	21.4	100.0	64.1	74.4	89.5
MINIMUM	17.1	20.0	17.1	2.9	1.4	2.9	16.7	7.4	31.8	33.3
RANGE	41.5	50.0	45.8	32.8	18.6	18.5	83.3	56.7	42.6	56.2
STD ERROR(SEM)	2.8	3.3	3.1	2.2	1.2	1.2	5.6	3.8	2.8	3.7
SEM %	7.3	8.0	7.6	14.7	17.9	13.0	8.3	10.4	5.3	6.2
STD DEV (SD)	10.7	12.9	11.8	8.5	4.8	4.8	21.5	14.6	11.0	14.5
SD % COEF VAR	28.3	30.9	29.3	57.0	69.1	50.2	32.1	40.3	20.7	24.0
>40% AVERAGE	35.4	40.5	29.8	10.6	4.8	8.3	54.5	35.1	52.8	61.7
SUM	318.6	364.3	268.5	95.6	42.8	74.3	490.4	316.2	475.6	554.9
SUMSQ	12581.6	16493.3	8461.0	1372.5	272.5	824.1	29514.3	13896.9	26520.2	36484.1
NUMBER (N)	9.0	9.0	9.0	9.0	9.0	9.0	9.0	9.0	9.0	9.0
SQRT OF N	3.0	3.0	3.0	3.0	3.0	3.0	3.0	3.0	3.0	3.0
MAXIMUM	55.7	70.0	38.6	21.4	10.0	15.7	75.0	57.7	69.2	89.5
MINIMUM	17.1	20.0	17.1	2.9	1.4	2.9	16.7	7.4	33.3	33.3
RANGE	38.6	50.0	21.5	18.5	8.6	12.8	58.3	50.3	35.9	56.2
STD ERROR(SEM)	4.3	5.6	2.4	2.1	1.0	1.4	6.5	5.6	4.0	6.2
SEM %	12.1	13.7	8.0	19.4	20.1	17.2	11.9	15.9	7.5	10.1
STD DEV (SD)	12.9	16.7	7.2	6.2	2.9	4.3	19.4	16.8	12.0	18.7
SD % COEF VAR	36.3	41.2	24.0	58.1	60.3	51.7	35.7	47.7	22.6	30.4
t-test<40/>40	N.S.	N.S.	P<0.0005	P<0.0125	P<0.025	N.S.	P<0.025	N.S.	N.S.	N.S.
<40% AVERAGE	41.5	43.6	56.2	21.2	10.2	11.4	86.0	38.2	53.7	58.8
SUM	248.7	261.5	337.2	127.2	61.4	68.5	515.8	228.9	322.1	352.6
SUMSQ	11044.2	11633.2	19035.7	3200.6	842.9	990.2	44686.2	10385.5	18679.6	21773.1
NUMBER (N)	6.0	6.0	6.0	6.0	6.0	6.0	6.0	6.0	6.0	6.0
SQRT OF N	2.4	2.4	2.4	2.4	2.4	2.4	2.4	2.4	2.4	2.4
MAXIMUM	58.6	52.9	62.9	35.7	20.0	21.4	100.0	64.1	74.4	76.9
MINIMUM	28.6	32.9	51.4	7.1	1.4	4.3	76.2	11.4	31.8	36.4
RANGE	30.0	20.0	11.5	28.6	18.6	17.1	23.8	52.7	42.6	40.5
STD ERROR(SEM)	5.0	3.3	1.9	4.8	3.1	2.9	4.0	8.8	7.1	6.8
SEM %	12.1	7.6	3.4	22.5	30.3	25.0	4.6	23.0	13.2	11.5
STD DEV (SD)	12.2	8.2	4.7	11.7	7.6	7.0	9.7	21.5	17.4	16.5
SD % COEF VAR	29.5	18.7	8.4	55.1	74.2	61.1	11.3	56.4	32.4	28.1

Of the per cent of child's positives an average of 53.2% matched the father's positives, 60.5% matched the mother's positives and 36.3% matched both parents. These data matches how the positives from each individual parent and that of both parents match up to the child's positives. This comparison is very specific and more directly reveals the familial origins of the test positive results in the children. Both parents (36.3+/-14.6 (40.3%CV) tested as significantly different at the 95 % or better confidence level compared to the averages for fathers (53.2+/-11.0 (20.7%CV) and mothers (60.5+/-14.5 (24.0%CV) with no significant difference between the fathers and mothers. It appears that the individual father and mother exert considerable influence over what is passed to the offspring. This does not change for the individual parents or for both parents together depending on whether the involvement of the child is less than or greater than 40% sensitivity, maybe because it's genetic!

REFERENCES

Black AP: A New Diagnostic method in allergy disease. Pediat 17: 717-724, 1956.

Bochner BS: Update on cells and cytokines cellular adhesion and its antagonism. J. Allergy Clin Immunol 100:581-5, 1997.

Bryan WTK and Bryan M: Cytotoxic Reactions in the diagnosis of food allergy. Otolarnyngol Clinics of N Am 4: 523-534, 1971.

Bryan WTK and Bryan M: Cytotoxic Reactions in the diagnosis of food allergy. Laryngoscope 79: 1453-1472, 1969.

Bryan WTK and Bryan M: The application of in vitro cytotoxic reactions to clinical diagnosis of food allergy. Laryngoscope 70: 810-824, 1960.

Ulett GA and Perry SG: Cytotoxic testing and leukocyte increase as an index to food sensitivity. II. Coffee and Tobacco. Annals Allergy 34: 150-160, 1975.

Ulett GA and Perry SG: Cytotoxic testing and leukocyte increase as an index to food sensitivity. Annals Allergy 33: 23-32, 1974.

Krupski WC: The peripheral vascular consequences of smoking. Ann. Vasc. Surg. 1991; 5:291-304

Lakier JB: Smoking and Cardiovascular Disease. Am J of Med 1992; 93(suppl 1A) 8s-12s

CHAPTER 6

Population Analysis

POPULATION DISTRIBUTION: IN this chapter we analyze 130 individuals (Figure 06.00) to reveal population trends that are related to age, male vs. female, and extent of sensitivity since we first published our findings (Ulett and Perry; 1974, 1975). **The data summaries from these studies (Table 06.01) reveal the following preliminary findings.** Over a broad range of individual total positive sensitivities to the food items tested for age statistically did not change. The age range studied span newborn to 82 years averaging 27.8+/- 7.2 SD (25.9%CV) with an average overall population sensitivity of 37.3%+/-5.9 SD (15.8%CV). The broad age range is reflected in the high coefficient of variation (CV); however, the average sensitivity has a reasonable CV (less than 20 %) for that same very age diverse population.

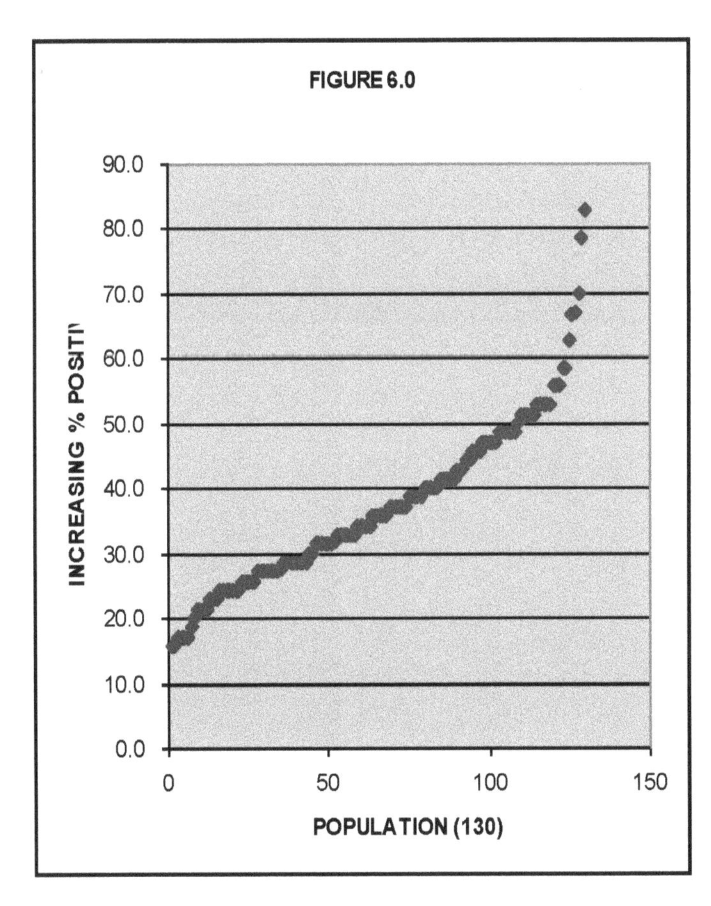

Arbitrarily analyzing those of the 130 individuals in this population (Table 06.01) with less than 40% sensitivity vs. those with greater than 40% sensitivity the age span for those with less than 40% sensitivity was newborn to 82 years averaging 28.8+/-9.2 SD (32.0%CV) with an overall sensitivity of 28.9%+/-2.6 SD(8.9%CV). The age span for those with greater than 40% sensitivity span 3 to 72 years averaging 26.2+/-9.7 SD(36.8%CV) with an overall sensitivity 50.4%+/-6.0 SD (11.9%CV). The difference in age between the two populations was not significant (P<0.25, Student-t-Test), however, the difference in sensitivity was highly significance (P<0.0005) revealing at least two distinctly separate populations based on the degree of sensitivity (percent positives). Overall sensitivity is not dependent on age. With regard to the genetic recognition and appearance of test positive food sensitivity it appears that most food positives are established inside the first two years according to maternal and family infant feeding schedules.

TABLE 06.01 ANALYSIS OF DISTRIBUTION OF RANKED CYTOTOXIC TEST POSITIVE FOOD SENSITIVITIES.

CASE	AGE	PERCENT (+)
AVERAGE	27.8	37.3
NUMBER (N)	130	130
SQRT OF N	11.4	11.4
SUMSQ	143723.0	203132.1
MAXIMUM	82.0	82.9
MINIMUM	0.0	15.7
RANGE	82.0	67.2
STD ERROR (SEM)	0.6	0.5
SEM %	2.3	1.4
STD DEV (SD)	7.2	5.9
SD % COEF VAR	25.9	15.8
> 40 % SENS		
AVERAGE	28.8	28.9
%POSITIVE(@SUM)	2277.0	2280.4
SUMSQ	91725.0	68868.6
NUMBER (N)	79.0	79.0
SQRT OF N	8.9	8.9
MAXIMUM	82.0	38.6
MINIMUM	0.0	15.7
RANGE	82.0	22.9
STD ERROR (SEM)	1.0	0.3
SEM %	3.6	1.0
STD DEV (SD)	9.2	2.6
SD % COEF VAR	32.0	8.9
< 40 % SENS		
AVERAGE	26.2	50.4
%POSITIVE(@SUM)	1338.0	2572.3
SUMSQ	51998.0	134263.5
NUMBER (N)	51.0	51.0
SQRT OF N	7.1	7.1
MAXIMUM	72.0	82.9
MINIMUM	3.0	40.0
RANGE	69.0	42.9
STD ERROR (SEM)	1.4	0.8
SEM %	5.2	1.7
STD DEV (SD)	9.7	6.0
SD % COEF VAR	36.8	11.9
>40% VS<40%	P < 0.25	P < 0.0005

The data analyzed in table 06.01 can be visualized in figure 06.00 tracking the ranked Cytotoxic Test positive food sensitivities against a background of age of the 130 individuals at the time of testing. The percent positive sensitivity ranges from 15.7 to 82.9 averaging 37.3+/-5.9SD (15.8%CV) for the 130 participants spanning newborn to 82 years. The sample shows relatively low percent sensitivity (less than 40%) compared to relatively high percent sensitivity (greater than 40%), that proved to be significantly different (P<0.0005) while the corresponding ages showed high variability and proved to not be significantly different (P<.025).

The following equations were used to make the students to test comparisons.

SP(SQ) =SUMXSQ-(SUMX)SQ/Nx + @SUMYSQ-(SUMY)SQ/Ny / Nx+Ny-2 (EQ 6.1)

SP=X(AVG)-Y(AVG)/SQRT1/Xn+1/Yn (EQ 6.2)

t = X(AVG)-Y(AVG) / SP * (SQRT I/Nx+1/Ny) (EQ6.3)

%POSITIVE >40% VS<40%	AGE >40% VS<40%	
3042.978	26095.519	SUMXSQ-@SUMXSQ/Nx
4523.749	16895.176	SUMYSQ-@SUMYSQ/Ny
59.115	335.865	SPSQ=SUM ABOVE/Nx+Ny-2
7.689	18.327	SP=SQRT ABOVE
15.567	0.790	
P< 0.0005	P < 0.25	

When male vs. female populations were analyzed 71 females age span newborn to 82 years in age averaging 25.2+/-9.7 SD (38.6%CV) with an overall sensitivity of 36.5+/-8.0 SD (21.8%CV). The 59 males age span newborn to 72 years in age averaging 30.9+/-9.4 SD (30.3%CV) with an overall sensitivity of 38.3+/-8.2 SD (21.4%CV). Although the difference in age between the males and female was not significant (P< 0.10) there was no significant difference (P<0.30) in the average sensitivity (percent positives) between the male and female populations. The data for male vs. female test positive sensitivities can be visualized below (Figure 6.1). It appears that sensitivities are manifest sequentially as antigen exposure triggers immune system prenatal and early postnatal genetic potential independent of sex or age. If you are not stung by a bee, "or a food", until late in life then the same fundamental process kicks in with a proportional reaction base on Phases I. II., and III of your genetic capacity.

We have proposed three phases that add up to an allergic manifestation. Phase I we identify (*in vitro* Cytotoxic Test or *in vivo* Food Challenge Test positive) as an immune cell recognition step, Phase II: initiated immune defense steps, and Phase III: end organ(s) allergic response (i.e., skin, gastrointestinal, cardiovascular, central nervous, etc. system(s). We are measuring Phase I. with the *in vitro* Cytotoxic

Test, or the *in vivo* Food Challenge reported here.

I am reminded, whenever statistics are used, that Oliver Lowry, M.D. told his students, "If you have to use statistics then you didn't design your experiment well", however, Dr. Lowry did arm his students with statistical short cuts (Lowry, OH and Passonneau, JV; 1972) to help validate and emphasize the messages hidden in results. SD = range/sq root of n, then if plus or minus 2 SD of compared results don't overlap the difference is significant at the 95% confidence level.

A rule of thumb example is given below used during a positive tobacco allergen and the central nervous system experiment (Lowry OH, and SG Perry; 1974, 1975). Note that with the NONSMOKERS 5.9+/-0.4 (+/-2SD 5.1-6.7)verses SMOKERS 10.6+/-0.75 (+/-2SD 9.1-12.1) , do not overlap and are statistically significant at the 95% confidence level. WBC normal range here is 4.8 -10.8 $10^3/mm^3$ and differs slightly laboratory to laboratory depending on method, reagents, and equipment employed.

RULE OR THUMB STATISTICS (LOWRY AND PASSONNEAU,1972).			
	NONSMOKER		SMOKER
	WBC		WBC
	4.8-10.8 $10^3/MM^3$		*4.8-10.8 $10^3/MM^3$*
	5.1		12
	6.6		11.5
	6.1		11
	6.5		9.5
	6		10.5
	5		11.9
	5.5		9.5
	5.7		11.5
	6.5		10.8
	5.3		9.5
	5.8		9.0
	6.7		12
	5.7		9.8
	6.7		10
	5.2		11.5
	5.3		10
SUM	93.7		170
n	16		16
AVERAGE	5.9		10.6
n SQ ROOT	4		4
RANGE	1.7		3
SD	0.6		1
VARIATION	10.20%		9.40%
	NOTE:		
	SD = RANGE/SQ ROOT OF n		
NONSMOKER	(SD = 1.7/16 SQ ROOT = 1.7/4 = 0.425)		
	5.9+/-0.4SD(7.2%CV) **+/-2SD 5.1 - 6.7**		
SMOKER	(SD = 3/16 SQ ROOT =3/4 = 0.75)		
	10.6+/-0.75SD(7.1%CV) **+/-2SD 9.1 - 12.1**		
IF +/- 2SD DON'T OVERLAP = STATISTICALLY UNIQUE AT 95% COFIDENCE LEVEL			

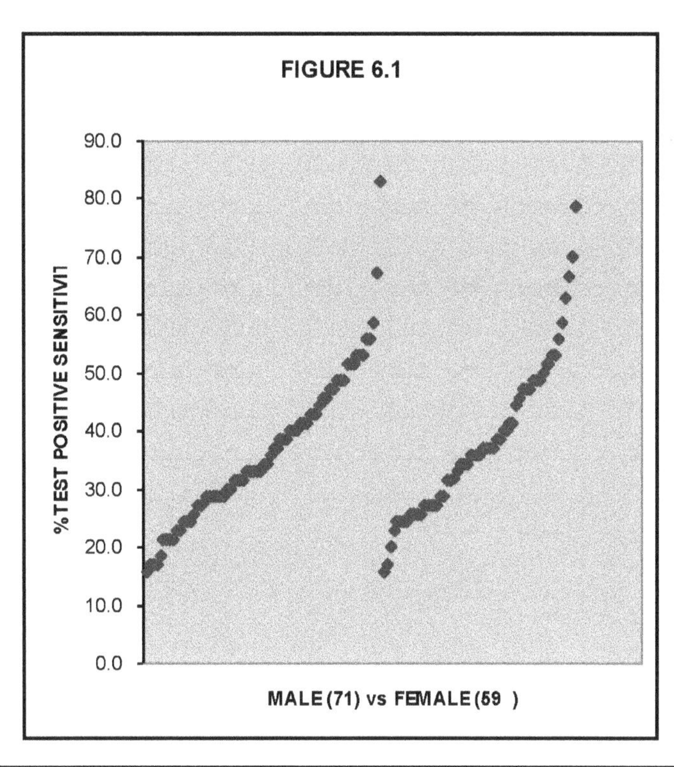

TABLE 06.02 SEXES: ANALYSIS OF POPULATION DISTRIBUTION OF RANKED CYTOTOXIC TEST POSITIVE FOOD SENSITIVITIES.

CASE	AGE	SEX	PERCENT(+)
AVERAGE	27.8		37.3
NUMBER (N)	130		130
SQRT OF N	11.4		11.4
MAXIMUM	82.0		82.9
MINIMUM	0.0		15.7
RANGE	82.0		67.2
STD ERROR (SEM)	0.6		0.5
SEM %	2.3		1.4
STD DEV (SD)	7.2		5.9
SD % COEF VAR	25.9		15.8
MALE AVERAGE	25.2		36.5
%POSITIVE(@SUM)	1790.0		2594.6
SUMSQ	70448.0		106364.2
NUMBER (N)	71.0		71.0
SQRT OF N	8.4		8.4
MAXIMUM	82.0		82.9
MINIMUM	0.0		15.7
RANGE	82.0		67.2
STD ERROR (SEM)	1.2		0.9
SEM %	4.6		2.6
STD DEV (SD)	9.7		8.0
SD % COEF VAR	38.6		21.8
t-TEST M vs F	P<0.10		P<0.30
FEMALE AVERAGE	30.9		38.3
%POSITIVE(@SUM)	1825.0		2258.1
SUMSQ	73275.0		96767.9
NUMBER (N)	59.0		59.0
SQRT OF N	7.7		7.7
MAXIMUM	72.0		78.6
MINIMUM	0.0		15.7
RANGE	72.0		62.9
STD ERROR (SEM)	1.2		1.1
SEM %	3.9		2.8
STD DEV (SD)	9.4		8.2
SD % COEF VAR	30.3		21.4

Age and Food Sensitivity: The 130 participants in this study, average age 27.8+/-7.2SD (25.9%CV) and an average percent positive food sensitivity of 37.3+/-5.9SD (15.8%CV), were further analyzed according to age in three different age groups (Table 06.03). The average percent positive food sensitivities for each age group; 9.9 years (37.6+/-9.5SD (25.3%CV), 29.1 years (36.7+/-7.6SD (20.9%CV), and 53.9 years (37.8+/-10.4SD (27.4%CV) was not significantly different indicating that age was not a discriminating factor for establishing the recognition phase for the onset of food sensitivities. This is visualized in figure 06.02 as age is ranked verses test positive sensitivity where sensitivity remains uniform over the entire range of age. As noted above, with the appearance of Phase I. test positive food sensitivity, it appears that most food positives are established inside the first two years as the maternal and family infant feeding schedules are established. High %CV indicate experiment design should be redesigned to achieve <20%CV.

TABLE 06.03 AGE: POPULATION DISTRIBUTION OF CYTOTOXIC POSITIVE FOOD SENSITIVITIES.

CASE	AGE	SEX	PERCENT (+)
AVERAGE	27.8		37.3
NUMBER (N)	130		130
SQRT OF N	11.4		11.4
MAXIMUM	82.0		82.9
MINIMUM	0.0		15.7
RANGE	82.0		67.2
STD ERROR (SEM)	0.6		0.5
SEM %	2.3		1.4
STD DEV (SD)	7.2		5.9
SD % COEF VAR	25.9		15.8
>21 AGE AVERAGE	9.9		37.6
%POSITIVE(@SUM)	494.0		1881.3
SUMSQ	6342.0		79556.5
NUMBER (N)>	50.0		50.0
SQRT OF N	7.1		7.1
MAXIMUM	20.0		82.9
MINIMUM	0.0		15.7
RANGE	20.0		67.2
STD ERROR (SEM)	0.4		1.3
SEM %	4.0		3.6
STD DEV (SD)	2.8		9.5
SD % COEF VAR	28.6		25.3
<20, > 40 AGE AVERAGE	29.1		36.7
%POSITIVE(@SUM)	1397.0		1762.9
SUMSQ	41689.0		72230.5
NUMBER (N)	48.0		48.0
SQRT OF N	6.9		6.9
MAXIMUM	39.0		70.0
MINIMUM	22.0		17.1
RANGE	17.0		52.9
STD ERROR (SEM)	0.4		1.1
SEM %	1.2		3.0
STD DEV (SD)	2.5		7.6
SD % COEF VAR	8.4		20.8
<39, > 82 AGE AVERAGE	53.9		37.8
%POSITIVE(@SUM)	1724.0		1208.5
SUMSQ	95692.0		51345.0
NUMBER (N)	32.0		32.0
SQRT OF N	5.7		5.7
MAXIMUM	82.0		78.6
MINIMUM	41.0		20.0
RANGE	41.0		58.6
STD ERROR (SEM)	1.3		1.8
SEM %	2.4		4.8
STD DEV (SD)	7.2		10.4
SD % COEF VAR	13.5		27.4

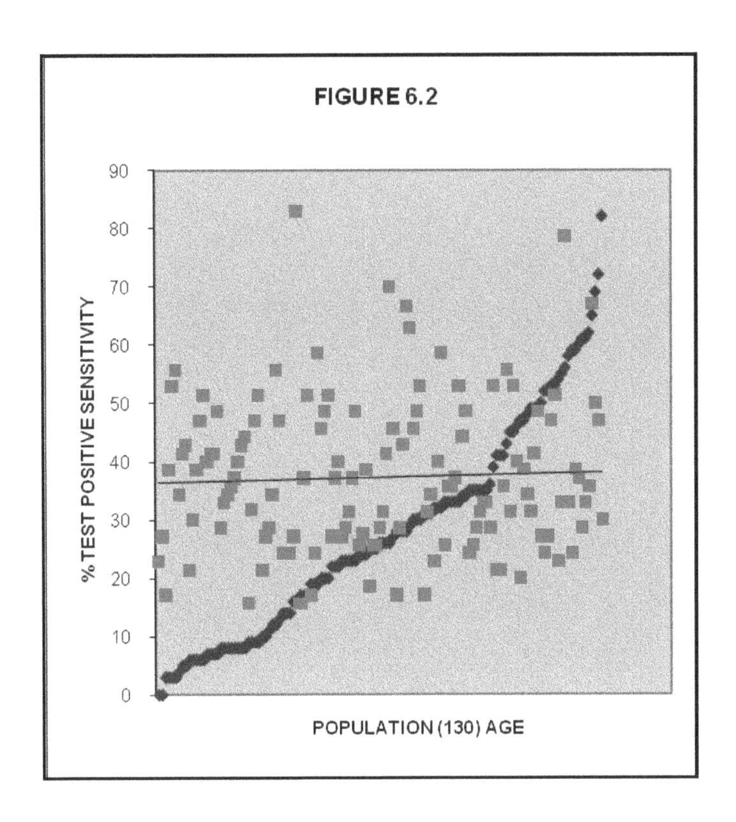

Familial Distribution of Food Sensitivity: In Chapter 5 we detailed several case studies. Here in Chapter 6, we will summarize our findings for the 15 offspring reported in these case studies (Table 06.04). The 15 offspring and both parents were collectively positive for 14.9% of the 70 items tested for. The father and child were positive for an additional 6.9% of the items tested for. This difference comparing the contribution of both parents against the contribution of a single parent, in the case of the father, tested out to be statistically significantly (P<0.005) for the 15 cases analyzed. The mother and the child exclusively matched on an additional 9.5% of the 70 items tested for, a difference that also proved to be statistically significant (P<0.05) when compared to the 14.9% for 15 offspring and both parents. The difference between the father and child (6.9%) and the mother and child (9.5%) was not statistically significant (P<0.15), that is, neither parent appeared to have a genetic advantage contributing individually to their offspring. Of the 70 items tested for 33.2% were not positive for any family members.

The right side of Table 06.01 evaluates the data as a percent of the 15 children's positives as opposed to a percentage of the 70 items tested for. Both parents and child accounted for 36.3% of the positives. The father and child accounted for an additional 17.3% of the children's positives which t-tested (P<0.0025) as separate and statistically distinct from the data for both parents and child. The mother and child accounted for yet an additional 24.7% of the children's positives, which also tested to be statistically (P<0.05) distinct and separate from both parents and child positives. The difference between the father and child (17.3%) compared to the mother and child (24.7%) did not prove statistically significant (P<0.01). 21.0% of the children's positives were exclusive to the child and did not appear in either parent. This accounts for 99.3% of the children's positives. The genetic picture painted here is very powerful. Both parents and the combination of each parent's individual contribution are statistically undeniable, especially considering this finding is emerging from this relatively small sample.

TABLE 06.04 ANALYSIS OF FAMILIAL CASE STUDIES OFFSPRING'S GENETIC DISTRIBUTIONS OF CYTOTOXIC TEST POSITIVE FOOD SENSITIVITIES: <40% vs. >40%.

CASE	AGE	SEX	PERCENT POSITIVE 70 FOODS	%POSITIVE PARENTS CHILD	%POSITIVE FATHER CHILD	%POSITIVE MOTHER CHILD	%NEGATIVE PARENTS CHILD	PERCENT OF CHILD POSITIVES: CHILD	PARENTS CHILD	FATHER CHILD	MOTHER CHILD
GPS	3	M	17.1	2.9	7.1	2.9	28.6	25.0	16.7	41.7	45.8
SSP	0	M	22.9	11.4	2.9	4.3	42.9	18.8	50.0	12.5	41.7
JLP	14	F	24.3	5.7	4.3	7.1	57.1	29.4	23.5	17.6	37.0
EJG	0	F	27.1	10.0	1.4	14.3	22.9	5.3	36.8	5.3	18.8
RRP	6	M	30.0	17.1	1.4	5.7	41.4	19.0	57.1	4.8	12.2
KKR	12	F	34.3	7.1	7.1	15.7	32.9	0.0	20.8	20.8	52.6
KES	17	F	37.1	17.1	5.7	7.1	32.9	19.2	46.2	15.4	24.3
GWS	22	M	37.1	21.4	2.9	2.9	30.0	26.9	57.7	7.7	25.0
SBS	3	M	38.6	2.9	10.0	14.3	50.0	29.6	7.4	25.9	7.7
JSR	6	M	51.4	12.9	11.4	21.4	31.4	11.1	25.0	22.2	7.7
BPP	3	M	52.9	22.9	1.4	12.9	31.4	31.4	43.2	2.7	29.4
MTZ	3	M	55.7	35.7	5.7	7.1	25.7	12.8	64.1	10.3	19.2
JES	12	F	55.7	25.7	14.3	4.3	28.6	20.5	46.2	25.6	16.7
EMS	19	F	58.6	22.9	8.6	7.1	20.0	34.1	39.0	14.6	12.8
MLK-W	29	F	62.9	7.1	20.0	15.7	22.9	31.8	11.4	31.8	19.0
AVERAGE	9.9		40.4	14.9	6.9	9.5	33.2	21.0	36.3	17.3	24.7
SUM	149.0		605.7	222.8	104.2	142.8	498.7	314.9	545.1	258.9	369.9
SUMSQ	2567.0		27496.7	4573.1	1115.4	1814.3	18096.8	8067.6	24282.4	6141.9	11845.4
NUMBER (N)	15.0		15.0	15.0	15.0	15.0	15.0	15.0	15.0	15.0	15.0
SQRT OF N	3.9		3.9	3.9	3.9	3.9	3.9	3.9	3.9	3.9	3.9
MAXIMUM	29.0		62.9	35.7	20.0	21.4	57.1	34.1	64.1	41.7	52.6
MINIMUM	0.0		17.1	2.9	1.4	2.9	20.0	0.0	7.4	2.7	7.7
RANGE	29.0		45.8	32.8	18.6	18.5	37.1	34.1	56.7	39.0	44.9
STDERROR(SEM)	1.9		3.1	2.2	1.2	1.2	2.5	2.3	3.8	2.6	3.0
SEM %	19.5		7.6	14.7	17.9	13.0	7.4	10.8	10.4	15.1	12.1
STD DEV (SD)	7.5		11.8	8.5	4.8	4.8	9.6	8.8	14.6	10.1	11.6
SD % COEF VAR	75.4		29.3	57.0	69.1	50.2	28.8	41.9	40.3	58.3	47.0
>40% AVERAGE	8.6		29.8	10.6	4.8	8.3	37.6	19.2	35.1	16.9	29.5
SUM	77.0		192.8	71.3	29.9	57.1	258.7	116.7	251.1	118.1	232.4
SUMSQ	1167.0		5594.6	906.1	164.1	611.2	10322.0	2600.5	10512.9	2925.8	9065.1
NUMBER (N)	9.0		9.0	9.0	9.0	9.0	9.0	9.0	9.0	9.0	9.0
SQRT OF N	3.0		3.0	3.0	3.0	3.0	3.0	3.0	3.0	3.0	3.0
MAXIMUM	22.0		38.6	21.4	10.0	15.7	57.1	29.6	57.7	41.7	52.6
MINIMUM	0.0		17.1	2.9	1.4	2.9	22.9	0.0	7.4	4.8	7.7
RANGE	22.0		21.5	18.5	8.6	12.8	34.2	29.6	50.3	36.9	44.9
STDERROR(SEM)	2.4		2.4	2.1	1.0	1.4	3.8	3.3	5.6	4.1	5.0
SEM %	28.6		8.0	19.4	20.1	17.2	10.1	17.1	15.9	24.3	16.9
STD DEV (SD)	7.3		7.2	6.2	2.9	4.3	11.4	9.9	16.8	12.3	15.0
SD % COEF VAR	85.7		24.0	58.1	60.3	51.7	30.3	51.3	47.7	73.0	50.8
t-Test<40/>40			P<0.0005	P<0.01	P<0.05	P<0.20					P<0.05
<40% AVERAGE	12.0		56.2	21.2	10.2	11.4	26.7	23.6	38.2	17.9	17.5
SUM	72.0		337.2	127.2	61.4	68.5	160.0	141.7	228.9	107.2	104.8
SUMSQ	1400.0		19035.7	3200.6	842.9	990.2	4374.8	3867.3	10385.5	2486.0	2096.0
NUMBER (N)	6.0		6.0	6.0	6.0	6.0	6.0	6.0	6.0	6.0	6.0
SQRT OF N	2.4		2.4	2.4	2.4	2.4	2.4	2.4	2.4	2.4	2.4
MAXIMUM	29.0		62.9	35.7	20.0	21.4	31.4	34.1	64.1	31.8	29.4
MINIMUM	3.0		51.4	7.1	1.4	4.3	20.0	11.1	11.4	2.7	7.7
RANGE	26.0		11.5	28.6	18.6	17.1	11.4	23.0	52.7	29.1	21.7
STDERROR(SEM)	4.3		1.9	4.8	3.1	2.9	1.9	3.8	8.8	4.9	3.6
SEM %	36.1		3.4	22.5	30.3	25.0	7.1	16.2	23.0	27.1	20.7
STD DEV (SD)	10.6		4.7	11.7	7.6	7.0	4.7	9.4	21.5	11.9	8.9
SD % COEF VAR	88.5		8.4	55.1	74.2	61.1	17.5	39.8	56.4	66.5	50.7

Degree of Sensitivity: The 15 children were ranked (table 6.1.1) according to the percent positive of the foods tested for. Arbitrarily those children with less than 40% test positive food sensitivities were compared to those with greater than 40% sensitivity. **This comparison was 29.8+/-7.2SD (24%CV) for >40% test positive food sensitivity and 56.2+/-4.7SD (8.4%CV) for <40% test positive food sensitivity, a difference that proved highly significant (P<0.0005)**

The average age was 8.6+/-7.3SD(85.7%) for children with >40% test positive food sensitivity, and 12+/-10.6SD(88.5%CV) for children with <40% test positive food sensitivity indicating that in this relatively small sample age was not a discriminating factor across the arbitrary boundary above and below 40% test positive food sensitivity.

Consistent with the arbitrary comparison above and below 40% test positive food sensitivity, the Father/ Mother-Child data, 21.2+/-11.7SD (55.1%CV) for greater than 40% sensitivity compared to 10.6+/-6.2SD (58.1%CV) as statistically significant (P<0.01) as did the data for Father-Child (P<0.05). The Mother-Child data comparison based on positives per 70 test positive foods tested did not prove statistically significant (P<0.20) among the 15 children and parents analyzed.

For the more specific comparison (Table 06.01) based on the percent of children's positives in the right side of the table the Father/Mother-Child and the Father-Child comparisons showed no significant differences. The Mother-Child comparison for > 40% test positive food sensitivity, 29.5+/-15SD (50.8%), proved to be significantly (P<0.05) different than the 17.5+/-8.9SD (50.7%CV) for the <40% test positive food sensitivity. In combination with the paternal contribution this is more consistent with a maternal role; however, the genetic contribution probably trumps all contributions. We have proposed three phases that add up to an allergic manifestation. Phase I we identify (*in vitro* Cytotoxic Test or *in vivo* Food Challenge Test positive) as an immune cell recognition step, Phase II: initiated immune defense steps, and Phase III: end organ(s) allergic response (i.e., skin, gastrointestinal, cardiovascular, central nervous, etc. system(s).

Female vs. Male Sensitivity: There were no significant differences between female, 42.9+/-14.6SD (34%CV), and male, 38.2+/-13.6 (35.7%CV), offspring (Table 06.05) regards the overall percent positive food sensitivity out of the 70 foods tested for. When analyzing female(13.7+/-7.6SD (55.4%CV) verses male (15.9+/-11.6SD (72.9%CV) offspring no significant difference was seen for Father/Mother-Child percent positive sensitivities out of 70 foods tested, as was the case for individual Father-Female Child (8.8+/-7.0SD (80.1%CV) verses Father-Male Child,(5.4+/-3.5SD (66.1%CV) or Mother-Female Child (10.2+/-4.3SD (42.3%CV) verses Mother-Male Child (8.9+/-6.5SD (73.2%CV) for the 70 foods tested for.

For the more specific comparison (Table 06.05) based on the percent of children's positives in the right side of the table the Father/Mother-Child and the Father-Child comparisons showed no significant differences. When analyzing female(32+/-13.2SD (41.1%CV) verses male (40.2+/-20.0SD (49.9%CV) offspring no significant difference was seen for Father/Mother-Child percent positive sensitivities out of the percent of child positive foods tested, as was the case for individual Father-Female Child (18.7+/-10.0SD (53.5%CV) and Father-Male Child,(16.0+/-13.8SD (86.3%CV) or Mother-Female Child (25.9+/-15.0SD (58.1%CV) and Mother-Male Child (23.6+/-13.5SD (57.1%CV). No significant difference for female (20.0+/-12.9SD (64.3%CV) or male (21.8+/-7.2SD (32.9%CV) percent of child only positives was seen either. **With regard to female verses male cell biology, it appears that we are observing cellular genetics and straight forward distribution of genetic material not driven by maternal or sex of the offspring factors.**

There is a significant age difference (P<0.05) based on analysis (Table 06.06) by age. Selecting for those ages 12 and above also selected for six females and one male, with the less than 12 age group the selection sorting out seven males and one female and no significant difference in food sensitivities.

TABLE 06.05 ANALYSIS OF CASE STUDIES OFFSPRING'S GENETIC DISTRIBUTIONS OF CYTOTOXIC POSITIVE FOOD SENSITIVITIES: FEMALE vs MALE.

			PERCENT	%POSITIVE	%POSITIVE	%POSITIVE	%NEGATIVE	PERCENT OF CHILD POSITIVES			
			POSITIVE	PARENTS	FATHER	MOTHER	PARENTS		PARENTS	FATHER	MOTHER
CASE	AGE	SEX	70 FOODS	CHILD	CHILD	CHILD	CHILD	CHILD	CHILD	CHILD	CHILD
EJG	0	F	27.1	10.0	1.4	14.3	22.9	5.3	36.8	5.3	18.8
KKR	12	F	34.3	7.1	7.1	15.7	32.9	0.0	20.8	20.8	52.6
JES	12	F	55.7	25.7	14.3	4.3	28.6	20.5	46.2	25.6	16.7
JLP	14	F	24.3	5.7	4.3	7.1	57.1	29.4	23.5	17.6	37.0
KES	17	F	37.1	17.1	5.7	7.1	32.9	19.2	46.2	15.4	24.3
LMS	19	F	58.6	22.9	8.6	7.1	20.0	34.1	39.0	14.6	12.8
MLK-W	29	F	62.9	7.1	20.0	15.7	22.9	31.8	11.4	31.8	19.0
SSP	0	M	22.9	11.4	2.9	4.3	42.9	18.8	50.0	12.5	41.7
GPS	3	M	17.1	2.9	7.1	2.9	28.6	25.0	16.7	41.7	45.8
SBS	3	M	38.6	2.9	10.0	14.3	50.0	29.6	7.4	25.9	7.7
BPP	3	M	52.9	22.9	1.4	12.9	31.4	31.4	43.2	2.7	29.4
MTZ	3	M	55.7	35.7	5.7	7.1	25.7	12.8	64.1	10.3	19.2
RRP	6	M	30.0	17.1	1.4	5.7	41.4	19.0	57.1	4.8	12.2
JSR	6	M	51.4	12.9	11.4	21.4	31.4	11.1	25.0	22.2	7.7
GWS	22	M	37.1	21.4	2.9	2.9	30.0	26.9	57.7	7.7	25.0
AVERAGE	9.9		40.4	14.9	6.9	9.5	33.2	21.0	36.3	17.3	24.7
SUM	149.0		605.7	222.8	104.2	142.8	498.7	314.9	545.1	258.9	369.9
SUMSQ	2567.0		27496.7	4573.1	1115.4	1814.3	18096.8	8067.6	24282.4	6141.9	11845.4
NUMBER (N)	15.0		15.0	15.0	15.0	15.0	15.0	15.0	15.0	15.0	15.0
SQRT OF N	3.9		3.9	3.9	3.9	3.9	3.9	3.9	3.9	3.9	3.9
MAXIMUM	29.0		62.9	35.7	20.0	21.4	57.1	34.1	64.1	41.7	52.6
MINIMUM	0.0		17.1	2.9	1.4	2.9	20.0	0.0	7.4	2.7	7.7
RANGE	29.0		45.8	32.8	18.6	18.5	37.1	34.1	56.7	39.0	44.9
STDERROR(SEM)	1.9		3.1	2.2	1.2	1.2	2.5	2.3	3.8	2.6	3.0
SEM %	19.5		7.6	14.7	17.9	13.0	7.4	10.8	10.4	15.1	12.1
STD DEV (SD)	7.5		11.8	8.5	4.8	4.8	9.6	8.8	14.6	10.1	11.6
SD % COEF VAR	75.4		29.3	57.0	69.1	50.2	28.8	41.9	40.3	58.3	47.0
FEMALE	14.7		42.9	13.7	8.8	10.2	31.0	20.0	32.0	18.7	25.9
SUM	103.0		300.0	95.6	61.4	71.3	217.3	140.3	223.9	131.1	181.2
SUMSQ	1975.0		14370.7	1710.6	781.8	867.2	7692.0	3855.4	8259.0	2887.4	5883.4
NUMBER (N)	7.0		7.0	7.0	7.0	7.0	7.0	7.0	7.0	7.0	7.0
SQRT OF N	2.6		2.6	2.6	2.6	2.6	2.6	2.6	2.6	2.6	2.6
MAXIMUM	29.0		62.9	25.7	20.0	15.7	57.1	34.1	46.2	31.8	52.6
MINIMUM	0.0		24.3	5.7	1.4	4.3	20.0	0.0	11.4	5.3	12.8
RANGE	29.0		38.6	20.0	18.6	11.4	37.1	34.1	34.8	26.5	39.8
STDERROR(SEM)	4.1		5.5	2.9	2.7	1.6	5.3	4.9	5.0	3.8	5.7
SEM %	28.2		12.9	20.9	30.3	16.0	17.1	24.3	15.5	20.2	22.0
STD DEV (SD)	11.0		14.6	7.6	7.0	4.3	14.0	12.9	13.2	10.0	15.0
SD % COEF VAR	74.5		34.0	55.4	80.1	42.3	45.2	64.3	41.1	53.5	58.1
MALE AVERAGE	5.8		38.2	15.9	5.4	8.9	35.2	21.8	40.2	16.0	23.6
SUM	46.0		305.7	127.2	42.8	71.5	281.4	174.6	321.2	127.8	188.7
SUMSQ	592.0		13126.1	2862.5	333.6	947.1	10404.7	4212.2	16023.4	3254.5	5962.0
NUMBER (N)	8.0		8.0	8.0	8.0	8.0	8.0	8.0	8.0	8.0	8.0
SQRT OF N	2.8		2.8	2.8	2.8	2.8	2.8	2.8	2.8	2.8	2.8
MAXIMUM	22.0		55.7	35.7	11.4	21.4	50.0	31.4	64.1	41.7	45.8
MINIMUM	0.0		17.1	2.9	1.4	2.9	25.7	11.1	7.4	2.7	7.7
RANGE	22.0		38.6	32.8	10.0	18.5	24.3	20.3	56.7	39.0	38.1
STDERROR(SEM)	2.8		4.8	4.1	1.3	2.3	3.0	2.5	7.1	4.9	4.8
SEM %	47.8		12.6	25.8	23.4	25.9	8.6	11.6	17.7	30.5	20.2
STD DEV (SD)	7.8		13.6	11.6	3.5	6.5	8.6	7.2	20.0	13.8	13.5
SD % COEF VAR	135.3		35.7	72.9	66.1	73.2	24.4	32.9	49.9	86.3	57.1

TABLE 06.06 FAMILIAL OFFSPRING'S GENETIC DISTRIBUTIONS OF CYTOTOXIC TEST POSITIVE FOODS: AGE <12 YEARS vs >12 YEARS.

			PERCENT	%POSITIVE	%POSITIVE	%POSITIVE	%NEGATIVE	PERCENT OF CHILD POSITIVES			
			POSITIVE	PARENTS	FATHER	MOTHER	PARENTS		PARENTS	FATHER	MOTHER
CASE	AGE	SEX	70 FOODS	CHILD	CHILD	CHILD	CHILD	CHILD	CHILD	CHILD	CHILD
SSP	0	M	22.9	11.4	2.9	4.3	42.9	18.8	50.0	12.5	18.8
EJG	0	F	27.1	10.0	1.4	14.3	22.9	5.3	36.8	5.3	52.6
GPS	3	M	17.1	2.9	7.1	2.9	28.6	25.0	16.7	41.7	16.7
SBS	3	M	38.6	2.9	10.0	14.3	50.0	29.6	7.4	25.9	37.0
BPP	3	M	52.9	22.9	1.4	12.9	31.4	31.4	43.2	2.7	24.3
MTZ	3	M	55.7	35.7	5.7	7.1	25.7	12.8	64.1	10.3	12.8
RRP	6	M	30.0	17.1	1.4	5.7	41.4	19.0	57.1	4.8	19.0
JSR	6	M	51.4	12.9	11.4	21.4	31.4	11.1	25.0	22.2	41.7
KKR	12	F	34.3	7.1	7.1	15.7	32.9	0.0	20.8	20.8	45.8
JES	12	F	55.7	25.7	14.3	4.3	28.6	20.5	46.2	25.6	7.7
JLP	14	F	24.3	5.7	4.3	7.1	57.1	29.4	23.5	17.6	29.4
KES	17	F	37.1	17.1	5.7	7.1	32.9	19.2	46.2	15.4	19.2
LMS	19	F	58.6	22.9	8.6	7.1	20.0	34.1	39.0	14.6	12.2
GWS	22	M	37.1	21.4	2.9	2.9	30.0	26.9	57.7	7.7	7.7
MLK-W	29	F	62.9	7.1	20.0	15.7	22.9	31.8	11.4	31.8	25.0
AVERAGE	9.9		40.4	14.9	6.9	9.5	33.2	21.0	36.3	17.3	24.7
SUM	149.0		605.7	222.8	104.2	142.8	498.7	314.9	545.1	258.9	369.9
SUMSQ	2567.0		27496.7	4573.1	1115.4	1814.3	18096.8	8067.6	24282.4	6141.9	11845.4
NUMBER (N)	15.0		15.0	15.0	15.0	15.0	15.0	15.0	15.0	15.0	15.0
SQRT OF N	3.9		3.9	3.9	3.9	3.9	3.9	3.9	3.9	3.9	3.9
MAXIMUM	29.0		62.9	35.7	20.0	21.4	57.1	34.1	64.1	41.7	52.6
MINIMUM	0.0		17.1	2.9	1.4	2.9	20.0	0.0	7.4	2.7	7.7
RANGE	29.0		45.8	32.8	18.6	18.5	37.1	34.1	56.7	39.0	44.9
STDRROR(SEM)	1.9		3.1	2.2	1.2	1.2	2.5	2.3	3.8	2.6	3.0
SEM %	19.5		7.6	14.7	17.9	13.0	7.4	10.8	10.4	15.1	12.1
STD DEV (SD)	7.5		11.8	8.5	4.8	4.8	9.6	8.8	14.6	10.1	11.6
SD % COEF VAR	75.4		29.3	57.0	69.1	50.2	28.8	41.9	40.3	58.3	47.0
AGE>12 AVG	3.0		37.0	14.5	5.2	10.4	34.3	19.1	37.5	15.7	27.9
SUM	24.0		295.7	115.8	41.3	82.9	274.3	153.0	300.3	125.4	222.9
SUMSQ	108.0		12484.1	2504.5	327.2	1143.2	10029.2	3516.7	14048.4	3223.3	7622.3
NUMBER (N)	8.0		8.0	8.0	8.0	8.0	8.0	8.0	8.0	8.0	8.0
SQRT OF N	2.8		2.8	2.8	2.8	2.8	2.8	2.8	2.8	2.8	2.8
MAXIMUM	6.0		55.7	35.7	11.4	21.4	50.0	31.4	64.1	41.7	52.6
MINIMUM	0.0		17.1	2.9	1.4	2.9	22.9	5.3	7.4	2.7	12.8
RANGE	6.0		38.6	32.8	10.0	18.5	27.1	26.1	56.7	39.0	39.8
STDERROR(SEM)	0.8		4.8	4.1	1.3	2.3	3.4	3.3	7.1	4.9	5.0
SEM %	25.0		13.1	28.3	24.2	22.3	9.9	17.1	18.9	31.1	17.9
STD DEV (SD)	2.1		13.6	11.6	3.5	6.5	9.6	9.2	20.0	13.8	14.1
SD % COEF VAR	70.7		36.9	80.1	68.5	63.1	27.9	48.2	53.4	88.0	50.5
t-test	P<0.05										
AGE<12 AVG	17.9		44.3	15.3	9.0	8.6	32.1	23.1	35.0	19.1	21.0
SUM	125.0		310.0	107.0	62.9	59.9	224.4	161.9	244.8	133.5	147.0
SUMSQ	2459.0		15012.7	2068.6	788.3	671.1	8067.6	4550.9	10234.0	2918.6	4223.1
NUMBER (N)	7.0		7.0	7.0	7.0	7.0	7.0	7.0	7.0	7.0	7.0
SQRT OF N	2.6		2.6	2.6	2.6	2.6	2.6	2.6	2.6	2.6	2.6
MAXIMUM	29.0		62.9	25.7	20.0	15.7	57.1	34.1	57.7	31.8	45.8
MINIMUM	12.0		24.3	5.7	2.9	2.9	20.0	0.0	11.4	7.7	7.7
RANGE	17.0		38.6	20.0	17.1	12.8	37.1	34.1	46.3	24.1	38.1
STDERROR(SEM)	2.4		5.5	2.9	2.4	1.8	5.3	4.9	6.6	3.4	5.4
SEM %	13.6		12.5	18.7	27.2	21.4	16.5	21.1	18.9	18.1	25.9
STD DEV (SD)	6.4		14.6	7.6	6.5	4.8	14.0	12.9	17.5	9.1	14.4
SD % COEF VAR	36.0		32.9	49.5	71.9	56.5	43.7	55.7	50.0	47.8	68.6

REFERENCES

Lowry, OH and Passonneau JV: A flexible system of enzymatic analysis. Academic Press. 1972.

Ulett GA and Perry SG: Cytotoxic testing and leukocyte increase as an index to food sensitivity. II. Coffee and Tobacco. Annals Allergy 34: 150-160, 1975.

Ulett GA and Perry SG: Cytotoxic testing and leukocyte increase as an index to food sensitivity. Annals Allergy 33: 23-32, 1974.

NOTES:

CHAPTER 7

Summary and Conclusions

SUBJECTS SUBMITTED THEMSELVES as volunteers for this study, filled out a medical history questionnaire and were scheduled for testing (Figure 07.01). All subjects were fasted 8 – 12 hours (overnight) before being examined using the Cytotoxic Test, essentially as described by Bryan and Bryan (1971). Based on the results, Cytotoxic Test positive challenge meals (3-4 items) were selected and follow up Cytotoxic Tests were scheduled for 8:00 AM to noon. At the zero hour a blood sample was taken for a second Cytotoxic Test and a WBC, and then the challenge meal (3-4 items) was consumed by 8:10 AM. The experimental challenge meal was followed at 30-minute intervals with a CBC till about a 90-minute-high point where a sample for a third Cytotoxic Test was performed and revealed some additional Cytotoxic Test positive food items as reported by Ulett and Perry (1974, 1975).

LABORATORY SET-UP

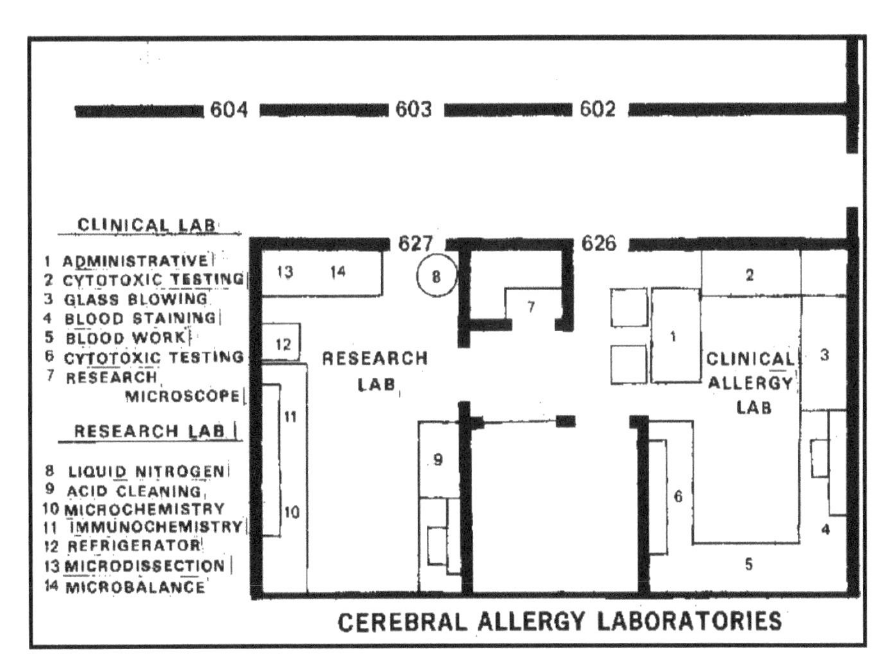

Figure 7.1

The Cytotoxic Test results were graded on the basis of the amount of *in vitro* leukocyte response. In the initial stage (Figure 7.2) the segmented lymphocytes (segs) begin to lose their amoeboid activity and platelets stick to their surface as compared to the control preparations. Subsequent stages (Ulett, Itil, and Perry; 1974) show marked deterioration compared to the control preparations. This led me to a series of hypothesis.

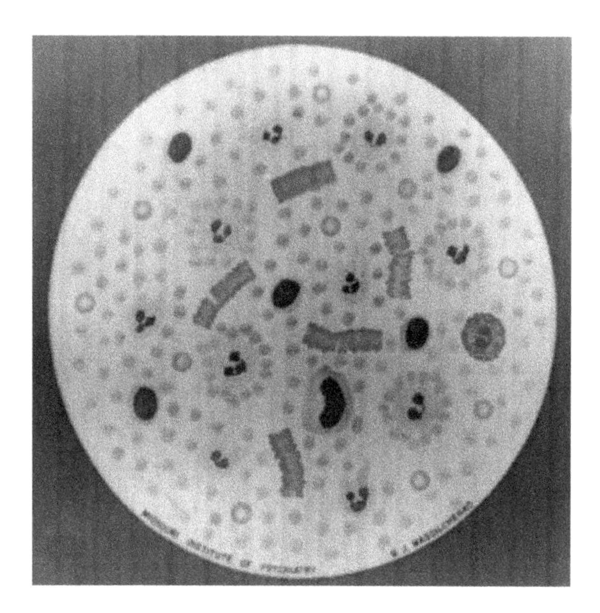

Figures 7.2

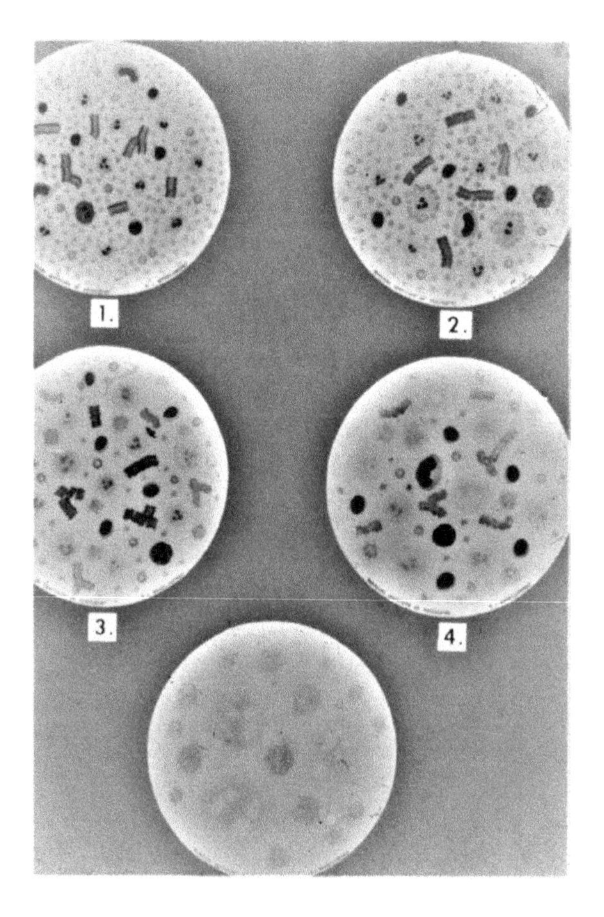

Hypothesis: If one consumes *in vitro* Cytotoxic Test positive foods and follows the CBC; then one might expect to see an *in vivo* WBC response that helps validate the results of *in vitro* Cytotoxic Testing. The very first experiment (Figure 07.03) showed the WBC to increase by 40% at 2.5 hrs. although consuming the test positive foods lasted 15 minutes. On repeat, consuming test positive foods were confined to ten minutes sharpening up the results to an 80% plus increase at 2.0 hrs. (Chapter 2.0). Conclusion: the hypothesis is on the right track, proceed with confirmatory tests.

Hypothesis: If one consumes the *in vitro* Cytotoxic Test negative foods and follows the CBC; then one might expect to see no *in vivo* WBC response. At first the observed results (Figure 07.03) were not zero (heavy bar–large circles), but the results were encouraging by the apparent separation of the experimental and control data. As described above, a repeat *in vitro* Cytotoxic Test panel performed on a blood sample at the peak of the WBC response revealed new test positive food items. Adjusting for these findings test negative foods, with "false negative foods" removed, compared to the zero-hour control. Now I had the framework to test realistic food allergen amounts against a known reliable clinical standard (CBC/WBC). Conclusion: the hypotheses are right on track.

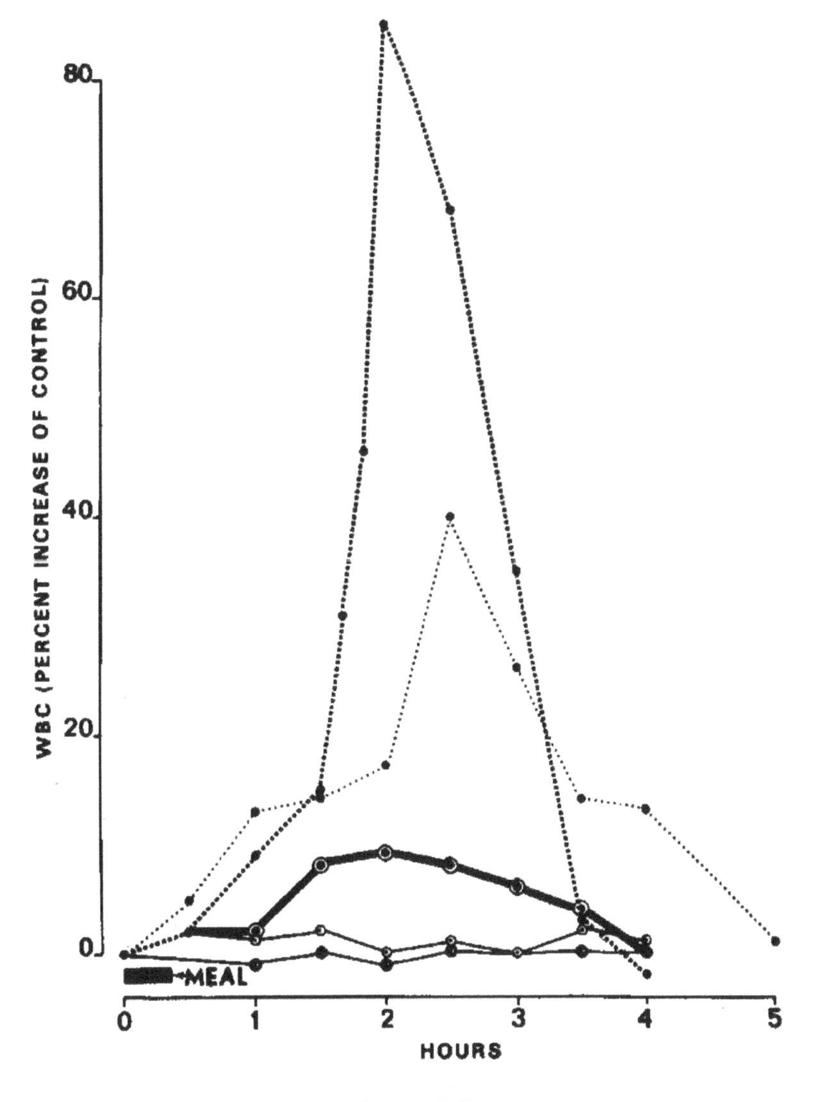

Figure 7.3

Hypothesis: If one consumes no foods and follows the CBC; then one might expect to see no *in vivo* increase to provide a base line control. The results (Figure 7.3) show a series of WBC counts that compare to the zero-hour controls. WBC published normal ranges without a requirement for fasting is not correct (Ulett and Perry 1974; 1975). Relative to known clinical values we hit the jackpot, extending the scope of investigation into possible immune antigen antibody models for allergy studies.

To confirm our initial observations nine additional volunteers were *in* vitro Cytotoxic Test and *in* vivo Food Allergen tested (Figure 7.4) showing remarkable agreement (see chapter 3.0), when human studies often require thousands of volunteers over sometimes many years to reach statistically significant conclusions. These ten subjects were tracked individually in the 0 – 3.0 hr. window (Figure 07.04), then picked up again as all ten subjects returned to control levels 3.5 – 5.0 hr. window. Hypothesizes confirmed, conclusions positive.

The Cytotoxic Test negative results showed zero increase and perfectly compared to the zero-time controls. A ten-minute meal triggers a highly variable period in the 1 - 0 – 3.0 hr. time frame, then it's time for lunch, then it's time for dinner, when does someone with food allergies get relief? The Drs Ulett's patient's fashioned diets of only Cytotoxic Test negative foods with remarkable results within days.

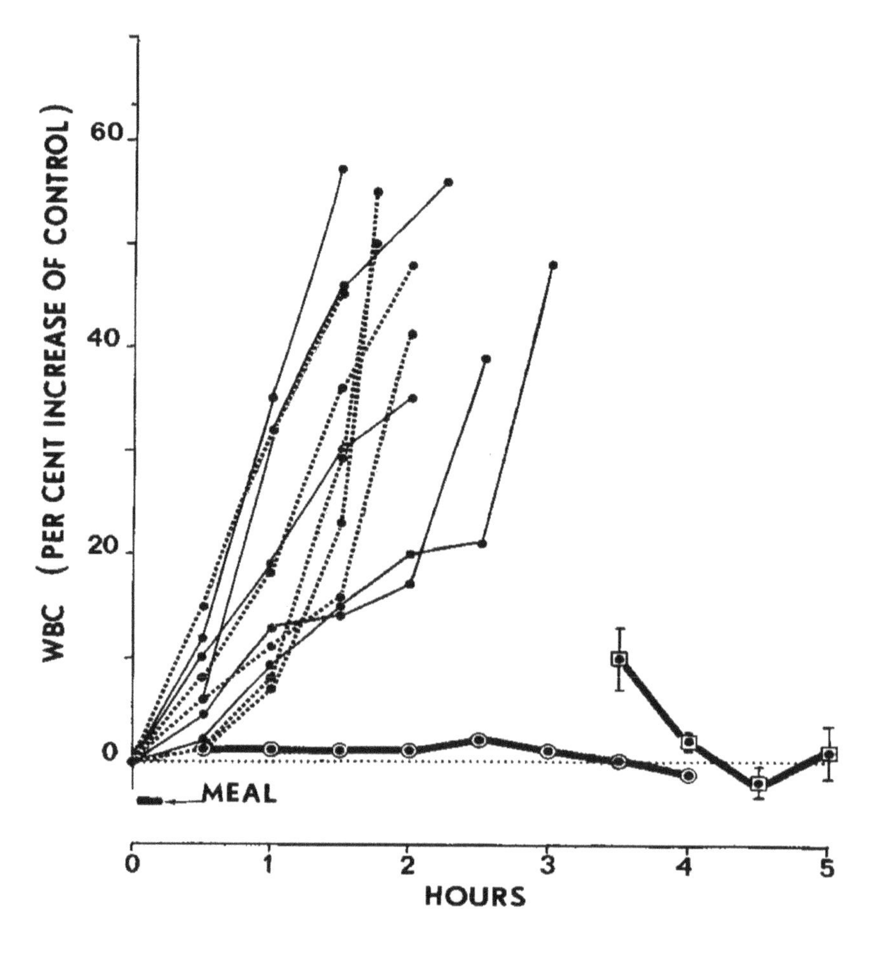

FIGURE 7.4

Hypothesis: If one consumes different amounts of a single Cytotoxic Test positive food and follows the CBC; then one might expect to see proportional *in vivo* responses to further validate the *in vivo* response results. Indeed, the results of randomized amounts of chocolate (circles), peanut (squares), and chicken (dots) showed classical sigmoid dose response curves (Figure 7.5) when plotted against the log of amount consumed vs. response. Varying amounts of Cytotoxic Test negative pork, cherry, and eggs (hatched line) showed no change

NOTES:

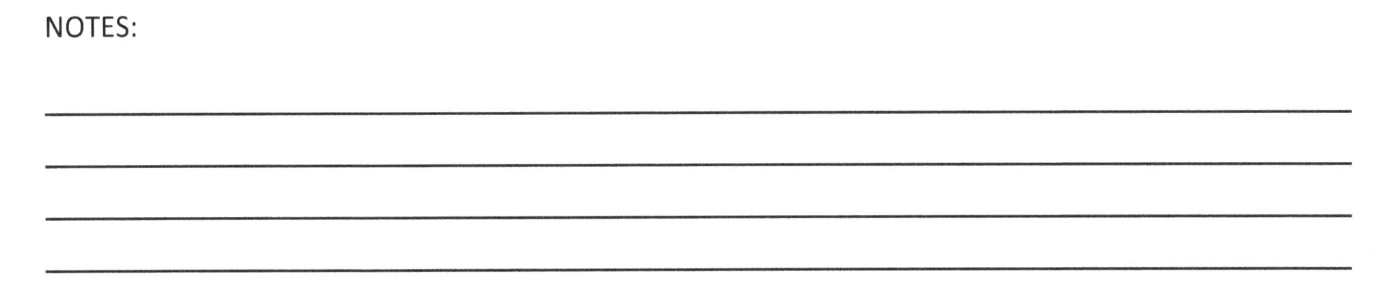

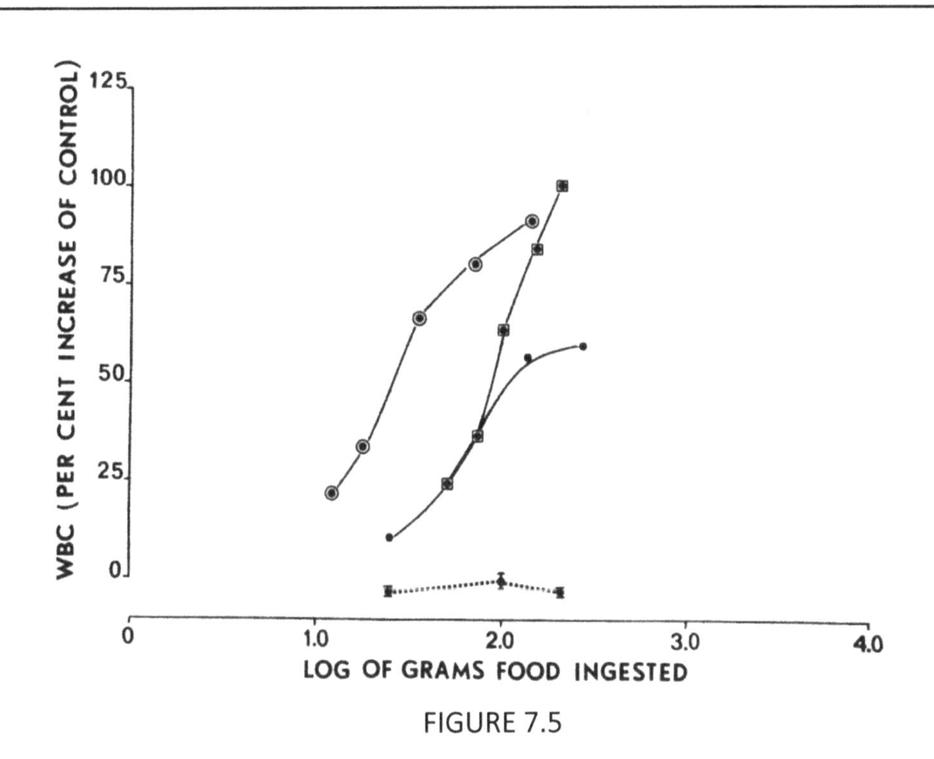

FIGURE 7.5

The sigmoid dose response has an accepted broad application in pharmaceutical, toxicology and clinical medicine as the 50% point on the curve determines the effective dose (ED50) or lethal dose (LD50) for medicines or poisons. For allergy, an (AD50) offers a two-dimensional new way to identify, quantitate and track the severity of specific allergens. Curves displaced to the left and up might be of more concern than curves displaced to the right and down. Treatment regimens might be tested for effectiveness in moving curves to the right and down.

Chocolate (circles) maximum was 4.0 (142.88 gms) Hershey's **milk** chocolate bars, then in descending sequence, 2.0 bars (71.44 gms), 1.0 bar (35.72 gms), 0.5 bar (17.86 gms), and 0.25 bar (8.93 gms). One chocolate bar triggers the AD50 in the range of 60% over the normal control WBC. What is happening within the immune-defense system? Is there an allergic effect on an end organ system (i.e., skin, cardiovascular, or CNS)? Each one of these amounts could collectively have its own magnitude of involvement

Peanut (squares) maximum was 200 gms, then in descending sequence, 150 gms, 100 gms, 75 gms, and 50 gms One bag of peanuts (75 gms) bar triggers the AD50 in the range of 60% over the normal control WBC. What allergen(s) is in peanut has been intensely researched primarily because of the number of severely allergic patients. Peanut has been purified to very specific peptide proteins that can trigger allergic reactions.

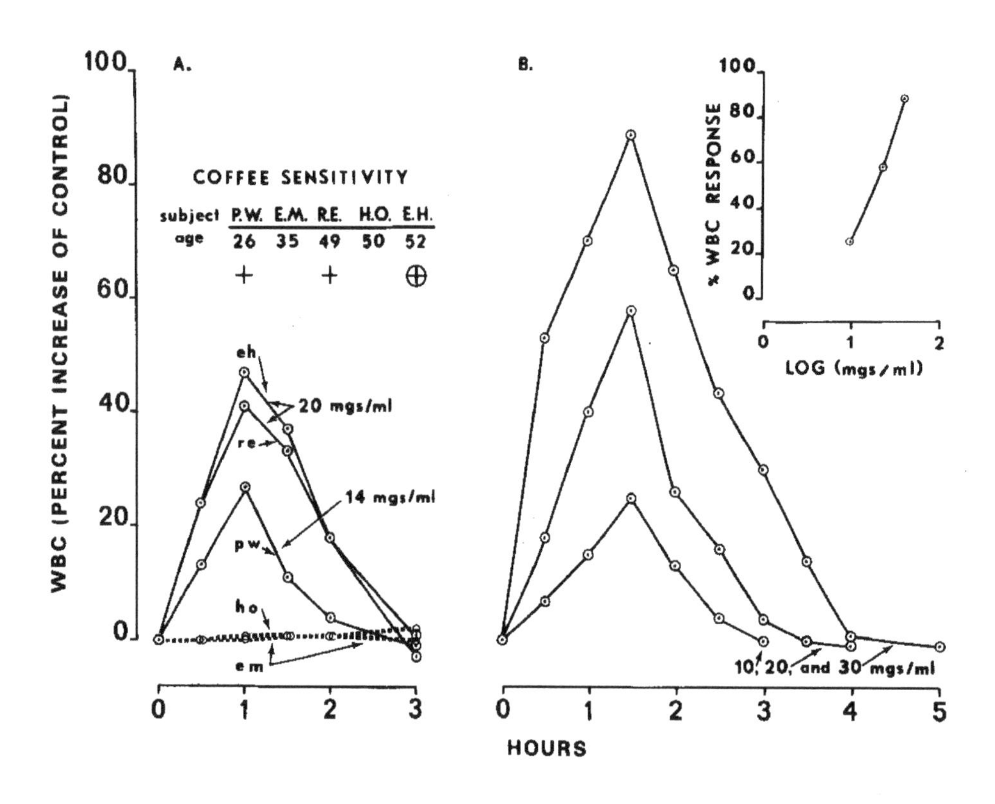

Figure 7.6

<u>Hypothesis</u>: <u>If</u> a subject consumes a food that they know is positive, <u>then</u> they will psychosomatically respond affecting the WBC response outcome of the experiment. Members of the Missouri Institute of Psychiatry staff imposed this challenge. Five members of the dietary department volunteered to test this challenge. The *in vitro* Cytotoxic Test results and the *in vivo* WBC response results conducted and posted before the subjects were informed of the results (Figure 7.6 A). The results silenced all critics. The conclusion favored trusting established quantitative clinical controls and normal ranges.

Testing the effect of varying concentrations of coffee (Figure 7.6 B) served to bracket the range that an individual might consume coffee. Coffee concentrations of 20 mgs/ml triggers the AD50 in the range of 60% over the zero-time control WBC (Figure 7.6 B inset). Coffee consumption can be habit forming, while some can take it or leave it. This research model presents an avenue to test several hypotheses for allergic and neurologic studies.

Hypothesis: If one uses the *in vitro* Cytotoxic Test positive tobacco and follows the CBC; then one might expect to use this alternative route for allergen entry to further validate the *in vivo* WBC response results. The results for the tobacco oral-respiratory route of entry (Figure 7.7), matched the pattern of results for the food oral-gastrointestinal route of entry (Figure 7.4). As one might expect the time frame for absorption was more immediate with effects apparent in the 15 min – 60 min window, and back to normal values by 2 – 3 hrs. As with coffee, tobacco can be habit forming concluding that this model might be used to test this possibility. When does one want the next "dose of tobacco"? Are those positive for tobacco allergens more prone for cancer or respiratory disease? Are there treatment regimens that might be specific for allergen prone patients?

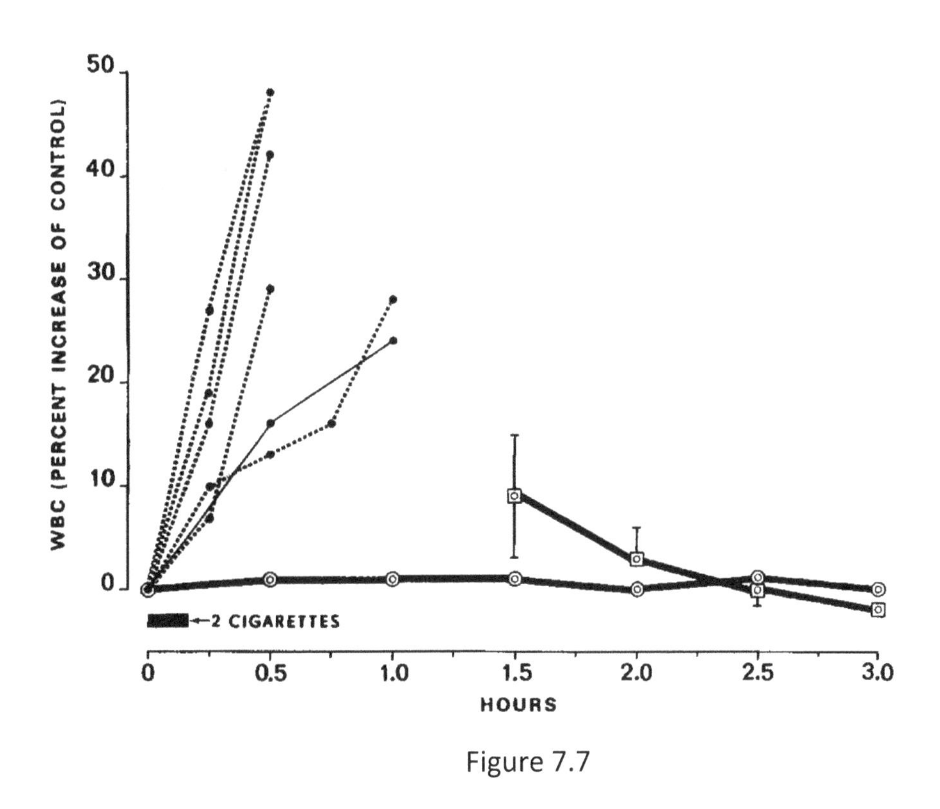

Figure 7.7

The previous experiments led me to an ultimate test, a test that would measure the potential involvement of allergy and an end organ in a rather spectacular way. The Missouri Institute of Psychiatry was conducting a very advanced worldwide recognized research program involving computer analyzed EEG's. From these two lines of research a pilot study was proposed.

Hypothesis: If a subject with an established *in vivo* WBC test positive allergen response repeats this test setup and concurrently monitors the EEG with computer analysis at 30 min intervals, then a specific end organ response might be recorded and correlated with the allergens. The first results detailed in Chapter 4 were very encouraging. The following observations detail parallel recorded quantitative data confirming the hypothesis and prompting continued research.

The electrodes were placed according to the EEG International 10/20 Electrode Placement System (10% or 20% of the total distance nasion to inion divides the head into eight left and right regions (Figure 7.8). Our initial placements: nasion ground, channel two frontal lobe (F3/F4), channel three central parietal

(CP3/CP4), channel four occipital (O1/O2), produced results between allergens and frontal vs. parietal brain regions that left us very excited.

NOTES:

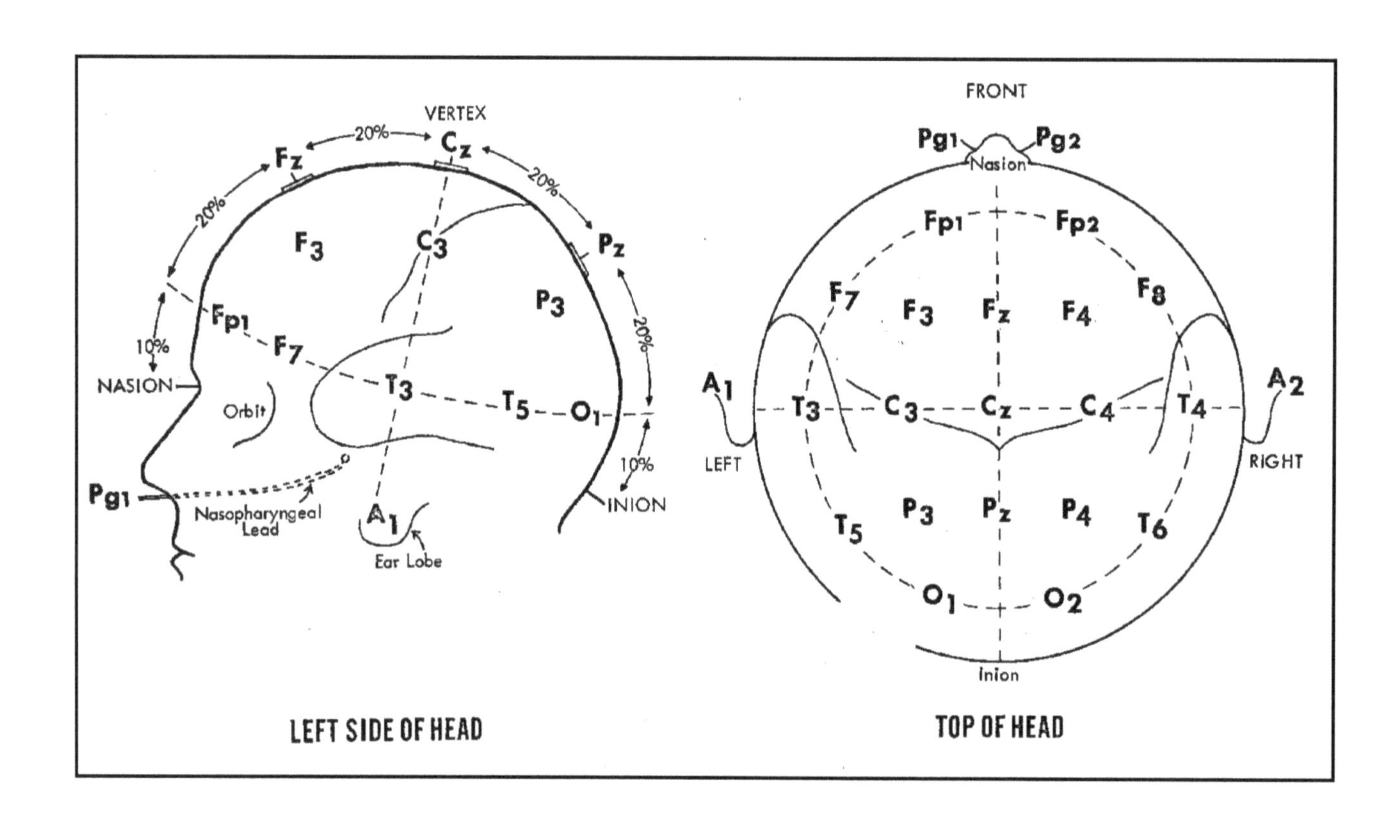

Figure 7.8

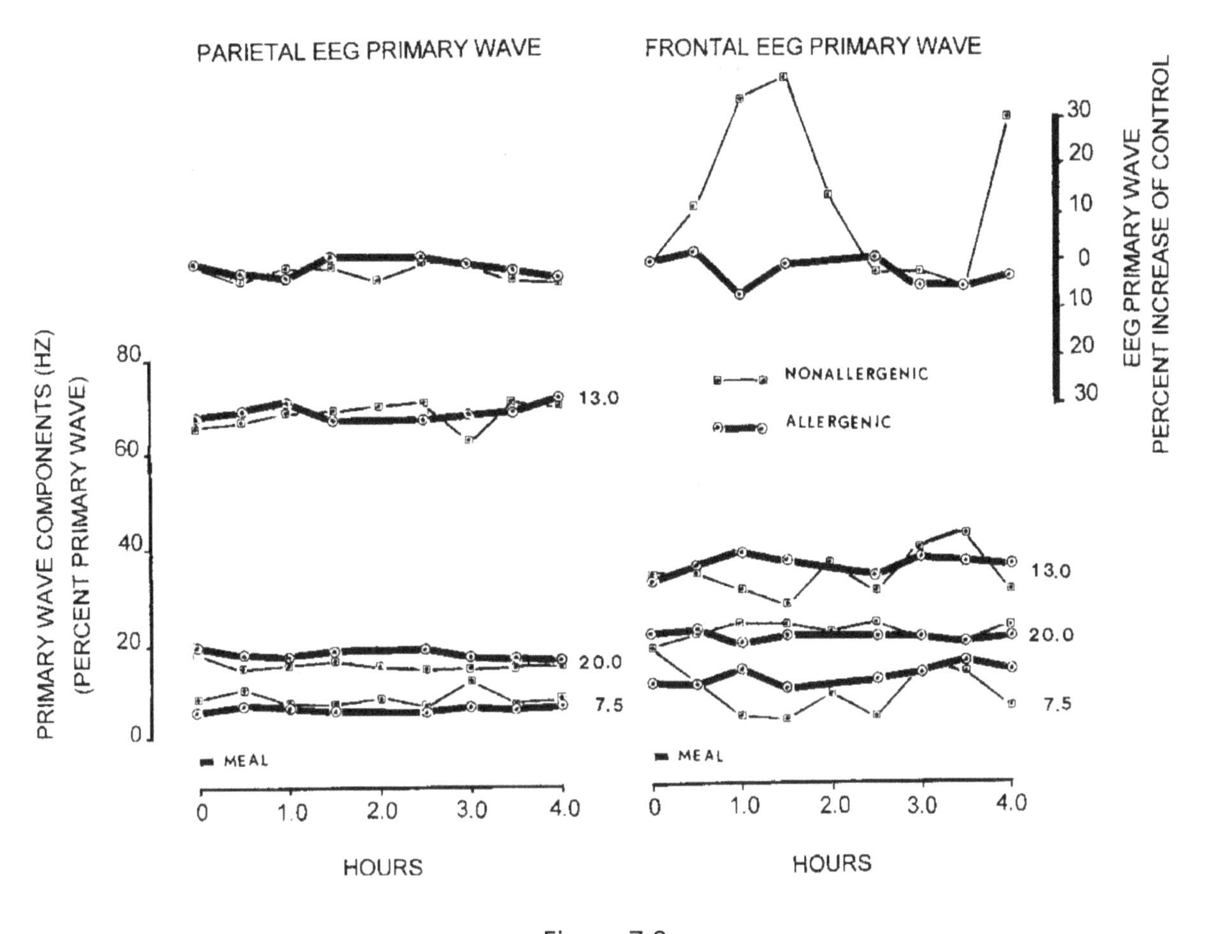

Figure 7.9

Initial attempts to sort out differences were not difficult. The frontal lobe primary wave nonallergenic test period displayed significantly (P<0.05) different from its 0.0-time control and the frontal lobe four-hour allergenic comparison period, or the allergenic and nonallergenic parietal lobe (internal control) results (Figure 7.9). The conclusions from this difference were not so apparent at this point beyond wild speculation, but there was a difference to explore.

The EEG primary wave can be further analyzed by computer to component waves (Figure 7.10), where the 0.0-hour primary and component waves are tabulated for the temporal lobe (F3, F4 10/20). All subsequent temporal lobe allergenic experimental readings have been compared to the 0.0-hour results in this table. Each experimental time frame gave us nine snap shots from which we calculated an average with reasonable SEM or SD that yielded results with acceptable variation. This provided a four-hour test period to analyze the EEG and compare it to the WBC response studies. These results are analyzed in detail in Chapter 4.

Key component waves are plotted (Figure 7.9) for the 0.0-hour control and subsequent four-hour experimental test periods. The parietal lobe 13 Hz component varied significantly (P<0.05) more of the primary wave than the temporal lobe 13 Hz component, with no separation of allergenic vs. nonallergenic test results for either brain region. Another result with conclusions pending, the "silver bullet" is still hiding.

TABLE 4.1 IN VIVO CYTOTOXIC TESTED FOOD (SENSITIZING) CHALLENGE WITH TEMPORAL ANALYSIS OF EEG RESULTS.

ANALYTE RANGE UNITS CASE/AGE/SEX	TIME HOUR	AVG PRIMARY WAVE	FRONTAL (F3,F4 10/20) EEG PRIMARY WAVE (320/SEC SAMPLING/20 SEC EPOCHS) 3.5 %	7.5 %	13.0 %	20.0 %	26.6 %	40.0 %	90.0 %	%
SGP/35/M	0.0	13.2	34.1	11.7	20.2	13.1	7.6	6.9	4.3	0.8
"ALLERGENIC"		19.0	0.0	12.8	33.0	22.4	11.2	13.2	6.6	0.6
MEAL(10 MIN)		19.5	0.0	10.8	34.8	21.7	12.1	12.1	7.3	0.8
(CHICKEN)		18.9	0.0	12.4	36.2	22.4	10.2	11.0	6.6	0.8
(CHOCOLATE)		18.7	0.0	8.0	38.9	25.4	10.0	10.5	6.3	0.7
(PEANUT)		18.8	0.0	11.8	34.2	24.7	10.5	11.7	6.3	0.7
(TEA)		20.2	0.0	11.5	30.2	21.8	12.2	16.4	7.1	0.6
5/30/73		20.7	0.0	10.0	29.2	25.0	13.2	13.7	7.8	1.0
9:30AM		19.4	0.8	12.1	33.5	23.2	11.5	11.1	7.0	0.7
========										
AVERAGE		**18.7**	**3.9**	**11.2**	**32.2**	**22.2**	**10.9**	**11.8**	**6.6**	**0.7**
SUM		168.4	34.9	101.1	290.2	199.7	98.5	106.6	59.3	6.7
SUM OF SQS		3188.7	1163.5	1153.0	9588.3	4539.8	1099.2	1316.3	398.5	5.1
NUMBER (N)		9.0	9.0	9.0	9.0	9.0	9.0	9.0	9.0	9.0
SQRT OF N		3.0	3.0	3.0	3.0	3.0	3.0	3.0	3.0	3.0
MAXIMUM		20.7	34.1	12.8	38.9	25.4	13.2	16.4	7.8	1.0
MINIMUM		13.2	0.0	8.0	20.2	13.1	7.6	6.9	4.3	0.6
RANGE		7.5	34.1	4.8	18.7	12.3	5.6	9.5	3.5	0.4
STD ERROR (SEM)		0.8	3.8	0.5	2.1	1.4	0.6	1.1	0.4	0.0
SEM %		4.5	97.7	4.7	6.4	6.2	5.7	8.9	5.9	6.0
STD DEV (SD)		**2.5**	**11.4**	**1.6**	**6.2**	**4.1**	**1.9**	**3.2**	**1.2**	**0.1**
SD % COEF VAR		**13.4**	**293.1**	**14.2**	**19.3**	**18.5**	**17.1**	**26.7**	**17.7**	**17.9**
EXPECTED-		13.7	-18.9	8.0	19.8	14.0	7.2	5.5	4.3	0.5
RANGE (2SD)		23.7	26.6	14.4	44.7	30.4	14.7	18.2	8.9	1.0

FIGURE 10.0

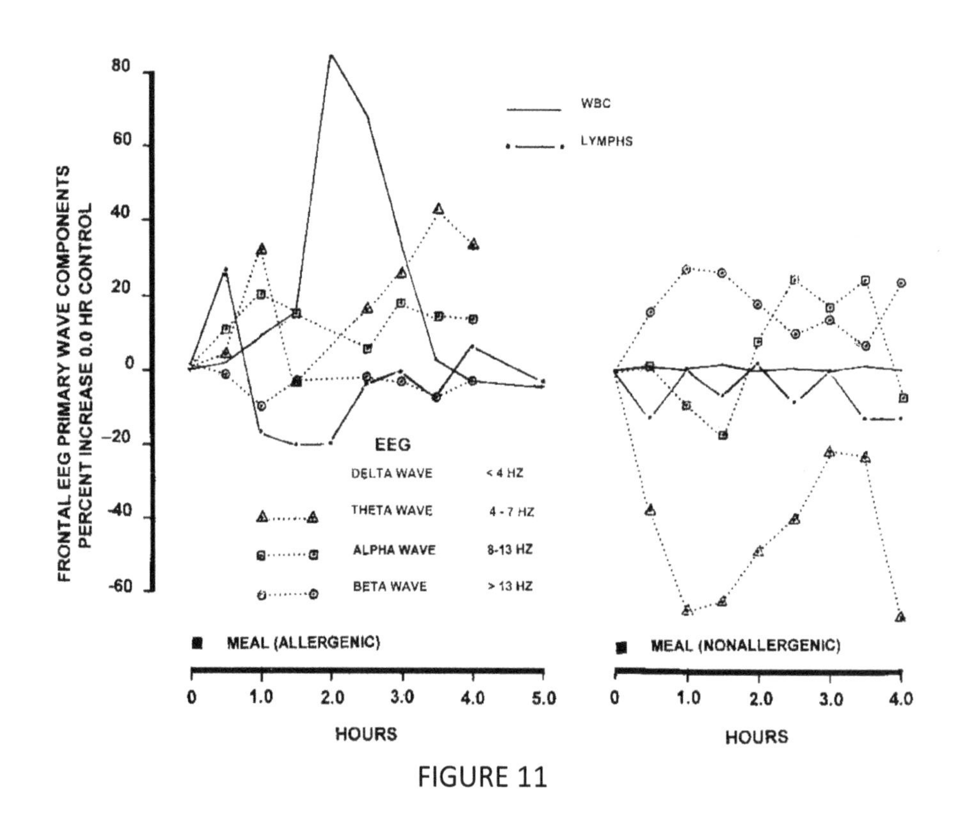

FIGURE 11

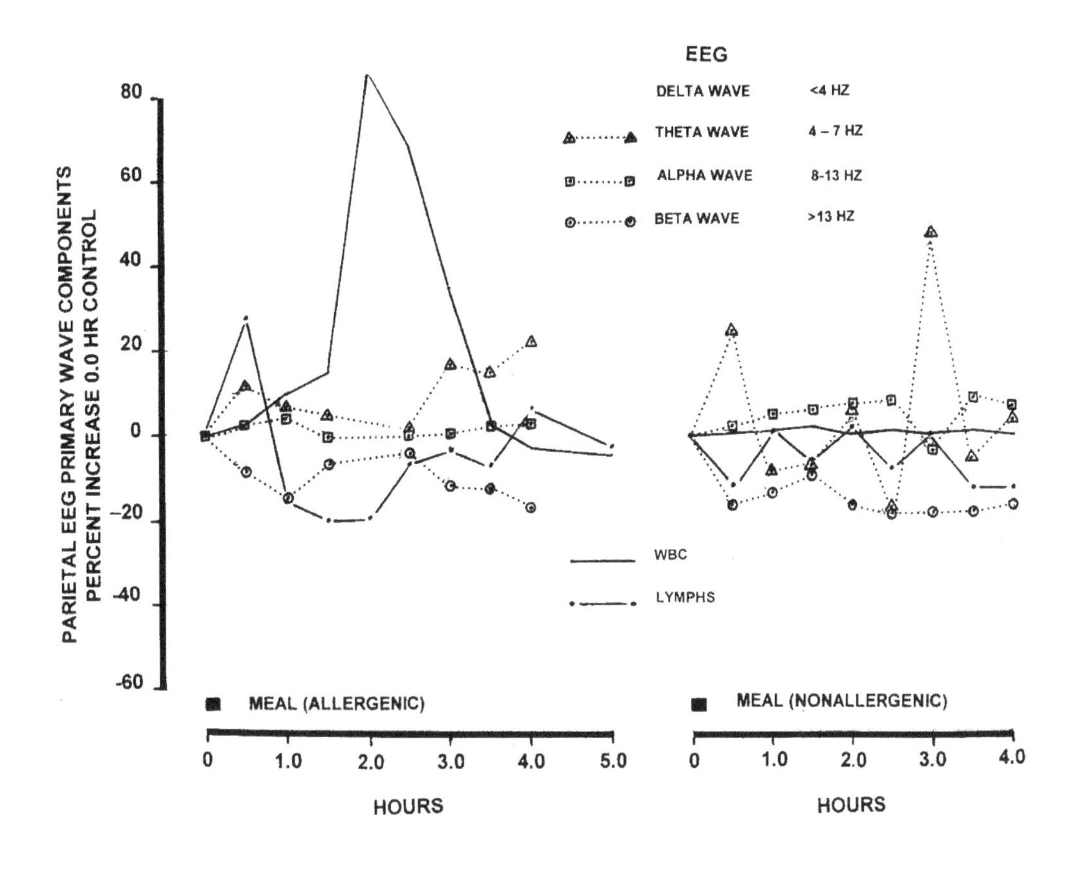

FIGURE 12

Below the surface of the WBC and the EEG primary wave some interesting relationships unfold during the temporal lobe and parietal lobe experimental test periods (Figure 7.11, 7.12). During the temporal lobe allergenic test period the lymphocytes rise at the 0.5-hour mark then fall during the 1.0-, 1.5-, and 2.0-hour marks before returning to 0.0 hr. control levels. The temporal lobe nonallergenic test period the lymphocytes do not vary much from the 0.0-time control values. Following the lymphocyte activity, the frontal lobe theta wave peaks at 1.0 hour, shows a parallel drop to 1.0 hour before there is a significant rise thru the 4.0-hour mark. The theta wave for the nonallergenic followed a much different pattern tracking well below the 0.0-hour control. The lymphocyte activity followed by frontal lobe component theta wave activity is followed at 2.5 hours by a significant rise in the WBC. As the WBC is returning to control levels the frontal lobe theta waves are rising sharply. The temporal lobe allergenic meal alpha component wave is consistently above the 0.0-hour control, while the nonallergenic meal shows more fluxuation. There appears to be something special happening in the "worry lobe" (frontal lobe).

During the parietal lobe allergenic test period the lymphocytes rise at the 0.5-hour mark then fall during the 1.0-, 1.5-, and 2.0-hour marks before returning to 0.0 hr. control levels. The nonallergenic test period the lymphocytes do not vary much from the 0.0-time control values. The parietal lobe theta wave increases a little above the control out to 2.5 hour, before there is a significant rise thru the 4.0-hour mark. The theta wave for the nonallergenic followed the 0.0-hour control with a positive spike at the 0.5- and 3.0-hour mark. The parietal lobe nonallergenic meal alpha component wave consistently compares to the 0.0-hour control, with more fluxuation.

Hypothesis: If the allergen (antigen) is triggering the immune defense system (antibodies), then classical immune-diffusion methods should permit detection of antigen/antibody relationships with the possibility that one might be vaccinated for severe allergies. The initial results produced a finger stick method and a Neurochemistry Lab (Figure 7.13) micro liter immune-diffusion system for antigen/antibody interactions that could detect several antigens with controls and standards (Figure 7.14).

Hypothesis: *If* food allergy is an immune disorder; then *in vitro* passive transfer techniques might be used to confirm the involvement of antigen/antibody mechanisms and Cytotoxic Test results. In the very first attempt to test this hypothesis the results summarized in Figure 14 show the results of three different combinations done in triplicate (Cases A, B and C) where buffy coat and plasma fractions were experimentally recombined then exposed to various food allergens based on *in vitro* Cytotoxic Test results. A whole new line of experimentation is revealed. In case A passive transfer of chocolate coffee, oat, tea, and watermelon between SGP and SSA jump out as examples of passive transfer that encouraged using immune system tags to monitor and possibly find a way to treat food allergy problems.

Hypothesis: If one knows that specific food antigen(s) are positive, then removing specific food antigen(s) from the diet makes more sense than removing "dairy product" or "green vegetables". The procedures described above provide specific ways to find the elusive needle in the haystack that may provide a key to wellness.

NOTES:

NEUROCHEMISTRY LAB

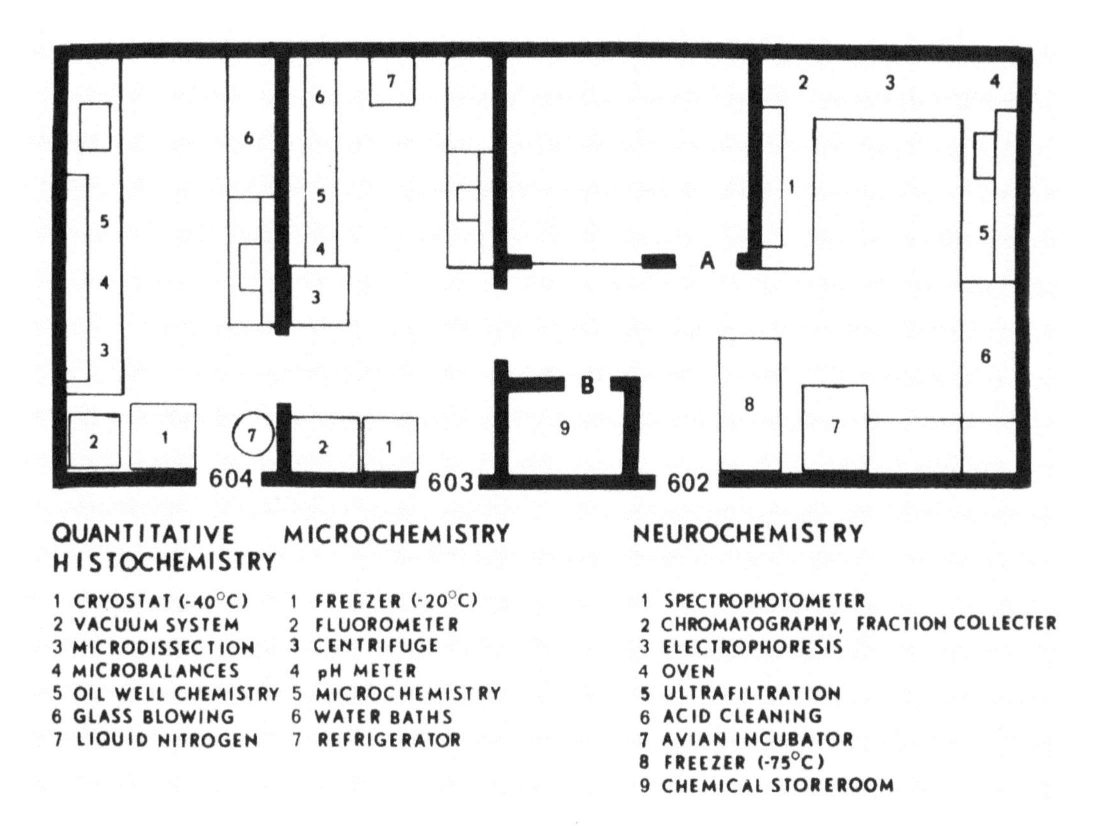

FIGURE 13

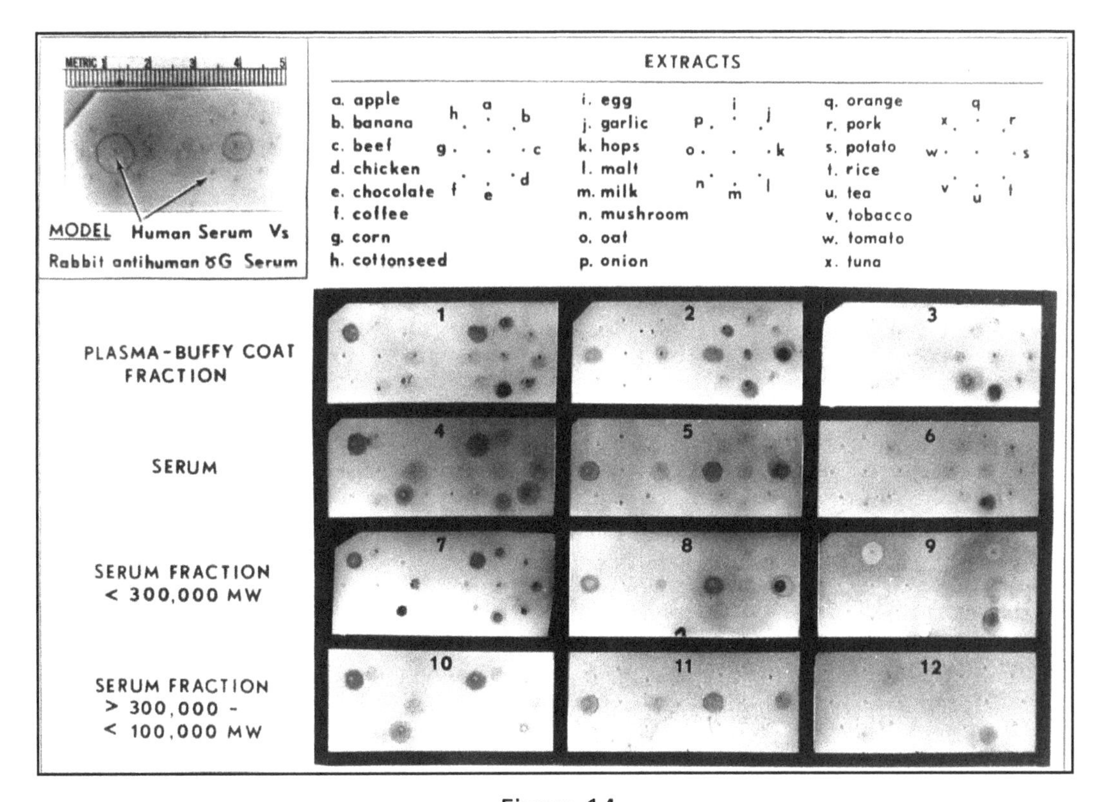

Figure 14

This was a first test for this procedure, optimizing conditions leads to a new allergy procedure.

In Case B, mustard, tomato, and watermelon sensitivities were transferred EJL to CAH. In return cabbage, cantaloupe, coffee, honeydew, and pea sensitivities were transferred from CAH to EJL. In Case C, chicken, chocolate, peanut, and string bean sensitivities transferred SGP to CAA, while in return egg, orange, pork, and cane sugar were transferred from CAA to SGP. This first series of experiments produced encouraging results (Figure 14) using classical immunological procedures to further advance our earlier studies. A new series of studies is suggested. Our very first attempt to analyze food allergy at the antigen/antibody micro-immunodiffusion (Figure15) level suggested further studies. The door is opened to the roles of the ratio of the immunoglobulin's, especially immunoglobulin G and E. We all know how proper and timely immunization helps some of us deal with rag weed and the flu each fall. There seemed to be no limit to the possibilities for these allergy and immunology studies to contribute to better understanding and treating neurological disorders.

TABLE 1 TRANSFER OF FOOD SENSITIVITY, AS DETERMINED BY CYTOTOXIC TESTING, FOLLOWING THE COMBINATION OF ONE SUBJECT'S BLOOD PLASMA TO A SECOND SUBECT'S BUFFY COAT[1]

A

Food Item[2]	SGP BUFFY COAT						SSA BUFFY COAT					
	SGP PLASMA			SSA PLASMA			SSA PLASMA			SGP PLASMA		
Apple			⊕									
Cantaloupe					⊞	⊞	+	+	+		+	+
Cherry	+		+			+				⊞		
Chocolate	+	+	+	+	+	+				⊞	⊞	⊞
Coffee	+	+	+	+	+	+				⊞	⊞	⊞
Cucumber	+	+	+			+						⊞
Hops	+	+	+			+						
Lobster	+	+	+	+	+	+					⊞	⊞
Milk			⊕			⊕						
Oat	+	+	+	+	+	+				⊞	⊞	⊞
Pea						⊞	+	+	+			+
Rice	+	+	+									
Stringbean	+	+	+	+	+	+					⊞	
Tea	+	+	+	+	+	+				⊞	⊞	⊞
Turpentine	+	+	+			+						
Vanilla			⊕									
Watermellon				⊞	⊞	⊞	+	+	+	+	+	+

B

Food Item	EJL BUFFY COAT						CAH BUFFY COAT					
	EJL PLASMA			CAH PLASMA			CAH PLASMA			EJL PLASMA		
Apple									⊕			⊕
Banana			⊕			⊕						
Barley									⊕			
Broccoli	+	+	+									
Cabbage				⊞	⊞		+	+	+	+	+	+
Cantaloupe				⊞	⊞	⊞	+	+	+	+	+	+
Coffee				⊞	⊞	⊞	+	+	+	+	+	+
Garlic			⊕			⊕						
Honeydew				⊞	⊞		+	+		+	+	
Mustard	+	+	+	+	+	+				⊞	⊞	⊞
Pea				⊞	⊞	⊞	+	+	+	+	+	+
Peanut			⊕			⊕						⊞
Soybean			+									
Spinach							+					
Tobacco						⊞			⊕			⊕
Tomato	+	+	+	+	+	+				⊞	⊞	⊞
Turpentine			⊕			⊕						
Watermelon	+	+	+	+	+	+				⊞	⊞	⊞

C

Food Item	SGP BUFFY COAT						CAA BUFFY COAT					
	SGP PLASMA			CAA PLASMA			CAA PLASMA			SGP PLASMA		
Cantaloupe									⊕			
Chicken	+	+	+[3]	+	+	+				⊞	⊞	⊞
Chocolate	+	+	+	+	+	+				⊞	⊞	⊞
Egg				⊞	⊞	⊞	+	+	+	+	+	+
Milk			+									
Orange					⊞	⊞	+	+	+	+	+	+
Peanut	+	+	+	+	+	+				⊞	⊞	⊞
Pork				⊞	⊞	⊞	+	+	+	+	+	+
Stringbean	+		+	+		+				⊞		⊞
Sugar, Cane				⊞	⊞	⊞	+	+	+	+	+	+

1) three experiments, 2) items exclusive to one or other individual based on cytotoxic testing, 3) subject was fasted prior to the third test, ⊞ cytotoxic test positive food items previously exclusive to the other subject, ⊕ cytotoxic test positive only at maximum leukocytosis.

Figure 15

A decision to reorganize the Missouri Institute of Psychiatry to Missouri Institute of Mental Health was underway. Key programs, Computer Analyzed EEGs, and Cerebral Allergy would be seriously affected.

sunday pictures

ST. LOUIS POST-DISPATCH
JANUARY 20, 1974

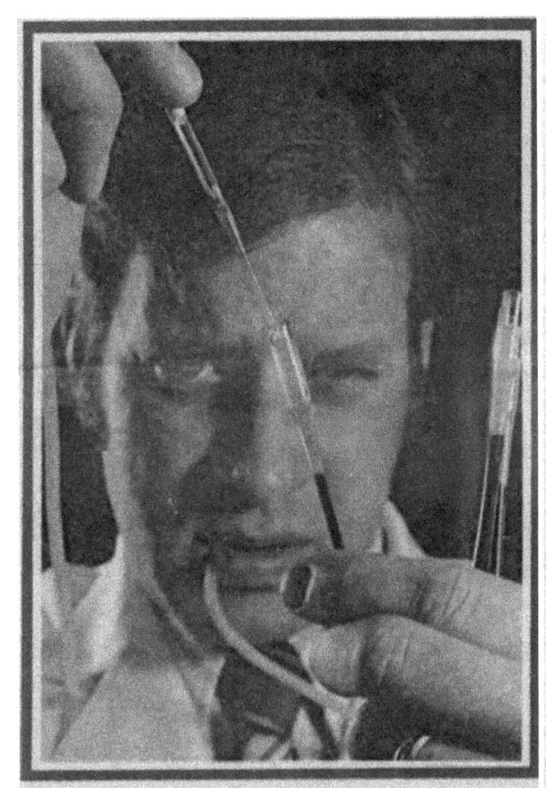

Crisis In Psychiatric Research

Stephen G. Perry, a biochemist, sat in his office at the institute and talked about allergies, which apparently have a relationship to mental disorders.

"Each individual's allergy profile is as distinct as his or her fringerprint," Perry said. "It is probable that each one of us is sensitive to many more foods than we think. It is this which could possibly alter the delicate chemical balance of our body and produce those unexplained headaches, a general feeling of malaise, perhaps depression."

Blood tests, which Perry said were easier to conduct than the conventional skin tests, are used to determine the allergy profile. They are done by introducing a substance — alcohol or wheat, for example — into the blood sample. Then the reaction of the blood's disease fighters is examined.

It is this reaction that determines the individual's degree of sensitivity to the substance being tested.

"A man or woman can rise in the morning, eat breakfast and begin to feel bad if they have a wide range of unidentified food allergies," Perry said. "They begin to pick up shortly before lunch as the stuff is emptied from their system. Then they eat lunch and poison themselves again."

Science already has found that hyperkinetic children on special diets seem to function better than those on regular diets. The Cerebral Allergy Laboratory, which Perry directs at the institute, has determined further that most alcoholics tested are sensitive (allergic) to alcohol, and that persons who suffer from schizophrenia have distinct allergy profiles, which, although different from each other, have enough similarities to identify schizophrenia.

Diet management may one day become as large a part of the treatment of mental disease as it is for a person suffering from diabetes.

Figure 7.16

My bucket list includes tracking biomarkers for cardiovascular disease (hs-CRP, $LpPLA_2$, Elisa analysis) in response to tobacco allergen sensitivities using the experimental model described by Ulett and Perry (1974; 1975). In these studies, I would use the Beckman Coulter UniCel DxH800 Cellular Analysis System to quantitate the hematology results obtained using Ulett, Perry (1974; 1975) model to analyze specific tobacco allergen sensitivities. This approach would extend and further standardize results providing three dimensional quantitative readings.

A hypothesis for a model using tobacco as an allergen affecting the intima of the cardiovascular system is proposed. The case has been made for coordinated hematological (Chaps 1 – 3) and electrophysiological (Chap 4) activities following synchronized exposure to known negative or positive allergens. If the allergens have a significant effect on the cardiovascular intima, then hematological markers should signal the onset of inflammation (hs-CRP) of the intima including potential evidence of plaque (macrophage) $LpPLA_2$ release. Key molecular and hematological markers could reflect the time frame of the tobacco challenge. The immune system can cause a welt in the skin as well as potential welts (plaques) in the intima of blood vessels.

Dr George A Ulett M.D., Ph.D. and I collaborated to publish our initial findings (Ulett and Perry, 1974; 1975). Dr. Ulett and his pediatric psychiatrist wife successfully continued to use these validated results to guide them in their clinical practice. I, (Stephen G Perry, PhD, Assistant Professor, University of Missouri Medical School), trained in NeuroAnatomy and Biochemistry to teach and do basic research in the medical sciences (University of Kansas Medical Center and Washington University Medical Center) looked to improve the testing method and analyze the results of this research. The summary above, using fundamental laboratory techniques, led to new findings and beg for more detailed analysis according to varying fields of expertise.

The contents of this book are dedicated to improving the health and the well-being of untold millions.

REFERENCES

Ulett GA and Perry SG: Cytotoxic testing and leukocyte increase as an index to food sensitivity. II. Coffee and Tobacco. Annals Allergy 34: 150-160, 1975.

Ulett GA and Perry SG: Cytotoxic testing and leukocyte increase as an index to food sensitivity. Annals Allergy 33: 23-32, 1974.

Ulett GA, Etil I and Perry SG: Cytotoxic food testing in Alcoholics. Quart J Stud Alc. 35: 930-942, 1974.

NOTES:

RESUME

STEPHEN G. PERRY, Ph.D.
1422 WALLACE DRIVE ALLEN TEXAS, 75013
sgpphd1@msn.com, 214-497-5336

2007-	AHS Substitute Teacher, Consultant, Author, Medical Research.
1998-2006	Allen School System, Allen Texas 7th Grade Science Teacher, Curtis/Ereckson
1990-1997	Laboratory Computer Consultant
1983-1989	Boeing/McDonnell Douglas Aerospace, Computer Data Management Division for Hospital Financials, plus Hospital Data Commutations (HDC), Tandem based HDC, and Departmental computer systems (**Lab**, Radiology, and Pharmacy).
1976-1983	Vice-President, and Director of Technical Operation, For the Central Region of SmithKline Clinical Labs St. Louis, Missouri.
1976	Chief: Radioimmunoassay/Immunochemistry, and Technical Administration SmithKline Clinical Labs, St. Louis, Missouri.
1975Chief:	Radioimmunoassay, SmithKline Clinical Labs, St. Louis, Missouri.
1971-1975	Assistant Professor Neurochemistry/Cerebral Allergy Units. Missouri Medical School/ Missouri Institute of Psychiatry (MIP) , St, Louis, Missouri.
EDUCATION	Ph.D., Neurochemistry/Biochemistry, Kansas University Medical School, 1968. B.A., Biology, Chemistry/History. Augustana College, Rock Island, Illinois, 1960.
1969-1971	Post-Doctoral: Oliver H. Lowry, M.D./Helen Burch, Ph.D., Biochemical Pharmacology, Washington University Medical School, St. Louis, Missouri.
1968-1969	Post-Doctoral: Victor Hamburger, Ph.D./Florence Moog, Ph.D., Developmental Biochemistry, Washington University Biology Department, St. Louis, Missouri.
1965	National Science Foundation Fellowship, Bermuda Biological Station, Experimental Embryology.
1963-1968	Research Assistant with Bryon Wenger, Ph.D., and his Lowry ultamicrochemistry laboratory in the Biochemistry Department, Kansas University, Lawrence, Kansas.
1960-1962	Teaching Assistantship Medical Anatomy, Irwin Baird, Ph.D., Kansas University, Medica School, Lawrence Kansas.

PUBLICATIONS

Perry, S.G. and B.S. Wenger (1967). A microchemical study of Glucose-6-Phosphate Dehydrogenase activity during the period of vertebral cartilage induction in the chick. J. Cell Biol., 35:102a.

Perry, S.G. (1968). Enzymatic differentiation Glucose-6-Phosphate Dehydrogenase activity in the components of the chick vertebral cartilage induction system. Ph.D. Thesis, University of Kansas, Lawrence Kansas.

Perry, S.G., (1970). Enzymatic differentiation of Glucose-6-Phosphate Dehydrogenase, NADP Isocitrate Dehydrogenase, and Phosphofructokinase in developing mouse duodenum. J. Cell Biol., 47:157a.

Burch, H.B., Perry, S.G., and O.H. Lowry (1973). Distribution of Glutamic pyruvic transaminase, L-Alanine and D-Amino Acid Oxidase in young and adult rat nephrons. Fed. Proc., 32:615.

Perry, S.G., Gouri, M.K.S.., and A.S. Marrazi (1973). Distinguishing properties of a stable preparation of cerebrally active human blood fractions. Fed. Proc., 32:476.

Burch, H.B., Lowry, O.H. Perry, S.G., Fan, L., and S. Fagioli (1974). Effect of age on Pyruvate Kinase and Lactic dehydrogenase distribution in rat kidneys. Amer. J. Physiol., 226: 1227-1231.

Ulett, Itil, E., and S.G. Perry (1974). Alcoholics increase their response to cytotoxic testing techniques. Quat. J. Studies Alcohol., 35: 930-932.

Ulett, G.A., and S.G. Perry (1974). Cytotoxic testing and leukocytes increase as an index to food sensitivity. Annals of Allergy, 33: 23-32.

Ulett, G.A. and S.G. Perry (1975). Cytotoxic testing leucocyte increase as an index to food sensitivity II.: Coffee and Tobacco. Annals of Allergy, 34:150-160.

Chan, A.W.K., Perry, S.G., Burch, H.B., Fagioli, S., Alvey, T.R., and Lowry, O.H. (1979). Distribution of Transaminase and D-Amino acid oxidase within the nephron of young and adult rats. J. Histochemical. and Cytochem., 27: 751-755.

BOOKS

Perry, S.G. (written, in review). GLUCOSE-6-PHOSPHATE DEHYDROGENASE ACTIVITY. IN COMPONENTS OF THE CHICK VERTEBRAL CARTILAGE INDUCTION SYSTEM.

Perry, S.G. (written, ready to publish). ALLERGY ANALYSIS: ALTERNATIVE METHODS

Perry, S.G., 2023. TYPE-2 DIABETES CASE STUDY, PUBLISHED with 45 BOOKS, OutSkirtspress.com

FAMILY

Widowed to Carol Ann Lollar-Perry ("Mother Superior") with three sons. Robert Roland, M.D., University of Texas, Houston, doubled boarded in Internal Medicine and Hospice and is a Family Practice Physician in Harker Heights (Fort Hood area) Texas. Brant Patrick M.S. Psychology, Ed.M. Administration, Ed.D. Administration. Greenville Texas H.S. Principal, Shawn Stephen, Ed.M. Administration, Ed.D. Administration, Frisco H.S Principal Frisco, Texas. Seven grandkids. Family fun: Camping, Golfing, Cruising, Youth Athletic Programs, Boy Scouts (Four EAGLES).

CPSIA information can be obtained
at www.ICGtesting.com
Printed in the USA
BVHW062104160623
666052BV00021B/1066